WILLIAM F. MAAG LIBRARY
YOUNGSTOWN STATE UNIVERSITY

Advances in
INORGANIC CHEMISTRY
AND
RADIOCHEMISTRY

———

Volume 28

CONTRIBUTORS TO THIS VOLUME

R. Bohra

G. A. Brinkman

N. M. N. Gowda

A. Haas

D. E. Hobart

John H. Holloway

David Laycock

Jack Martin Miller

S. B. Naikar

J. R. Peterson

G. K. N. Reddy

H. W. Roesky

Ralf Steudel

Eva-Maria Strauss

Advances in
INORGANIC CHEMISTRY
AND
RADIOCHEMISTRY

EDITORS

H. J. EMELÉUS

A. G. SHARPE

University Chemical Laboratory
Cambridge, England

VOLUME 28

1984

ACADEMIC PRESS, INC.
(*Harcourt Brace Jovanovich, Publishers*)

Orlando San Diego New York London
Toronto Montreal Sydney Tokyo

COPYRIGHT © 1984, BY ACADEMIC PRESS, INC.
ALL RIGHTS RESERVED.
NO PART OF THIS PUBLICATION MAY BE REPRODUCED OR
TRANSMITTED IN ANY FORM OR BY ANY MEANS, ELECTRONIC
OR MECHANICAL, INCLUDING PHOTOCOPY, RECORDING, OR ANY
INFORMATION STORAGE AND RETRIEVAL SYSTEM, WITHOUT
PERMISSION IN WRITING FROM THE PUBLISHER.

ACADEMIC PRESS, INC.
Orlando, Florida 32887

United Kingdom Edition published by
ACADEMIC PRESS, INC. (LONDON) LTD.
24/28 Oval Road, London NW1 7DX

LIBRARY OF CONGRESS CATALOG CARD NUMBER: 59-7692

ISBN 0-12-023628-1

PRINTED IN THE UNITED STATES OF AMERICA

84 85 86 87   9 8 7 6 5 4 3 2 1

CONTENTS

CONTRIBUTORS ix

Fast-Atom Bombardment Mass Spectrometry and Related Techniques

JACK MARTIN MILLER

I. Introduction	1
II. The Fast-Atom Bombardment Method	2
III. Applications to Organometallic Compounds	7
IV. Applications to Coordination Compounds	13
V. Applications to Other Inorganic Systems	17
VI. Applications of Secondary Ion Mass Spectrometry	19
VII. Applications of Other Related Techniques	21
VIII. Summary	22
IX. Addendum	22
References	22

The Chemistry of Berkelium

J. R. PETERSON AND D. E. HOBART

I. Introduction	29
II. Nuclear Properties, Availability, and Applications	30
III. Separation and Purification	32
IV. Atomic Properties	34
V. Metallic State	41
VI. Compounds	47
VII. Solution Chemistry	55
VIII. Concluding Remarks	63
References	64

Preparations and Reactions of Oxide Fluorides of the Transition Metals, the Lanthanides, and the Actinides

JOHN H. HOLLOWAY AND DAVID LAYCOCK

I. Introduction	73
II. The Oxide Fluorides of the Transition Metals	74

III.	The Oxide Fluorides of the Lanthanide Elements	85
IV.	The Oxide Fluorides of the Actinide Elements	86
	References	91

Chemical Effects of Nuclear Transformations

G. A. Brinkman

I.	Introduction	101
II.	Reactions of Thermalized Recoil Atoms	102
III.	Stereochemistry in Substitution Reactions	112
IV.	Muonium Chemistry	119
	References	130

Homocyclic Selenium Molecules and Related Cations

Ralf Steudel and Eva-Maria Strauss

I.	General	135
II.	Neutral Selenium Ring Molecules	136
III.	Homocyclic Selenium Cations	152
IV.	Selenium Iodide Cations, Se_nI^+	161
V.	Conclusions and Outlook	162
	References	163

The Element Displacement Principle: A New Guide in p-Block Element Chemistry

A. Haas

I.	Introduction	168
II.	The Element Displacement Principle	169
III.	Application of the Element Displacement Principle to Recent Chemical Problems	172
IV.	Halogen-Like Properties of Parahalogens of Different Derivation	176
V.	Interchangeability of Elements and Paraelements in Known Structures	194
VI.	Electronegativities	197
	References	199

Compounds of Pentacoordinated Arsenic(V)

R. Bohra and H. W. Roesky

I.	Introduction	203
II.	As—C-Containing Compounds	204
III.	As—N-Containing Compounds	216
IV.	As—O- and As—S-Containing Compounds	223
V.	As—Halogen-Containing Compounds	238
VI.	Conclusion	248
	References	249

Perchlorate Ion Complexes

N. M. N. Gowda, S. B. Naikar, and G. K. N. Reddy

I.	Introduction	255
II.	Perchlorate Ion Coordination and Methods of Identification	256
III.	Complexes of Early Transition Metals	260
IV.	Complexes of Iron Group Metals	263
V.	Complexes of Cobalt Group Metals	265
VI.	Complexes of Nickel Group Metals	268
VII.	Complexes of Copper Group Metals	273
VIII.	Complexes of Zinc Group Metals	283
IX.	Complexes of Lanthanides	287
X.	Complexes of Other Metals	288
XI.	Conclusion	290
XII.	Abbreviations	291
	References	294

Index	301
Contents of Previous Volumes	305

CONTRIBUTORS

Numbers in parentheses indicate the pages on which the authors' contributions begin.

R. BOHRA (203), *Department of Chemistry, University of Rajasthan, Jaipur 302004, India*

G. A. BRINKMAN (101), *Department of Chemistry, National Institute of Nuclear Physics and High-Energy Physics, 1009 AJ Amsterdam, The Netherlands*

N. M. N. GOWDA (255), *Department of Chemistry, Central College, Bangalore University, Bangalore 560001, India*

A. HAAS (167), *Lehrstuhl für Anorganische Chemie II, Ruhr-Universität Bochum, 4630 Bochum 1, Federal Republic of Germany*

D. E. HOBART (29), *Isotope and Nuclear Chemistry Division, Los Alamos National Laboratory, Los Alamos, New Mexico 87545*

JOHN H. HOLLOWAY (73), *Department of Chemistry, The University, Leicester LE1 7RH, England*

DAVID LAYCOCK[1] (73), *Department of Chemistry, The University, Leicester LE1 7RH, England*

JACK MARTIN MILLER (1), *Department of Chemistry, Brock University, St. Catharines, Ontario L2S 3A1, Canada*

S. B. NAIKAR (255), *Department of Chemistry, Central College, Bangalore University, Bangalore 560001, India*

J. R. PETERSON (29), *Department of Chemistry, The University of Tennessee, Knoxville, Tennessee 37996, and Transuranium Research Laboratory, Oak Ridge National Laboratory, Oak Ridge, Tennessee 37831*

G. K. N. REDDY (255), *Department of Chemistry, Central College, Bangalore University, Bangalore 560001, India*

H. W. ROESKY (203), *Institute of Inorganic Chemistry, University of Göttingen, D-3400 Göttingen, Federal Republic of Germany*

RALF STEUDEL (135), *Institut für Anorganische und Analytische Chemie, Technische Universität Berlin, D-1000 Berlin 12, Federal Republic of Germany*

EVA-MARIA STRAUSS (135), *Institut für Anorganische und Analytische Chemie, Technische Universität Berlin, D-1000 Berlin 12, Federal Republic of Germany*

---

[1] Present address: Mobil Oil Company, Ltd., Research and Technical Service Laboratory, The Manorway, Coryton, Stanford-le-Hope, Essex SS17 9LN, England.

FAST-ATOM BOMBARDMENT MASS SPECTROMETRY AND RELATED TECHNIQUES

JACK MARTIN MILLER

Department of Chemistry, Brock University, St. Catharines, Ontario, Canada

I.	Introduction	1
II.	The Fast-Atom Bombardment Method	2
	A. Basic Methods	2
	B. Instrumentation	3
	C. Matrix Liquids	5
	D. Relation to Other "Soft" Ionization Techniques	6
III.	Applications to Organometallic Compounds	7
	A. Main Group Examples	10
	B. Transition Metal Examples	11
IV.	Applications to Coordination Compounds	13
V.	Applications to Other Inorganic Systems	17
VI.	Applications of Secondary Ion Mass Spectrometry	19
VII.	Applications of Other Related Techniques	21
VIII.	Summary	22
IX.	Addendum	22
	References	22

I. Introduction

Mass spectrometry has become a very important tool for the inorganic chemist, having been extensively applied particularly in the fields of organometallic and coordination chemistry (1–3). However, for the inorganic chemist dealing with ionic materials, thermally labile systems, or polymeric systems not readily volatilized in the "organic" mass spectrometer (1), mass spectrometry remained a dream for a long time. The exceptions were those with access to "inorganic" mass spectrometers, with thermionic and spark sources, designed primarily for elemental analysis or isotope ratio determination and the surface scientists with access to secondary ion mass spectrometers (SIMS), ion microprobes, and related devices. Most of these were low-mass, low-resolution devices, not particularly adaptable to the problems of the inorganic chemist, i.e., problems analogous to those for which the organic chemists have so widely applied mass spectrometry. In addition, there was the widely held belief that SIMS was only applicable to conducting samples and that surface charging effects precluded application to neutral molecules. Early attempts at molecular SIMS suggested that the sample lifetime would also be exceedingly short.

This article deals primarily with fast-atom bombardment (FAB) mass spectrometry. It is basically an outgrowth of SIMS, except that it uses an energetic beam of fast neutral atoms of an inert gas, rather than 2- to 8-keV ions, as the means of sputtering charged species from a surface. This technique, which was readily adapted to high-mass, high-resolution (organic) mass spectrometers, together with the discovery that dissolving the sample in a matrix liquid, such as glycerol, gave long-lived signals, burst onto the mass spectrometry scene in 1980–1981 *(4–9)*—initially finding its greatest application to biological molecules that were polar and thermally labile. However, from the earliest papers on FAB, the inorganic applications were apparent. Since FAB mass spectra were able to show a parent quasi-molecular ion for vitamin B_{12} (m/z 1355) and its coenzyme (m/z 1579) *(4, 5, 8)*, almost everyone with early access to the technique tried these cobalamins as ideal model compounds *(10, 11)*. The observation of vitamin B_{12} by mass spectrometry has been described as a "milestone" *(12)*. Many people, myself included, saw or used commercial prototype FAB sources before the first paper on the subject *(5)* appeared in print.

Although the growth in applications of FAB has been explosive, inorganic applications have been slower in coming for several reasons. First, as Wilson and I suggested in an earlier review *(1)*, the instrumentation seldom belonged to the inorganic chemist and our metals may have been feared by traditional mass spectroscopists. More important, the matrix liquid has been a greater problem in inorganic chemistry, than for the polar organic and bioorganic molecules for which FAB (using a glycerol matrix) had proved so successful. I hope, however, to show that FAB offers great potential for solving problems in inorganic chemistry.

II. The Fast-Atom Bombardment Method

A. BASIC METHODS

The basic method involves the addition of a source of "fast atoms" to a conventional mass spectrometer. This neutral atom beam is directed onto a sample deposited on the metal tip of a direct-insertion probe such that the sputtered ion beam is directed onto the entrance slits of a conventional mass spectrometer. The atom beam usually is directed onto a bevelled probe tip so that it makes an angle of about 60–70° with the surface normal (13) and an angle of about 90° with the direction of the spectrometer ion beam. No new instruments have had to be designed for the technique. FAB sources have been retrofitted to both new mass spectrometer designs and to instruments designed more than 20 years earlier. As long as the basic pumping speed of the ion source is fast enough to remove the inert gas coming from the atom gun, and as long as there is a suitable port onto which the gun can be mounted while still providing for a direct insertion probe, any mass spectrometer, including quadrupole mass analyzers, can be retrofitted for FAB, although new source components are required to extract the sputtered ions efficiently. Martin et al. (13) have published a useful summary of the optimization of the FAB experiment, though Campana (14) has reminded the FAB community that most of the factors that have been found optimal for SIMS by surface chemists also apply to FAB, thus demonstrating the similarities of the techniques. Rollgen and Geismann (118) have replied, suggesting that things were less obvious than suggested by Campana.

The factor that makes FAB–MS so different from EI–MS is that, in its usual form, the sample coating the probe tip consists of a solution or suspension in a relatively nonvolatile matrix liquid such as glycerol. This provides for a continually renewed surface exposed to the atom beam and thus spectra that are stable over a period of many minutes. No heating of the sample is required other than the localized energy implanted in the sample by the atom beam. Although complications may result from interactions with the matrix liquid, they are often less than, or certainly no worse than, such complications as thermal decomposition or ion molecule reactions, involved in other techniques for sample volatilization. In addition, FAB–MS is looking at condensed-phase systems similar to those investigated by NMR or IR. Thus perhaps the data are easier to correlate. Several reviews or introductions to the method have appeared (4, 7–9, 13, 15–22).

B. Instrumentation

Figure 1 shows schematically the important parts of a FAB source. The atom guns typically can handle argon or xenon, though they have also been used with heavy metal atoms such as mercury (23). The heavier the bombarding atom, the greater the sensitivity, which is roughly proportional to atomic weight (24). Thus argon is used for

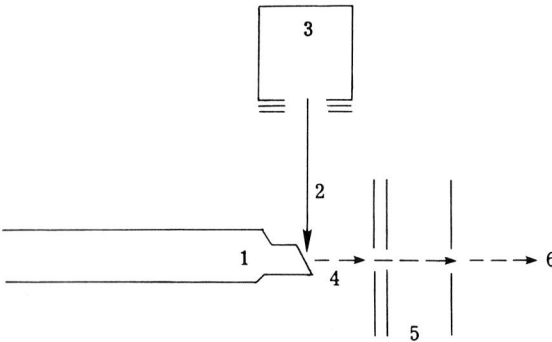

FIG. 1. Schematic diagram of a simplified FAB source: (1) probe, (2) fast-atom beam, (3) atom gun, (4) sputtered secondary ion beam, (5) source ion optics, (6) to mass analyzer.

routine work where sensitivity is good, and xenon for situations where sensitivity is a problem. However, gas consumption is low enough so that, if care is taken, xenon can be used at all times without excessive cost. The guns used are usually commercial models, typically a saddle-field source [e.g., Ion Tech. (28)] or Capillaratron [Phasor (26)] though home-built systems are also described [e.g., (26, 27)]. There are also guns designed to produce beams of Cs ions (26). A power supply is needed for the source to provide for voltages of 2–10 keV for acceleration of the ions produced, which are then neutralized by collision to produce the fast atoms. The probe tip is usually metal; copper [which occasionally has been observed to form cluster ions with the sample or matrix (13)] or stainless steel are the most commonly used materials. In a magnetic analyzer instrument, the repeller voltage is simply applied to the probe tip. The collimating and focusing electrodes are similar to those in an EI source but with wider spacings, allowing for the greater dispersion of the sputtered ions.

FAB sources readily produce high-mass cluster ions and ions corresponding to high-molecular-weight compounds not previously amenable to study by mass spectrometry. As a result, new magnetic analyzers have been developed that, through the use of nonhomogeneous magnetic fields, can attain masses as high as 15,000 daltons at full accelerating voltage and hence full sensitivity. New detectors for these high-mass ions have also been developed. Data systems can be used at these high masses, but the problems of calibration, mass defect (often greater than one nominal mass for very large ions) and digitization rate have to be overcome. Older data systems may not be adaptable to work at very high masses without both software and hardware modification.

C. MATRIX LIQUIDS

Glycerol, the most popular matrix liquid for bioorganic applications of FAB, is still the most popular starting point for choosing a matrix liquid. It is clear that even in cases where one might suspect that "protic" glycerol would react with the complex being studied, good spectra can be obtained in that matrix, e.g., hydrates and hydrogen-bonded solvates of tetraalkylammonium fluorides (28). In some cases, to clear up overlap caused by matrix ions, thioglycerol has been used. DMSO has been used as a matrix liquid (29), but we have found that many organometallic and coordination compounds are decomposed in it. Sulfolane is more promising, although it doesn't produce spectra that last as long. Another successful matrix liquid for coordination compounds and large macrocycle complexes is di-tert-amylphenol. Crown ethers, such as 18-crown-6, with 10% tetraglyme to depress the melting point, have given good results for organometallic compounds (30). We have recently found HMPA to give good results as a matrix liquid for fluoride systems, though its toxic nature makes one wary of its regular use. Poly(phenyl ether) and various other pump oils, including silicones, have been used with some success (16). Triton X100 has also been recommended as a solubilizing reagent (1%), though pure Triton X100 has been used for porphyrins (16). Doping the matrix liquid with a trace of acid or base often helps in obtaining positive- or negative-ion spectra. The addition of sodium or chloride salts can have the same effect. Many other viscous, relatively nonvolatile organic liquids have been tried. Neat liquid samples also work well if sufficient material is available, and slightly impure materials that give an oil rather than crystals can often serve as their own matrix. Do not hesitate to try new substances as matrix liquids appropriate to the nature of the compounds in which you are interested.

The study of various doping materials in a FAB matrix can, in fact, lead to the determination of chemical equilibrium constants as demonstrated by Caprioli (31) who was able to measure the pK_a values of a series of acids.

It is also possible to do the FAB experiment on solid materials without the presence of a matrix liquid. In such cases, the lifetime of the observed spectra is very much shorter but often adequate to characterize the material of interest. We have had considerable success with metal fluorides and metal fluorides coated on solid supports such as silica or alumina. The powdered sample is simply held on the probe tip with double-sided tape. This method can also be used for obtaining FAB spectra from unextracted TLC spots (116).

A new and promising matrix for less polar organometallics and coordination compounds is p-nitrophenyl octyl ether.

D. RELATION TO OTHER "SOFT" IONIZATION TECHNIQUES

As Campana has suggested (14), FAB and SIMS spectra are very similar in nature. This is perhaps in part due to the fact that the so-called fast-atom beam may well be a mixture of both ions and neutral particles (32). Since the deflection of the "low-energy" charged species in the ion gun does not remove the very energetic ions formed by the acceleration of multiply charged ions, they are then only partially neutralized by collision. It seems more a matter of nomenclature, with FAB sources being found on instruments designed as organic mass spectrometers, and SIMS being more likely on "surface science" instruments, often with quadrupole analyzers.

FAB is most often compared to the soft ionization method known as field desorption (FD) mass spectrometry, a technique in which the sample, deposited on an emitter wire coated with microcrystalline carbon needles, is desorbed under the influence of a high electric field gradient. As usual, bioorganic systems are best represented by both techniques (21, 33). Though FAB is the easier of the two, they are complementary, FAB being particularly suited for the case of extreme thermal lability and FD for the case of chemical lability or matrix interference. Cerny et al. (33) compare the two techniques for the study of coordination complexes and conclude FD is better for molecular-ion determination, while FAB provides better fragmentation information, which is useful in elucidating structures.

Laser desorption (LD) mass spectrometry is at an even less developed stage for organometallic and coordination compounds than is FAB (19, 34–35), though pure inorganic application preceded organic applications. Typically, pulsed lasers are used with time-of-flight (or quadrupole) mass analyzers. Thermal degradation seems to be greater than for other desorption methods. Similarly, plasma desorption (PD), or ^{252}Cf desorption, is based on the use of the californium fission fragments to energize and desorb samples from thin foils. Very high-molecular-weight species are detected (4000–10,000 daltons) using time-of-flight mass analyzers, which are, unfortunately, low resolution devices. It has been suggested (34) that all the desorption ionization techniques (DI) should be classed together and should include, in addition to the above, thermal desorption and electrohydrodynamic desorption, all being relatively "soft" and capable of yielding high-mass spectra.

III. Applications to Organometallic Compounds

Organometallic compounds were featured in the first published FAB paper by Barber et al. (5), and have been the subject of a brief review (22). Most exciting was the report that FAB yielded good spectra for vitamin B_{12} and its coenzyme. More details appeared in the proceedings of a conference held the previous year (4) and in a report in *Nature* (8). Prior to learning of these results, among others, we had also looked at vitamin B_{12} (11), using a Kratos demonstrator instrument (10). Most manufacturers offering FAB retrofits used vitamin B_{12} spectra in their promotional literature.

Barber's report on the FAB spectra of cobalamins is the most complete (36). The basic cobalamin structure is shown in Fig. 2, with the molecular weight ranging from 1343 to 1578 depending on the axial ligand R. In all cases, a reasonably intense quasi-molecular ion (i.e., M + 1), is observed in the positive-ion spectra obtained from a glycerol matrix, with an intense peak at m/z 1329, corresponding to the loss of the axial ligand, CN, CH_3, the coenzyme side chain, or OH. They reported the relative intensities of the M + 1 peak as 1355 > 1344 > 1579 > 1346. Major losses then occur from the 1329 peak, as illustrated on Fig. 2. Although some acetamide is lost, major fragmentation occurs from the other axial chain. Dimethylbenzimidazole is lost, followed by the sugar and phosphate, yielding the base peak at m/z 971. The negative-ion spectra show M − H ions as well as ions for the attachment of a chloride ion [e]. These spectra are relatively clean, long-lasting, and free from thermal decomposition.

McLafferty et al. (37) studied the collisionally activated dissociation (CAD) of the ions resulting from the fast-atom bombardment ionization of cyano- and methylcobalamins, using a tandem mass spectrometer. They showed that m/z 1329 is produced only from the quasi-molecular ion. Similar CAD spectra are produced by both the quasi-molecular ions and the 1329 peak, indicating that essentially all fragmentation arises from this latter ion. CAD spectra produced for the 1329 ion derived from methylcobalamin or vitamin B_{12} coenzyme are also similar. This is consistent with the work of Taylor (10), who used linked-scan metastable ion techniques in a conventional double-focusing mass spectrometer to show the same thing. The sequential

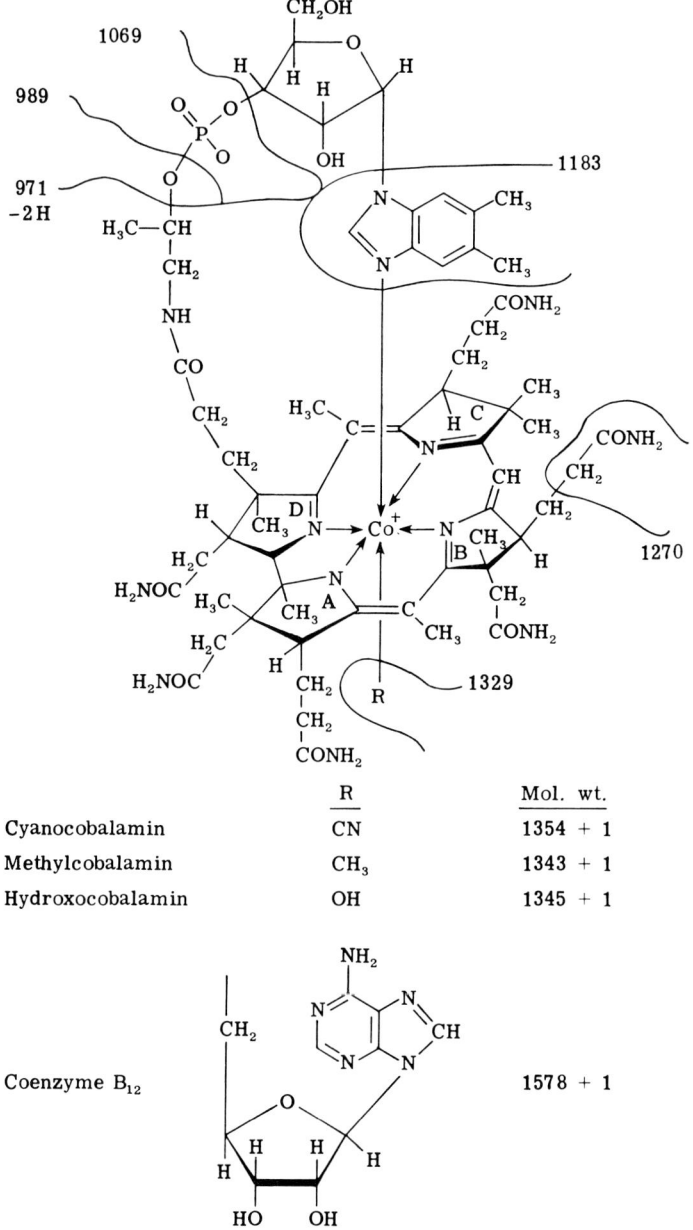

FIG. 2. Vitamin B_{12} FAB–MS fragmentation. [Reproduced in part with permission (36).]

losses from the axial ligand suggested by Barber are verified by both the tandem MS (37) and linked-scan (10) studies, though McLafferty warned that if the structure of the vitamin were not already known, then alternate fragmentation of the parts of the molecule could cause confusion in the assignment of the structure.

Taylor (10) also showed that high resolution was possible with FAB and that it could be applied to ions as large as these vitamins. However, higher resolution at high masses becomes essential, not for empirical formula determination (which for metal containing systems is often done better by a fit to the isotopic abundance patterns), but simply to be sure of resolving nominal masses as the molecules studied get heavier.

Schwartz et al. (38) reported on the use of FAB for "unequivocal and rapid molecular weight determination of corrins," with two cyano groups as the axial ligands. They reported no M^+ or $(M + H)^+$ ions, but rather the loss of both axial CNs, corresponding to reduction of the metal from Co(III) to Co(I), analogous to the EI behavior. However, negative-ion spectra gave very strong $(M - H)^-$ ions ranging from 20 to 100% relative intensity. Field desorption MS of cobalamin derivatives has been compared to FAB (39) with the suggestion that, for the type of compound discussed above, FD gives ions characteristic of the thermal degradation process, while FAB produces ions more characteristic of impact-ionization processes. However, because only positive-ion FAB was employed in this latter study, they did not observe the negative-ion spectra of the essentially intact parent molecule less a hydrogen.

Meili and Seibl (40) have investigated matrix effects in the FAB analysis of cobalamins to determine if changing from a glycerol matrix would improve the chances of observing molecular ions for such systems, which so readily suffer cleavage of the axial ligands, as discussed above. For cobyrinic acid derivatives, they achieved promising results with the ethyl and butyl esters of citric acid (TEC, TBC), benzoic acid benzyl ester (bzbz), and 2-nitrophenyloctyl ether (NPOE), all of these providing information on one or both axial ligands. Ion intensities are $(M - 2CN)^+ > (M - CN)^+ > (M)^+ \cdot$ (bzbz only). The $(M - CN)^+$ ion dominates the spectrum in NPOE. They suggest that the slightly oxidizing matrix prevents the metal valence change from Co(III) to Co(II). The abundance of the molecular ion versus $(M - 2CN)^+$ changes from 120–150% immediately after insertion to 20–30% after a few minutes. These results suggest that if any metal-containing system appears to give, as the highest mass ion in FAB, a species corresponding to the reduction of the metal, then a weakly oxidizing matrix such as NPOE

TABLE I

COMPARISON OF EI AND FAB DATA[a]

O			S	
EI	FAB	Ion$^+$	EI	FAB
0.3	—	M	0.1	—
—	0.4	(M + 2Gl − 2H)[b]	—	—
25.9	49.3	Ph$_3$Sn	25.5	56.6
3.5	1	Ph$_2$Sn	2.4	—
8.9	22	PhSn	9.9	24.0
8.4	6.5	Sn	7.9	13.3
—	3.2	(Sn + Gl − H)[b]	—	—
3.3	2.5	C$_6$F$_5$	—	5.9

[a] (C$_6$H$_5$)$_3$SnEC$_6$F$_5$, E = O, S.
[b] Gl, glycerol.

is advisable. One might similarly predict that in negative-ion spectra, where species appear to be undergoing unwanted oxidations, a reducing matrix should be considered.

A. MAIN GROUP EXAMPLES

We have studied the FAB–MS of some phenyl derivatives of the Group IV elements (41). In Table I, a comparison of the EI and FAB (glycerol) spectra of (C$_6$H$_5$)$_3$SnEC$_6$F$_5$ (E = O or S) is made. Unfortunately, in this case FAB does not give a more intense molecular ion than does EI, though this might not be considered surprising since Group IV organometallics are known for their weak or absent parent ions. There is only slight evidence for complexation of the sample with the glycerol matrix.

Triphenylgermanium and tin and lead halides were examined in both glycerol and sulfolane matrices. In the case of sulfolane, ions were sometimes observed corresponding to the coordination of a sulfolane molecule to the organometallic halide, but simple parent ions were not seen. In contrast, Pang and Costello reported the FD spectra of some organogermanes with good molecular-ion intensities (42). In a rather different FAB use of Group IV organometallics, Wong et al. (43) reported on the use of a silicone pump fluid as the bombarding particle.

FAB and FD–MS have also been used for the determination of organoarsenicals such as arsenobetaine, (CH$_3$)$_3$As$^+$CH$_2$COO$^-$, in biological systems such as seafoods (44, 45). EI gives no molecular ion for this compound. The FD spectra showed decarboxylation–protonation to give $(CH_3)_4As^+$ (the base peak) as well as dimeric and trimeric species. Arsenocholine bromide was debrominated to give the base peak. Impurities in the marine extracts did, however, complicate the spectra. The FAB spectrum of arsenobetaine in glycerol was similar to the FD observations, with extra peaks to provide more structural information. In actual marine extracts, the quantities of arsenicals were so low that glycerol dominated the spectrum, but under high resolution conditions the protonated arsenobetaine at m/z 179.0053 $(AsC_5H_{12}O_2)$ was measured with reference to the glycerol $(2M + H)^+$ ion, using a multichannel analyzer for signal averaging. Thus though the spectra of the real extracts were less clear than those from FD, in all cases with the long-lived spectra, high resolution permitted the detection of the arsenical in all samples examined.

Barber's group also investigated the application of FAB to transition-metal organometallics (4, 8). The silylcyclopentadiene–cyclooctadienerhodium compound (I) was both thermally labile and moisture

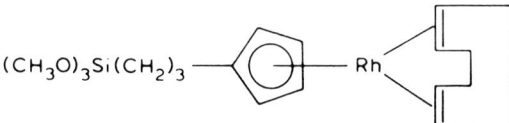

sensitive and gave no worthwhile mass spectrum, using other means of ionization. It was an oil, and its FAB spectrum was obtained as the neat liquid, without the use of a matrix liquid. The compound, being moisture sensitive, is likely to have reacted with glycerol. A strong parent ion was observed at m/z 438. The other important high-mass ions corresponded to the loss of neutral C_8H_4, trimethoxysilane, trimethoxysilylethylene, and the corresponding propylene. The rhodium COT bonds break readily, while rhodium remains strongly attached to the cyclopentadienyl fragment. Prominent silicon-containing positive ions also include $(CH_3O)_3Si^+$ and $(CH_3O)_2SiH^+$.

B. Transition Metal Examples

They also looked at the spectra of ionic complexes such as $(Ph_3P)_4$-$Au^+ClO_4^-$, which gave a simple positive-ion spectrum essentially corresponding to that of the cation and a successive loss of ligand. Loss of biphenyl was also observed. The negative-ion spectrum was that of the perchlorate anion.

Minard and Geoffroy (30) used 18-crown-6 (with 10% tetraglyme to depress the melting point) as their matrix liquid for organometallics to avoid the stability problems that glycerol or other protic solvents might have presented. They looked at two rhodium compounds, $(Ph_3P)_3RhCl$, i.e., Wilkinson's catalyst, and $C_8H_{12}Rh(PPh)(Cl)RhC_8H_{12}$, and a cobalt cluster compound, $(Ph_2P)_3$—$Co_3(CO)_6$. Molecular or overlapping molecular and quasi-molecular ions were observed. Successive loss of carbonyl was observed, just as would be expected from EI spectra.

Davis et al. (46) studied a series of organometallic complexes of rhodium, ruthenium, rhenium, palladium, and platinum, including metal cluster systems of iron and osmium. They used a series of matrix liquids including diamylphenol (DAP), glycerol, and carbowax 200. Wilkinson's catalyst gave similar results from the DAP to Minard's from a crown–glyme matrix. They show that FAB gives good spectra for triphenylphosphine complexes whose EI spectra are dominated by ions arising from the Ph_3P, produced by thermal decomposition of the complex. $ReCl(CO)_3[P(C_6H_4F\text{-}p)_3]_2$ and the corresponding ONO_2 compound having a p-methyl group in place of the p-F group both showed, as the highest-mass species, the ion $[(M + H)X]^+$. No evidence for the loss of X in solution has been reported. The Re—X bond may be particularly labile after ionization. These compounds did not give negative-ion spectra, suggesting that ionization occurs in the matrix, either via dissociation and loss of X or by protonation, to give a quasi-molecular ion, which may then fragment. The cluster compound $Fe_4(H)C(CO)_{12}AuPPh_3$ gave quasi-molecular ions in both positive- and negative-ion modes. The addition of a proton rather than its loss is unexpected in the negative mode. The salt $[Os_4H_3(CO)_{12}(MeCN)_2]PF_6$ gave both positive- and negative-ion spectra. The former gives the expected cation as the highest mass species, though the most intense ion corresponds to the additional loss of the acetonitriles. This latter ion is also seen in the negative-mode spectra, and it corresponds to species reported in the electrochemical reduction. Neither mode shows the ion pair of the parent molecule.

Sharp et al. (47, 48) used FAB–MS (with the crown–glyme matrix discussed above) to examine a series of rhodium, iridium, and platinum organometallic complexes containing either σ- or π-bonded cumulene ligands. For both rhodium and platinum compounds, the metal–bis(triphenylphosphine) cation is the base peak, while for iridium a similar ion is obtained, but with one carbonyl ligand remaining as well. Platinum complexes gave the most intense molecular ions (up to 35% relative intensity), while the rhodium molecular ions were least intense. Fragmentation occurred at the metal–ligand bond, the cumulene always being lost intact. The FAB spectra of rhodiumtris(triphenylphosphine) chloride, Wilkinson's catalyst, also reported by

Minard (30), and by Davis (46), who used a diamylphenol–acetonitrile matrix, did not show any parent molecular ion or quasi-molecular ion as the others had reported. Ir[P(C$_6$H$_5$)$_3$]$_2$COCl (Vaska's compound), its corresponding rhodium derivative, and Pt[(C$_6$H$_5$)$_3$]$_2$(C$_2$H$_4$) were also obtained for comparison. It is clear that greater intensities of high-mass fragments and molecular ions were observed when heavier bombarding particles were used. It is suggested that reproducibility between laboratories is equivalent to that for EI spectra. They also showed that isotopic cluster patterns are adequate to identify species without resorting to exact mass measurement. Sharp suggests that the Minard crown ether matrix will become the standard for organometallics.

Tkatchenko et al. (49) have compared the FAB and FD spectra of cationic η^3-allylic complexes of palladium and nickel. They found an intense cation and few fragments from FD data, while FAB is reported to give a smaller cation peak and numerous fragments. Since they used glycerol or glycerol–DMF matrices, they were perhaps not working under optimal conditions found by others such as Minard or Davis above. They were also able to observe species resulting from ligand exchange when they had two species present in the matrix liquid.

Kalinoski et al. (50) studied palladium-mediated reactions of organomercurials with glycals. If a heteroarylmercuric acetate, a chiral pyranoid or furanoid glycal, and palladium(II) acetate were added in equimolar quantities to a glycerol droplet, the FAB spectrum showed both the heteroarylpalladium species, formed by transmetallation from the mercurial, and the adduct formed by insertion of the glycal double bond into the palladium–heteroaryl carbon bond. Thus FAB has been used to observe labile organometallics not previously reported.

We can thus conclude that, though the applications of FAB to organometallics is still in its infancy, the technique shows great promise. It certainly works least well where EI works best, as we found for symmetrical nonpolar tetraaryls of Group IV elements. It obviously works best for those situations for which EI is of no value. In the intermediate case, there appears to be good evidence that FAB is complementary to FD. One will not replace the other, and at this stage it is difficult to say which will be the more successful in the long term, though FAB is easier to use.

IV. Applications to Coordination Compounds

We have studied some coordination compounds involving main-group Lewis acids (51a). In collaboration with Steve Hartman (51b) we have used FAB to detect the bis(quinuclidine)difluoroboron cation. The base peak in glycerol is the $D_2BF_2^+$ parent ion, with loss of quinuclidine giving the next most abundant ion. Group IV organometallic halides act as sufficiently strong Lewis acids to complex the glycerol or sulfolane matrix liquids, which are incorporated in some of the ions observed—sometimes displacing the halide. Triphenyltin bromide, for example, shows ions corresponding to coordination of sulfolane, as well as the corresponding ion that has lost either a phenyl group or bromine (51). We are not yet able to predict unequivocally the behavior of a weak Lewis acid when placed in a matrix liquid. There is clear evidence for ligand-exchange reactions in solution, and we are investigating the use of FAB as a complement to NMR for study of ligand-exchange reactions. Phosphonium salts, on the other hand, are somewhat more regular, showing the expected cation as well as weaker peaks for (cation)$_2$anion species. When we move to transition-metal complexes (51), the bistriphenylphospine–palladium halide system showed ions corresponding to one, two, three, and four halides.

Cerny *et al.* (33) have carried out extensive comparisons of FAB and FD mass spectrometry of neutral and 1+ and 2+ cationic transition-metal coordination complexes with glycerol used as the matrix liquid and with the sample solutions coated on aluminum foil to avoid memory effects. They concluded that FAB is preferable for the 1+ complexes, producing both parent ions and fragments. As one might have expected, FD gave better molecular-weight information on the neutral complexes, while FAB gave the fragments useful in elucidating structures. This may well be a case for further study of appropriate matrix liquids. Dicationic systems did not yield to either method, though some fragmentation information was obtained. Spectra were dominated by 1+ ions from reduction of the metal. They also found that for ligand losses and reductions, the FAB results paralleled the solution chemistry and thus could be useful in predicting solution chemistry. This would suggest that the observation of Meili and Seibl (40), that weakly oxidizing matrix liquids can prevent reduction of the metals, should receive wider attention. Among the metals included in the study by Bursey's group (33) were Re(III), Re(I), Ag(I), Rh(I), Os(II), Os(III), Os(VI), and Mo(VI). They also observed that in FAB, monodentate ligands are lost in preference to bidentate ligands, which are only lost from fragment ions. They report bipyridyl and phenanthroline-type ligands to be particularly tenaciously held. Fragments correspond either to reduction of the metal, or if that is an unfavorable process, by simple ligand loss. If there is a "redox accessible zero oxidation state," the redox process will be observed. For a series of Re complexes, they found chloride retained in preference to the π-acceptor ligand for Re(III), with the reverse observed for Re(I). Finally, they concluded that since the FAB and solution chemistries are so similar "that there is no evidence yet to discount a hypothesis that the energy-transfer process from the primary atom beam to the metal complex–glycerol matrix produces the initially formed parent ion in excited vibrational states of the ground state, which also can be populated by conventional thermolysis of the complex in a polar alcoholic solvent." In another work, the same group studied the FAB spectra of a series of copper complexes of biological interest (52).

Chan and Cook (53) studied the factors affecting the mass spectral sensitivity for ions sampled by field evaporation from a liquid matrix. These results, including those on metal complexes, show some of the types of interactions expected for ions in solutions, and they obtain results consistent with FAB–MS.

Barber's group has also compared positive- and negative-ion FAB and FD (54) for a series of iron(III) complexes, with hydroxamate-containing siderophores as the ligands. They propose these soft ionization techniques as a way to characterize siderophores (low molecular weight chelating agents with a high affinity for ferric iron, which are secreted by a wide range of microorganisms) and their metabolites. They found FD quite difficult, but useful results could be obtained with good technique. However, useful results were also obtained by FAB much more easily. FAB is recommended as the preliminary screening technique for potentially novel siderophores, with FD then being used to determine possible heterogeneity in the ferrichrome skeleton. Sensitivity was such that microgram quantities were adequate for identification of metabolites.

Johnstone et al. (55, 56) have extensively studied crown-ether complexes of metallic cations, using FAB. Metallic salts gave molecular ions of the type [crown + M^{n+} + $A^{(n-1)-}$]$^+$, where A is the anion. In particular, they looked at chlorides, acetates, and nitrates of Groups I and II, along with Cu, Hg, La, Ce, Th, Co, etc. They all gave molecular ions, but with a facility that reflected the complexation of the cations found in solution; e.g., K > Cs > Na for 18-crown-6, other crowns giving different results. The method is proposed as being suitable for analyzing mixtures of metal cations in solutions at concentrations as low as 10^{-10} M.

They also proposed the method as a means of measuring stability constants for the complexes. As well, they analyzed the thermodynamics of the desorption process and were able to calculate the localized "temperature" which must be created in a small volume around the impacting fast atom for desorption of the complex to occur. Only species near the surface will be desorbed, since further into the bulk matrix excess translational energy is dissipated by multiple collisions. A temperature of about 860 K is calculated for the crown complexes, which compares to the 900 K reported by Stoll (57) for desorption from a heated filament of a benzo-15-crown-5–sodium ion complex. They again confirm that the best results for FAB are for preformed ions. Thus the matrix chemistry is critical for obtaining good spectra of neutral species. We have also (51) observed results similar to those of Johnstone for $HgCl_2$–18-crown-6 in a diamylphenol matrix. In glycerol, we saw the expected peaks for NaF–18-crown-6, except that weak peaks are also seen for a species with two crown molecules.

Kurlansik et al. (58) have used FAB to study disproportionation and recombination of synthetic porphyrin in the matrix. They observed reactions in the sample matrix that yielded new compounds from recombinations of molecular fragments, and if FAB is to be used extensively for peptide sequencing, warned of the importance of understanding these processes involving the formation and rupture of amides or peptide bonds. Bazzaz et al. (59) have used FAB to detect a new naturally occurring chlorophyll from a mutant maize plant. Glycerol did not yield a spectrum, but a mixture of chloroform and trigol as solvent and matrix gave molecular and quasi-molecular ions. However, "in beam" EI had to be used to determine the structure of the hydrocarbon tail. Leznoff and I (60) have successfully used DAP as a matrix for studying a series of synthetic porphyrins and their related metal complexes, including structures with two porphyrin rings and metals.

Barber et al. (61) have used FAB to study the antibiotics bleomycin A_2 and B_2 and their metal complexes. In particular, the ferrous sulfate complex of bleomycin A_2 shows a pseudomolecular ion at m/z 1566, corresponding to the salt ($A_2 - H^+ + FeSO_4$) since the amide hydrogen of the histidine residue is lost on complexation. An ion at m/z 472 corresponds to the ferrous ion complexed to the surrounding ligands but lacking the disaccharide groups and the peptide chain.

Another biomedical area of interest in metal complexes concerns various anticancer drugs. Puzo et al. (117) and Theodoropoulos (63) have used FAB to study cisplatin [i.e., cis-$Pt(NH_3)_2Cl_2$] and its complexes and related compounds. For the 1:2 complex with guanosine, peaks due to the complex less a Cl, or less Cl and HCl, along with their corresponding glycerol adducts were observed. Less clear results were observed for other amino acid complexes (63). Cohen et al. (62) and Costello et al. (64) have used FAB to study the complexes of technitium(III) and technitium(I) and -(V), respectively, these compounds being important radioactive materials for diagnostic imaging. Cohen favored monothioglycerol as the matrix agent. For monovalent cations of technitium(III), he obtained intense cation spectra and distinctive fragmentation information. Costello has identified the probable species present in some of the commercially available Tc imaging agents, using FAB with various mixed matrices involving glycerol. Hansen *et al.* (*65*) report FD to be more successful than FAB in identifying another Tc complex, pointing out the difficulty of predicting success of the method for any one type of compound.

V. Applications to Other Inorganic Systems

FAB has also been used for basic surface analysis of inorganic solids such as glasses (*66, 67*) and zeolites (*68, 69*). In this respect, the use of FAB closely parallels that of SIMS. Reproducible results are obtained, which can be quantified, while organic materials absorbed on the surfaces can also be determined. It also proved possible to use FAB to determine a depth-profile analysis of these samples. No charging effects, such as those that may cause difficulties with SIMS, were observed. Dwyer *et al.* (*68*), in their work on zeolites, correlate the surface Si^+/Al^+ ratio with the bulk Si/Al ratio and show the ion ratio to be dependent on the bulk atom ratio rather than on the zeolite structure. Negative-ion FAB studies were used to investigate the presence of paired aluminum species in both zeolites and aluminas (*69*). We have found that FAB is an ideal way of measuring the nature of inorganic or quaternary ammonium fluorides supported on alumina or silica, in particular where the presence or absence of bifluoride ion is important (*28, 51*). Coverdale *et al.* (*70*) have used negative-ion FAB to investigate the structure of rhenium(VII) oxide on alumina, and they obtained spectra similar to those of the bulk unsupported material. Similarly, supported NH_4ReO_4 shows quantitatively comparable spectra to be bulk crystalline material.

FAB–MS has been compared to the analytical use of negative-ion FD in the determination of fluoroborate anion in water at the ppb level (*71*). Positive FAB is however only sensitive down to the microgram level in glycerol and other matrix liquids, but negative FAB results were not reported. Smith (*72*) has reported the use of FAB for isotope-ratio analysis of such biologically important metals as calcium, magnesium, zinc, and iron. The long-term stability of spectra are an advantage in achieving statistically good results. Other researchers have evaluated the usefulness of adding inorganic salts, such as KCl, NaCl, or $AgClO_4$, to solutions of the analyte in glycerol (*73, 74*). Silver ion, in particular, adds an isotopic pattern to the spectra, which aids in picking out the ions of interest from a background that may well be very complex.

Javanaud and Eagles (*75*) have studied in detail the positive and negative FAB spectra of simple inorganic salts involving both various metal chlorides and sodium salts of some complex anions. They report that, in general, the FAB data more closely resemble FD rather that EI mass spectral data. Various polymeric species are also observed, as will be discussed below. Teeter (*76*) has similarly studied anionic surfactants and other sulfonate salts. The spectra were all relatively easy to interpret "knowing in advance the structures involved." He suggests that, for these systems, the application of FAB to quantitative analysis is not easy, with glycerol certainly complicating the spectra. Spectra were not additive with respect to the condensed phase. Teeter used a computer program, written for use on either a personal computer or a larger system, to work out all the possible combinations present.

We have looked at a simple inorganic material, stannous chloride, using positive and negative FAB in glycerol and in other matrix liquids (*41*). The negative-ion spectrum in HCl-doped glycerol showed ions corresponding to $SnCl_2$, $SnCl_3$, and $SnCl_3 \cdot H_2O$. The latter diminishes as HCl concentration increases, as does the $SnCl_2OH$ intensity. The base peak is the chloride ion, followed in intensity by the tin trichloride anion. Glycerol-containing ions were weak except for the chloride adduct. In the positive-ion spectrum, species such as $(SnCl)_2$, glycerol-H, and SnCl with one and two molecules of coordinated glycerol were observed.

The most extensively studied of the simple inorganic systems have been the alkali metal halides (*75, 77–80*). KI and CsI have been used as calibration compounds for high-molecular-weight studies, since clusters of the type M_nX_{n-1} are readily observed in the positive-ion spectra with n values of 50 or higher [values greater than 200 are now observed, i.e., masses exceeding $m/z = 50,000$ (*79, 80*)] being reported. Campana and co-workers have shown that there are two major reaction pathways for decomposition of the cluster cations, involving losses of either MX or of $(MX)_2$. In the negative-ion spectra, the corresponding anions contain an extra halide rather than the extra metal cation, but the behavior is otherwise the same. Interestingly, the intensity of the cluster species does not fall off regularly with n. Rather the data show regions of extra stability, and hence spectral enhancement, centered on $n = 13, 22$, and corresponding regions of low abundance for clusters having n greater than the stable points by one or two. Collisionally activated successive reactions of these clusters have also been reported (*75, 81*). Campana also reports (*80*) that "these extended mass spectra are fit to the distributions predicted by a model based on random bond breaking . . . when Bethe lattices of the appropriate coordination number are used," i.e., these clusters from solution mimic the solid state structure. We (*51*) have looked at the corresponding glycerol–MF systems, which have as an additional factor the possibility of strong H bonding between the fluoride ion and the OH proton of the matrix. Clusters are observed, which are similar to those seen for the other alkali metal halides, but complexation with the glycerol is more common until very high MF concentrations are reached, i.e., in the region of a 1:1 mol ratio. Most astonishing, however, is the observation in the positive-ion spectrum of a weak series of glycerol · $(MF_nF^+$. At first glance one might attribute this behavior to the presence of the bifluoride ion. However, this is ruled out by the negative-ion spectrum, which is dominated by the simple fluoride ion with no evidence for bifluoride. Bifluoride is seen in systems intentionally doped with HF_2^-, or simply with water, leading to its formation. This we take to be the first example of a mass spectrum of a strong H bond to fluoride.

VI. Applications of Secondary Ion Mass Spectrometry

A complete review of the inorganic and related applications of SIMS is beyond the scope of this article, since the technique is so much more mature than FAB, although for a long time it has been confined chiefly to surface chemistry studies. It is only with the development of the liquid-matrix technique that "organic" SIMS became popular. However, there are a great many similarities between the two techniques, and as mentioned earlier, SIMS experts feel that the FAB workers are basically "reinventing the wheel." Although this may be the case, and the similarities between the two techniques are great, the availability of FAB on the organic mass spectrometers available to most inorganic, organometallic, and coordination chemists has resulted in a burgeoning of interest. Typical of the recent reviews of SIMS is the work of Winograd (*17, 82*). We shall mention briefly some of the SIMS work that closely parallels the FAB results described above.

The applications of SIMS to organometallic and coordination compounds have been pioneered by Cooks and Walton and their co-workers (*83–90*). They have used solid matrices, such as ammonium or sodium chlorides, to decrease the abundance of ions resulting from intermolecular processes while increasing the abundance of ions that are structurally informative. Their results on triphenyltin chloride (83) are very similar to our FAB results. Interestingly, they report that neither iron nor manganese trisacac gives a parent ion in SIMS, while weak ones are observed in simple EI spectra. They also report on the similarity of SIMS to laser microprobe mass analysis (LAMMA) and the distinctions between them and EI. In a study of metal complexes of β-diketones (84), bimetallic species, such as $Co_2(acac)_4^+$ and the corresponding trisacac cation, are reported. These have, however, been reported in the past for EI spectra as well (91). In an approach that is the opposite to the derivatization used in EIMS (i.e., the production of volatile derivatives from polar or ionic species), Cooks (85) has reported the derivatization of neutral samples to produce nonvolatile ionic species via Bronsted or Lewis acids or bases, which enhance the sensitivity of the SIMS technique. Other work from these groups includes the study of silver complexes (86), nitrosylisocyanide complexes of chromium and molybdenum (87), the distinction between phosphines and phosphonium ions (88), the characterization of nickel complexes on oxide supports (89), and a study of phosphonium salts (90).

Another area of SIMS that has received a great deal of attention is the study of simple inorganic salts. Marien and De Pauw (92) have looked at the ion-beam-induced effects in SIMS spectra. It is clear that under high ion fluxes significant damage to the surface occurs, with the resulting complication of the spectrum. FAB, or liquid-matrix SIMS, is free of this effect but has the equivalent complication of reactions with the matrix, as discussed previously.

Campana's group has pioneered the development of high-performance SIMS instrumentation (93), in particular for studies at very high masses. They have investigated metal salt clusters in a manner analogous to the FAB results described in the previous section (94–100). For cubic-like structures, the CsI spectrum shows stable structures for species of the type $[Cs(CsI)_n]^+$ where $n = 13, 22, 31, 37, 52$, and 62 (94) with results now surpassing $n = 100$ (95).

Other examples of SIMS for surface analyses are studies of Cu_2S–CdS solar-cell samples (101) and the study of chemisorbed species on inorganic substrates such as methanol on Cu (100) and titania (102). De Pauw's studies of such adsorbed systems may prove to be valuable in determining the mechanism of catalytic reactions on surfaces (103). Winograd and co-workers (104–6) have studied chemisorption on metal surfaces, using SIMS. In a related study (107), Unger et al. have used molecular SIMS to study the reactions of thiophene on a silver surface. They observed the self-hydrogenation of thiophene on the surface of silver foils. Such studies are beginning to appear from the groups working with FAB sources.

VII. Applications of Other Related Techniques

Among the other "soft" ionization techniques is laser microprobe mass spectrometry (LAMMA) in which a laser pulse is used to vaporize a small amount of sample, as discussed in a 1982 review (108). Of interest to us is the application to the study of some cobalamins (109). (M + H) and (M − H) ions were observed in the positive and negative ion modes, respectively. However, there were few other high-mass fragments that could be used to impart structural information.

Schulten (110, 111) has used laser-assisted field desorption mass spectrometry to study some inorganic and organometallic systems. This method is intermediate between LAMMA and simple FD. Metal cations predominate from inorganic salts. The technique also showed clusters of the type reported from both FAB and SIMS studies. By carefully controlling the laser, a chlorophyll molecular ion could be obtained as well as fragments relating to its structure.

The most established of these other techniques is field desorption, but it has never been particularly widely used for inorganic, organometallic, or coordination compounds. In a 1982 review, Costello discussed in detail some of these applications (108)—note that only 16 references are quoted going back as far as 1972. She is the most active worker in the field having published papers on organogermanium, tin, and silicon compounds (112) and a technetate(V) complex used as a radiopharmaceutical (113). Staal et al. (114) have used FD to characterize various metal carbonyl 1,4-diazabutadiene compounds and report that in most cases only molecular ions are generated, though some loss of a single carbonyl is also observed. Since FAB is becoming more readily available than FD, is easier to use, and provides molecular weight and fragmentation information, though perhaps no less a "black art" than FD, it is unlikely that FD will establish as strong a position in the inorganic and related fields as will FAB.

The rarest of these specialized techniques, californium-252 plasma desorption mass spectrometry, has not been applied extensively to inorganic systems, though in a 1983 review (115) Macfarlane quotes several examples, such as polymeric "platinum blue," with molecular ions extending to m/z 3000. At present there are only 10 functioning systems and this certainly limits its growth.

VIII. Summary

In conclusion, I hope that I have shown that fast-atom bombardment mass spectrometry is a potentially useful tool for the synthetic chemist working in many areas of inorganic, organometallic, and coordination chemistry. In addition, as further fundamental research is done in the field with these applications in mind, the technique should become as routine as IR or NMR. Combined with developments in high-resolution NMR of solids, FAB should provide particularly useful data on supported catalytic reactions. Certainly FAB and the other complementary mass spectrometric techniques, mentioned in less detail, constitute a major way of quickly characterizing new compounds.

IX. Addendum

In the 5 months since this article was written, literature relevant to it has increased by one-third. At the end of the original 117 references I have added important new references, along with key words for each that will relate them to the text. They are presented topically in approximately the same order as material is discussed in the text, and I hope they will be of value.

REFERENCES

1. Miller, J. M., and Wilson, G. L., *Adv. Inorg. Chem. Radiochem.* **18**, 229 (1976).
2. Litzow, M. R., and Spalding, T. R., "Mass Spectrometry of Inorganic and Organometallic Compounds." Elsevier, Amsterdam, 1973.
3. Charalambous, J., ed., "Mass Spectrometry of Metal Compounds." Butterworths, London, 1975.
4. Barber, M., Bardoli, R. S., and Sedgwick, R. D., *in* "Soft Ionization Biological Mass Spectrometry" (H. R. Morris, ed.), p. 137. Heyden, London, 1981.
5. Barber, M., Bardoli, R. S., Sedgwick, R. D., and Tyler, A. N., *J. Chem. Soc. Chem. Commun.* 325, (1981).
6. Surman, D. J., and Vickerman, J. C., *J. Chem. Soc. Chem. Commun.* 324, (1982).
7. Williams, D. H., Bradley, C., Bojesen, G., Santikarn, S., and Taylor, L. C. E., *Am. Chem. Soc.* **103**, 5700 (1981).
8. Barber, M., Bordoli, R. S., Sedgwick, R. D., and Tyler, A. N., *Nature (London)* **293**, 270 (1981).
9. Taylor, L. C. E., *Ind. Res. Dev.* **124**, Sept. (1981).
10. Taylor, L. C. E., *Annu. Conf. Mass Spectrom. (ASMS), 29th, Minneapolis,* June (1981).
11. Miller, J. M., *ACS/CIC Symp. Bloomington,* June (1982).
12. Schulten, H. R., and Schiebel, H. M., *Naturwissenschaften* **65**, 223 (1978).

13. Martin, S. A., Costello, C. E., and Biemann, K., *Anal. Chem.* **54**, 2362 (1982).
14. Campana, J. E., *Int. J. Mass Spectrom. Ion Phys.* **51**, 133 (1983).
15. McNeal, C. J., *Anal. Chem.* **54**, 43A (1982).
16. Barber, M., Bardoli, R. S., Elliott, G. J., Sedgwick, R. D., and Tyler, A. N., *Anal. Chem.* **54**, 645A (1982).
17. Garrison, B. J., and Winograd, N., *Science* **216**, 805 (1982).
18. Rinehart, K. L., *Science* **218**, 254 (1982).
19. MacFarlane, R. D., *Acc. Chem. Res.* **15**, 268 (1982).
20. Fenselau, C., Cotter, R. J., Heller, D., and Yergey, J., *Int. J. Chromatogr.* **271**, 3 (1983).
21. Przybylski, M., and Fresenius, Z., *Anal. Chem.* **315**, 402 (1983).
22. Miller, J. M., *J. Organomet. Chem.* **249**, 299 (1983).
23. Stoll, R., Schade, U., Rollgen, F. W., Giessmann, U., and Barofsky, D. F., *Int. J. Mass Spectrom. Ion Phys.* **43**, 227 (1982).
24. Morris, H. R., Panico, M., and Haskins, N. J., *Int. J. Mass Spectrom. Ion Phys.* **26**, 363 (1983).
25. Franks, J., *Int. J. Mass Spectrom. Ion Phys.* **46** 343-6 (1983).
26. Rudat, N. A., and McEwen, C. N., *Int. J. Mass Spectrom. Ion Phys.* **26**, 351 (1983).
27. Hogg, A. M., *Int. J. Mass Spectrom. Ion Phys.* **49** 25 (1983).
28. Miller, J. M., and Clark, J. H., unpublished observations (1983).
29. De Pauw, E., *Anal. Chem.,* in press (1983).
30. Minard, R. D., and Geoffroy, G. L., *Abstr. Annu. Conf. Mass Spectrom. Allied Top., 30th, Honolulu* 321 (1982).
31. Caprioli, R. M., *Abstr. Conf. ASMS, 31st, Boston,* 69 (1983); *Analyt. Chem.* **55**, 2387 (1983).
32. Ligon, W. V., Jr., *Int. J. Mass Spectrom. Ion Phys.* **41**, 205 (1982).
33. Cerny, R. L., Sullivan, B. P., Bursey, M. M., and Meyers, T. J., *Anal. Chem.* **55**, 1954 (1983).
34. Busch, K. L., and Cooks, R. G., *Science* **218**, 247 (1982).
35. Cotter, R. J., Van Breemen, R. N., Yergey, J., and Heller, D., *Int. J. Mass Spectrom. Ion Phys.* **46**, 395 (1983).
36. Barber, M., Bardoli, R. S., Sedgwick, R. D., and Tyler, A. N., *Biomed. Mass Spectrom.* **8**, 492 (1981).
37. Amster, I. J., Baldwin, M. A., Cheng, M. T., Proctor, C. J., and McLafferty, F. W., *J. Am. Chem. Soc.* **105**, 1054 (1983).
38. Schwartz, H., Eckart, K., and Taylor, L. C. E., *Org. Mass. Spectrom.* **17**, 459 (1982).
39. Schiebel, H. M., and Schulten, H.-R., *Biomed. Mass Spectrom.* **9**, 354 (1982).
40. Meili, J., and Seibl, J., *Annu. Conf. Mass Spectrom. (ASMS), 31st, Boston* May (1983).
41. Miller, J. M., *Int. Conf. Organomet. Coordi. Chem. Montreal* August, (1983).
42. Pang, H., and Costello, C. E., *Abstr. Conf. ASMS, 31st Boston* 379 (1983).
43. Wong, S. S., Stoll, R., and Rollgen, F. W., *Z. Naturforsch.* **37a**, 718 (1982).
44. van der Greef, J., and ten Noever de Brauw, M. C., *Int. J. Mass Spectrom. Ion Phys.* **46**, 379 (1983).
45. Luten, J. B., Riekwel-Booy, G., v.d. Greef, J., and ten Noever de Brauw, M. C., *Chemosphere* **12**, 131 (1983).
46. Davis, R., Groves, I. F., Durrant, J. L. A., Brooks, P., and Lewis, I., *J. Organomet. Chem.* **241**, C27 (1983).
47. Sharp, T. R., White, M. R., Davis, J. F., and Stang, P. J., *Abstr. Conf. ASMS, 31st, Boston* 371 (1983).

48. Sharp, T. R., White, M. R., Davis, J. F., and Stang, P. J., *Org. Mass Spectrom.* **19**, 107 (1984).
49. Tkatchenko, I., Neibecker, D., Fraisse, D., Gomez, F., and Farofsky, D. F., *Int. J. Mass Spectrom. Ion Phys.* **46**, 499 (1983).
50. Kalinoski, H. T., Hacksell, U., Barofsky, D. F., Barofsky, E., and Daves, G. D., Jr., personal communication.
51a. Miller, J. M., *Annu. Conf. Chem. Inst. Canada, Calgary* June (1983).
51b. Farquharson, M. J., and Hartman, J. S., *J. Chem. Soc. Chem. Commun.* (in press).
52. Cerny, R. L., Bursey, M. M., and Hass, J. R., *Abstr. Conf. ASMS, 31st, Boston* 369 (1983).
53. Chan, K. W. S., and Cook, K. D., *Anal. Chem.* **55**, 1306 (1983).
54. Dell, A., Hideer, R. C., Barber, M., Bordoli, R. S., Sedgwick, R. D., and Tayler, A. N., *Biomed. Mass Spectrom.* **9**, 158 (1982).
55. Johnstone, R. A. W., and Lewis, I. A. S., *Int. J. Mass Spectrom. Ion Phys.* **26**, 451 (1983).
56a. Johnstone, R. A. W., Lewis, I. A. S., and Rose, M. E., *Tetrahedron* **39**, 1597 (1983).
56b. Johnstone, R. A. W., and Rose, M. E., *J. Chem. Soc. Chem. Commun.* 1268 (1983).
57. Stoll, R., and Rollgen, F. W., *Org. Mass Spectrom.* **16**, 2296 (1981).
58. Kurlansik, L., Williams, T. J., Campana, J. E., Green, B. N., Anderson, L. W., and Strong, J. M., *Biochem. Biophys. Res. Commun.* April (1983).
59a. Bazzaz, M. B., Bradley, C. V., and Brereton, R. G., *Tetrahedron Lett.* **23**, 1121 (1982).
59b. Brereton, R. G., Bazzaz, M. B., Santikarn, S., and Williams, D. H., *Tetrahedron Lett.* **24**, 5775 (1983).
60. Miller, J. M., and Leznoff, C. C., unpublished observations.
61. Barber, M., Bardoli, R. S., Sedgwick, R. D., and Tyler, A. N., *Biochem. Biophys. Res. Commun.* **101**, 632 (1981).
62. Cohen, A. I., Glavan, K. A., and Kronauge, J. F., *Biomed. Mass Spectrom.* **10**, 287 (1983).
63. Theodoropoulos, D., Dalietos, D., Furst, A., Flessel, P., Guirguis, G., and Lee, T., *Abstr. ASMS Conf., 31st, Boston* 64 (1983).
64. Costello, C. E., Pang, H., Davidson, A., and Jones, A. G., *Conf. ASMS, 31st, Boston* 377 (1983).
65. Hansen, G., Heller, D., Yergey, J., Cotter, R. J., and Fenselau, C., *Chem. Biomed. Environ. Instrum.* **12**, 275 (1982–1983).
66. Surman, D. J., and Vickerman, J. C., *Appl. Surf. Sci.* **9**, 108 (1981).
67. Wakefield, C. J., Haselby, D., Taylor, L. C. E., and Evans, S., *Int. J. Mass Spectrom. Ion Phys.* **46**, 491 (1983).
68. Dwyer, J., Fitch, F. R., Qin, G., and Vickerman, J. C., *J. Phys. Chem.* **86**, 4574 (1982).
69. Dwyer, J., Elliott, I. S., Fitch, F. R., van den Berg, J. A., and Vickerman, J. C., *Zeolites* **3**, 97 (1983).
70. Coverdale, A. K., Dearing, P. F., and Ellison, A., *J. Chem. Soc. Chem. Commun.* 567 (1983).
71. van der Greef, J., Ten Noever de Brauw, M. C., Zwinselman, J. J., and Nibbering, N. M. M., *Biomed. Mass Spectrom.* **9**, 330 (1982).
72. Smith, D. L., *Abstr. Conf. ASMS, 31st, Boston* 248 (1983).
73. Allison, J., Musselman, B., Dolnikowski, G., and Watson, J., *Abstr. Conf. ASMS, 31st, Boston* 728 (1983).

74. Walther, H.-J., Parker, C. E., Harvan, D. J., Voyksner, R. D., Hernadez, O., Hagler, W. M., Hamilton, P. B., and Hass, J. R., *J. Agric. Food Chem.* **31**, 168 (1983).
75. Javanaud, C., and Eagles, J., *Org. Mass Spectrom.* **18**, 93 (1983).
76. Teeter, R. M., *ASMS Meet., Boston,* May (1983); and personal communication.
77. Taylor, L. C. E., Evans, S., and Wakefield, C. J., *ASMS Meet., Boston* May (1983).
78. Dunlap, B. I., Campana, J. E., Green, B. N., and Bateman, R. H., *J. Vac. Sci. Technol.,* in press (1983).
79. Campana, J. E., Coulton, R. J., Wyatt, J. R., Bateman, R. H., and Green, B. N., Submitted (1983).
80. Campana, J. E., and Dunlap, B. I., *Int. J. Mass Spectrom. Ion Proc.* **57**, 103 (1984); and Campana, J. E. Pittsburgh Conference, Atlantic City, March (1984).
81. Campana, J. E., Wyatt, J. R., Bateman, R. H., and Green, B. N., *Abstr. Conf. ASMS, 31st, Boston* 132 (1983); and Campana, J. E., and Green, B. N. *J. Am. Chem. Soc.* **106**, 531 (1984).
82. Winograd, N., *Prog. Solid State Chem.* **13**, 285 (1982).
83. Pierce, J., Busch, K. L., Walton, R. A., and Cooks, R. G., *J. Am. Chem. Soc.* **103**, 2583 (1981).
84. Pierce, J. L., Busch, K. L., Cooks, R. G., and Walton, R. A., *Inorg. Chem.* **21**, 2597 (1982).
85. Busch, K. L., Unger, S. E., Vincze, A., Cooks, R. G., and Keough, T., *J. Am. Chem. Soc.* **104**, 1507 (1982).
86. Pierce, J. L., Busch, K. L., Cooks, R. G., and Walton, R. A., *Inorg. Chem.* **22**, 2492 (1983).
87. Unger, S. E., Day, R. J., and Cooks, R. G., *Int. J. Mass Spectrom. Ion Phys.* **39**, 231 (1981).
88. Pierce, J. L., DeMarco, D., and Walton, R. A., *Inorg. Chem.* **22**, 9 (1983).
89. Pierce, J. L., and Walton, R. A., *J. Catal.* **81**, 375 (1983).
90. Ba-isa, A., Busch, K. L., Cooks, R. G., Vincze, A., and Granoth, I., *Tetrahedron* **39**, 591 (1983).
91. Miller, J. M., unpublished observations.
92. Marien, J., and De Pauw, E., *Int. J. Mass Spectrom. Ion Phys.* **43**, 233 (1982).
93. Colton, R. J., Campana, J. E., Bariak, T. M., DeCorpo, J. J., and Wyatt, J. R., *Rev. Sci. Instrum.* **51**, 1685 (1980).
94. Colton, R. J., Bariak, T. M., Wyatt, J. R., DeCorpo, J. J., and Campana, J. E., *J. Vac. Sci. Technol.* **20**, 421 (1982).
95. Barlak, T. M., Campana, J. E., Wyatt, J. R., Dunlap, B. I., and Colton, R. J., *Int. J. Mass Spectrom. Ion Phys.* **46**, 523 (1983).
96. Barlak, T. M., Campana, J. E., Wyatt, J. R., and Colton, R. J., *J. Phys. Chem.,* in press (1983).
97. Barlak, T. M., Campana, J. E., Colton, R. J., DeCorpo, J. J., and Wyatt, J. R., *J. Phys. Chem.* **85**, 3840 (1981).
98. Barlak, T. M., Wyatt, J. R., Colton, R. J., DeCorpo, J. J., and Campana, J. E., *J. Am. Chem. Soc.* **104**, 1212 (1982).
99. Barlak, T. M., Wyatt, J. R., Colton, R. J., DeCorpo, J. J., and Campana, J. E., *J. Am. Chem. Soc.* **104**, 1212 (1982).
100. Barlak, T. M., Campana, J. E., Colton, R. J., DeCorpo, J. J., and Wyatt, J. R., *J. Phys. Chem.* **85**, 3840 (1981).
101. Satkiewicz, F. G., and Charles, H. K., *Annu. Conf. Mass Spectrom., 29th, San Diego* 307 (1976).

102. Marien, J., De Pauw, E., and Pelzer, G., personal communication from E. De Pauw (1983).
103. De Pauw, E., and Marien, J., *Int. J. Mass Spectrom. Ion Phys.* **46**, 519 (1983).
104. Kimock, F. M., Baxter, J. P., and Winograd, N., *Surf. Sci.* **124**, L41 (1983).
105. Moon, D. W., and Winograd, N., personal communication (1983).
106. Moon, D. W., Bleiler, R. J., Karwacki, E. J., and Winograd, N., personal communication.
107. Unger, S. E., Cooks, R. G., Steinmetz, B. J., and Delgass, W. N., *Surf. Sci.* **116**, L211 (1982).
108. Costello, C. E., *Spectra* **8**, 28 (1982).
109. Graham, S. W., Dowd, P., and Hercules, D. M., *Anal. Chem.* **54**, 649 (1982).
110. Schulten, H.-R., Monkhouse, P. B., and Muller, R., *Anal. Chem.* **54**, 654 (1982).
111. Schulten, H.-R., Mueller, R., Haaks, D., and Fresenius, Z. *Anal. Chem.* **304**, 15 (1980).
112. Pang, H., and Costello, C. E., personal communication (1983).
113. Jones, A. G., Davison, A., LaTegola, M. R., Brodack, J. W., Orvig, C., Sohn, M., Toothaker, A. K., Lock, C. J. L., Franklin, K. J., Costello, C. E., Carr, S. A., Biemann, K., and Kaplan, M. L., *J. Nucl. Med.* **23**, 801 (1982).
114. Staal, L. H., Van Koten, G., Fokkens, R. H., and Nibbering, N. M. M., *Inorg. Chim. Acta* **50**, 205 (1981).
115. Macfarlane, R. D., *Anal. Chem.* **55**, 1248A (1983).
116. Chang, T. T., Lay, Jr., J. O., and Francel, R. J., *Anal. Chem.* **56**, 109 (1984).
117. Puzo, G., Prome, J. C., Macquet, J. P., and Lewis, I. A. S., *Biomed. Mass Spectrom.* **9**, 552 (1982).
118. Rollgen, F. W., and Giessmann, U., *Int. J. Mass Spectrom. Ion Proc.* **56**, 229 (1984).

FAB METHODOLOGY

Mahoney, J., Perel, J., and Taylor, S., *Am. Lab.* 92, March (1984).
Ross, M. M., Wyatt, J. R., Colton, R. J., and Campana, J. E., *Int. J. Mass Spectrom. Ion Proc.* **54**, 237 (1983).
Cook, K. D., and Chan, K. W. S., *Int. J. Mass Spectrom. Ion Proc.* **54**, 135 (1983).
Clay, K. L., Wahlin, L., and Murphy, R. C., *Biomed. Mass Spectrom.* **10**, 489 (1983).
Beckner, C. F., and Caprioli, R. M., *Biomed. Mass Spectrom.* **11**, 60 (1984).
Musselman, B. D., and Watson, J. T., *ASMS Meet., San Antonio, Texas,* May (1984).

ORGANOMETALLICS BY FAB

Grotjahn, L., Koppenhagen, V. B., and Ernst, L., *Z. Naturforsch.* **39b**, 248 (1984).
Livingston, D. A., Pfaltz, A., Schreiber, J., Eschenmoser, A., Ankel-Fuchs, D., Moll, J., Jaenchen, R., and Thauer, R. K., *Helv. Chim. Acta* **67**, 334 (1984).
Gregory, B., Jablonski, C. R., and Wang, Y-P., *J. Organomet. Chem.* **269**, 75 (1984).
Tolun, E., Proctor, C. J., Todd, J. F. J., Walshe, J. M. A., and Connor, J. A., *Org. Mass Spectrom.* **19**, 294 (1984).
Tolun, E., Proctor, C. J., Heaton, B. T., and Todd, J. F. J., 13th *Br. Mass Spectrom. Symp.* (1983).

COORDINATION COMPOUNDS BY FAB

Unger, S. E., *Anal. Chem.* **56,** 363 (1984).
Costello, C. E., Brodack, J. W., Jones, A. G., Davison, A., Johnson, D. L., Kasina, S., and Fritzberg, A. R., *J. Nucl. Med.* **24,** 353 (1983).

OTHER APPLICATIONS OF FAB

Baldwin, M. A., Proctor, C. J., Amster, I. J., and McLafferty, F. W., *Int. J. Mass Spectrom. Ion Proc.* **54,** 97 (1983).
Smith, D. L., *Anal. Chem.* **55,** 2391 (1983).
Lyon, P. A., Stebbings, W. L., Crow, F. W., Tomer, K. B., Lippstreu, D. L., and Gross, M. L., *Anal. Chem.* **56,** 8 (1984).
Dolnikowski, G. G., Watson, J. T., and Allison, J., *Anal. Chem.* **56,** 197 (1984).

SIMS

Cox, III, X. B., Linton, R. W., and Bursey, M. M., *Int. J. Mass Spectrom. Ion Proc.* **55,** 281 (1983/1984).
Suib, S. L., Coughlin, D. F., Otter, F. A., and Conopask, L. F., *J. Catal.* **84,** 410 (1983).
Long, J. V. P., and Hinton, R. W., *Int. J. Mass Spectrom. Ion Proc.* **55,** 307 (1983/1984).
Ramseyer, G. O., Brenna, J. T., Morrison, G. H., and Schwartz, R., *Anal. Chem.* **56,** 402 (1984).
Pimminger, M., Grasserbauer, M., Schroll, E., and Cerny, I., *Anal. Chem.* **56,** 407 (1984).

MISCELLANEOUS

Castro, M. E., and Russel, D. H., *Anal. Chem.* **56,** 578 (1984).
Schueler, B., Feigl, P. K. D., and Krueger, F. R., *Z. Naturforsch.* **38a,** 1078 (1983).
Michiels, E., and Gijbels, R., *Mikrochim. Acta* **III,** 277 (1983).
Chait, B. T., and Field, F. H., *J. Am. Chem. Soc.* **106,** 1931 (1984).
Blumenthal, T., Bruce, M. I., Shawataly, U. B., Green, B. N., and Lewis, I., *J. Organomet. Chem.* **269,** C10 (1984).
Lehmann, W. D., Kessler, M., and Konig, W. A. *Biomed. Mass Spectrom.* **11,** 217 (1984).

THE CHEMISTRY OF BERKELIUM

J. R. PETERSON* and D. E. HOBART**

*Department of Chemistry, The University of Tennessee, Knoxville, Tennessee,
*Transuranium Research Laboratory, Oak Ridge National Laboratory, Oak Ridge,
Tennessee, and **Isotope and Nuclear Chemistry Division, Los Alamos National
Laboratory, Los Alamos, New Mexico

I. Introduction	29
II. Nuclear Properties, Availability, and Applications	30
III. Separation and Purification	32
IV. Atomic Properties	34
A. Electronic Energies	34
B. Emission Spectra	35
C. Solution Absorption Spectra	36
D. Solid State Absorption and Raman Spectra	38
V. Metallic State	41
A. Preparation	41
B. Physical Properties	42
C. Chemical Properties	45
D. Theoretical Treatment	46
VI. Compounds	47
A. General Summary	47
B. Oxides	50
C. Halides and Oxyhalides	51
D. Pnictides, Chalcogenides, and Other Compounds	53
E. Magnetic Behavior of Berkelium Ions	54
VII. Solution Chemistry	55
A. Ionic Species	55
B. Thermodynamic Quantities	56
C. Stability Constants and Other Properties	56
D. Oxidation–Reduction Behavior and Potentials	59
VIII. Concluding Remarks	63
References	64

I. Introduction

As was the case for the previously discovered transuranium elements, element 97 was first produced via a nuclear bombardment reaction. In December, 1949, ion-exchange separation of the products formed by the bombardment of ^{241}Am with accelerated α particles provided a new electron-capture activity eluting just ahead of curium (1, 2). This activity was assigned to an isotope (mass number 243) of element 97. The new element was named berkelium after Berkeley, California, the city of its discovery, in a manner parallel to the naming of its lanthanide analog terbium after Ytterby, Sweden. The initial investigations of the chemical properties of berkelium were limited to tracer experiments (ion exchange and coprecipitation), but these were sufficient to establish the stability of Bk(III) and the accessibility of Bk(IV) in aqueous solution and to estimate the electrochemical potential of the Bk(IV)–Bk(III) couple (2, 3).

Since a complete study of the chemistry of an element is not possible by tracer methods alone, a program for long-term neutron irradiation of about 8 g of ^{239}Pu was initiated in 1952 in the Materials Testing Reactor (Arco, Idaho) to provide macro amounts of berkelium (4). In 1958 about 0.6 μg of ^{249}Bk was separated, purified, and used in experiments to determine the absorption spectrum of Bk(III) in aqueous solution and to measure the magnetic susceptibility of Bk(III) (4). No Bk(III) absorption was observed over the wavelength range 450–750 nm, but an upper limit of about 20 was set for the molar absorptivity of any absorption by Bk(III) in this wavelength region. The magnetic susceptibility, measured from 77 to 298 K with the Bk(III) ions sorbed in a single bead of cation-exchange resin, was found to conform to the Curie–Weiss law with an effective moment of 8.7 Bohr magnetons (μ_B), suggesting a 5f^8 electronic configuration for the Bk(III) ion.

The first structure determination of a compound of berkelium, the dioxide, was carried out in 1962 (5). Four X-ray diffraction lines were obtained from 4 ng of BkO$_2$ and indexed on the basis of a face-centered cubic structure with $a_0 = 0.533 \pm 0.001$ nm.

II. Nuclear Properties, Availability, and Applications

Selected nuclear properties of the principal isotopes of berkelium are listed in Table I (6). In addition to these isotopes, ranging from mass numbers 240 to 251, there are spontaneously fissioning isomers known for berkelium mass numbers 242, 243, 244, and 245, all with half-lives of less than 1 μsec. Only ^{249}Bk is available in bulk quantities for chemical studies, as a result of prolonged neutron irradiation of Pu, Am, or Cm (7). About 0.66 g of this isotope has been isolated from

TABLE I

NUCLEAR PROPERTIES OF BERKELIUM ISOTOPES (6)

Mass number	Half-life	Decay mode[a]	Decay energy[b]	Ground-state spin, parity[c]	Representative production method
240	5.0 min	EC			^{232}Th(^{14}N,6n)
242	7.0 min	EC			^{235}U(^{11}B,4n)
243	4.5 hr	EC(99.85%)		(3/2−)	^{241}Am(α,2n)
		α(0.15%)	6.758 g, 6.574		
244	4.35 hr	EC(99+%)			^{243}Am(α,3n)
		α(0.006%)	6.667, 6.625		
245	4.90 days	EC(99.88%)		3/2−	^{243}Am(α,2n)
		α(0.12%)	6.349 g, 6.145		
246	1.80 days	EC		(2−)	^{244}Cm(α,pn)
247	1380 years	α	5.794 g, 5.531	(3/2−)	^{244}Cm(α,p)
					^{247}Cf$\xrightarrow{EC}$
248	23.5 hr	β^-(70%)	0.65 g	(1−)	^{247}Bk(n,γ)
		EC(30%)			
248	>9 years			(6+)	^{246}Cm(α,pn)
	>1 × 10^4 years (β^-)				
249	325 days	β^-(99+%)	0.125 g	7/2+	^{248}Cm(n,γ)$\xrightarrow{\beta^-}$
	1.87 × 10^9 years (SF)	α(0.00145%)	5.437 g, 5.417		
250	3.22 hr	β^-	1.76 g, 0.725	2−	^{249}Bk(n,γ)
					^{254}Es$\xrightarrow{\alpha}$
251	55.6 min	β^-	~1.0, ~0.5	(3/2−)	^{255}Es$\xrightarrow{\alpha}$

[a] EC, Electron capture; $\alpha = {}_2^4\text{He}^{2+}$; $\beta^- = {}_{-1}^{0}e^1$; percentage of decays via particular mode given in parentheses.

[b] Energy in MeV of ground state (g) and most intense transitions.

[c] Values in parentheses are tentative.

target rods irradiated with neutrons in the High Flux Isotope Reactor (Oak Ridge, Tennessee) over the period 1967 through 1983 (8–10). The relative atomic mass of berkelium-249 is given as 249.075 by the International Union of Pure and Applied Chemistry (IUPAC) (11).

Besides the research use of ^{249}Bk for the characterization of the chemical and physical properties of element 97, its relatively rapid decay to ^{249}Cf (0.2% per day) makes it a valuable source of this important isotope of californium for chemical study. This genetic relationship has been exploited in studies of the chemical consequences of beta (β^-) decay in the bulk-phase solid state (12, 13).

There have been no reports of practical applications for any of the isotopes of berkelium.

III. Separation and Purification

Berkelium may be purified by many methods that are also applicable to other actinide elements. Therefore, only those methods that specifically apply to berkelium separation and purification will be treated here.

Since berkelium can be readily oxidized to Bk(IV), it can be separated from other, nonoxidizable transplutonium elements by combining oxidation–reduction (redox) methods with other separation techniques. The first application of this approach was performed by oxidizing Bk(III) with BrO_3^- in nitric acid solution (14). The resultant Bk(IV) was then extracted with hydrogen di(2-ethylhexyl)orthophosphoric acid (HDEHP) in heptane followed by back-extraction with nitric acid containing H_2O_2 as a reducing agent. In addition to other reports of the use of BrO_3^- as an oxidizing agent in berkelium purification procedures (15–19), the use of CrO_4^{2-} (16, 20), $Cr_2O_7^{2-}$ (20–22), $Ag(I)-S_2O_8^{2-}$ (20, 23), PbO_2 (21, 24, 25), BiO_3^- (21), O_3 (24), and photochemical oxidation (24) has also been reported. Separation of the oxidized berkelium has been accomplished by the use of (1) liquid–liquid extraction with HDEHP (14, 16, 19, 26), trioctylphosphine oxide (27), alkylpyrocatechol (28), 2-thenoyltrifluoroacetone (22), primary, tertiary, or quaternary amines (20, 23, 29–31), or tributyl phosphate (32, 33); (2) extraction chromatography with HDEHP (18, 34–36), zirconium phosphate adsorbant (21, 24, 25); (3) precipitation of the iodate (15, 17); or (4) ion exchange (18, 37, 38), applied separately or in combination with one another.

The purification procedures outlined above provide separation of berkelium from all trivalent lanthanides and actinides with the notable exception of cerium. Since berkelium and cerium exhibit nearly identical redox behavior, most redox separation procedures include a Bk–Ce separation step (21, 27, 37–41). Separation of Bk(III) from Ce(III) and other trivalent lanthanide and actinide elements can also be accomplished without the use of redox procedures (37, 39–47).

Personnel in the Transuranium Processing Plant (TRU) at the Oak Ridge National Laboratory have isolated and purified 0.66 g of ^{249}Bk during the period 1967 through 1983 (9, 10), using the procedure outlined in Fig. 1 (48). The transcurium elements, partitioned by LiCl-based anion exchange, are precipitated as hydroxides, filtered, and dissolved in nitric acid. Initial isolation is accomplished by high-pressure elution from cation-exchange resins with α-hydroxyisobutyrate (BUT) solution. The berkelium fraction is oxidized and extracted into HDEHP–dodecane from HNO_3–$NaBrO_3$ solution. The organic fraction

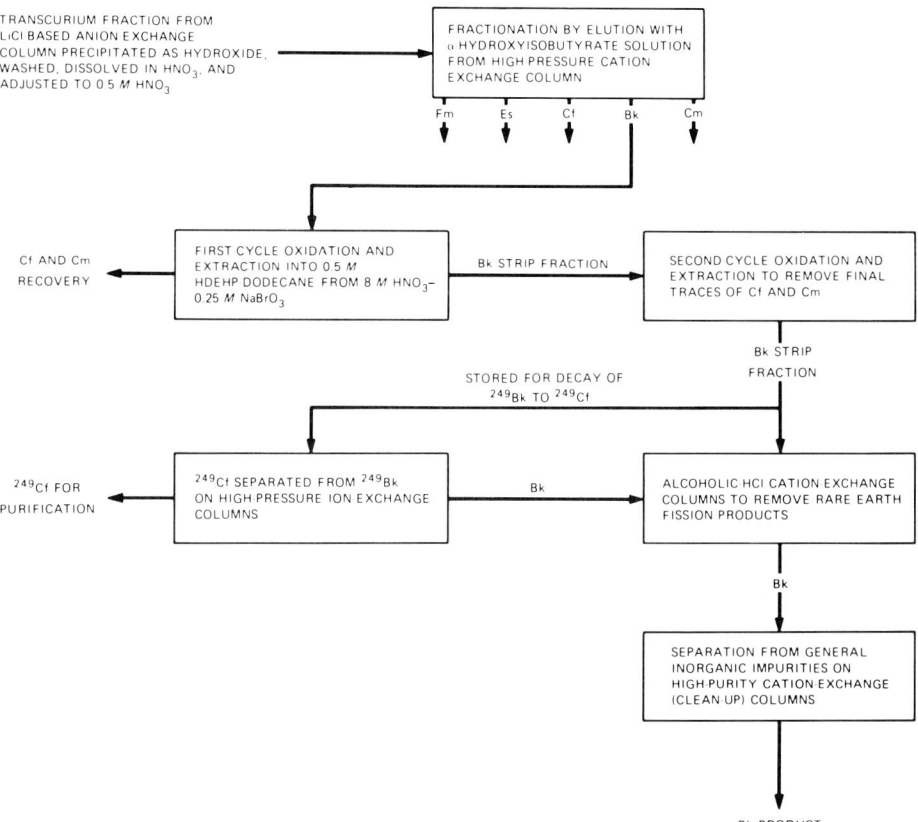

FIG. 1. Schematic diagram of procedures used in the final isolation and purification of berkelium in the Transuranium Processing Plant (TRU) at the Oak Ridge National Laboratory (ORNL). Adapted from (48).

containing Bk(IV) is treated with 2,5-di-*tert*-butylhydroquinone (DBHQ) to reduce the Bk(IV) to Bk(III) before back-extracting (stripping) it into HNO_3–H_2O_2 solution. Then another oxidation–extraction and another reduction–back-extraction cycle are carried out. The solution at this point is radiochemically pure except for fission-product cerium. After solvent cleanup and evaporation to dryness, the berkelium is dissolved in 0.1 M HCl for final ion-exchange purification steps including alcoholic–HCl elution from cation-exchange resin and cation cleanup columns (48).

An innovative procedure for the rapid separation of berkelium from other actinides, lanthanides, and fission products has been reported

(*49*). Expediency was required in order to measure the decay properties of short-lived isotopes. Berkelium and cerium were first separated from the other elements via solvent extraction of Bk(IV) and Ce(IV), using HDEHP in the usual way. Then the two-element mixture [as Bk(III) and Ce(III)] was separated in the heated cation-exchange column of a high-pressure liquid chromatograph (HPLC), using α-hydroxyisobutyrate as the eluant. The total separation time for berkelium from the initial complex mixture was about 8 min, half of the time being required for the operation of the HPLC column (*49*). Procedures for fast separations of berkelium from beryllium foil targets and gold catcher foils resulting from nuclear bombardment reactions have also been published (*50*).

For additional discussion of berkelium separation procedures, the reader is referred to several reviews and comprehensive texts on the subject (*51–58*).

IV. Atomic Properties

A. Electronic Energies

The ionization potential of neutral berkelium ($5f^9 7s^2$) has been derived from spectroscopic data to be 6.229 ± 0.025 eV (*59*). The changes in entropy associated with the stepwise ionization of gaseous berkelium atoms have also been calculated (*60*). The energy interval between the ground ($^7H_8$) and first excited level ($^5H_7$) of singly ionized berkelium was determined to be 1.48752×10^5 m^{-1} from measurements on plates taken with a high-resolution emission spectrograph (*61*). Several authors have calculated the energies of, and energy intervals between, the lowest-lying levels of the various electronic configurations of neutral berkelium (*62–65*) and of singly, doubly, and triply ionized berkelium (*65, 66*).

From measurements of the energies of a number of internal conversion lines in berkelium-249 (produced by the α decay of einsteinium-253), the atomic electron binding energies in berkelium were calculated for the K through O shells (*67*). The K-series X-ray energies and intensities of berkelium were later measured, and the K-shell electron binding energy calculated (*68*). The measured energies and relative transition probabilities agreed well with theoretical predictions (*69, 70*). A crystal spectrometer has been used to measure precisely the berkelium K-series X-ray energies, natural line widths, and relative intensities (*71*).

Also available are the results of relativistic relaxed-orbital *ab initio* calculations of L-shell Coster–Kronig transition energies for all possible transitions in berkelium atoms (*72*), relativistic relaxed-orbital Hartree–Fock–Slater calculations of the neutral-atom electron binding energies in berkelium (*73*), refined K- and L-shell electron binding energies for berkelium (*74*), calculations and compilations of the electron binding energies in berkelium (*75, 76*), and single-zeta, double-zeta, and extended (triple-zeta-valence) Slater-type basis sets for berkelium in the $5f^86d7s^2$ configuration (*77*).

Relativistic Hartree–Slater values of the X-ray emission rates for the filling of K- and L-shell vacancies in berkelium have been tabulated (*78*). X-ray emission rates for the filling of all possible single inner-shell vacancies in berkelium by electric dipole transitions have been calculated, using nonrelativistic Hartree–Slater wavefunctions (*79*).

B. EMISSION SPECTRA

Twenty emission lines, produced from 0.2 µg of berkelium in a high-voltage spark, were reported in 1965 (*80*). In 1967 between 3000 and 5000 lines were recorded in the wavelength region 250–900 nm from 38 µg of ^{249}Bk in an electrodeless discharge lamp (*81*). Many of the emission lines exhibited well-resolved eight-component hyperfine structure, which established the nuclear spin of ^{249}Bk to be 7/2 (*81*). This value is in agreement with that derived from nuclear decay systematics.

The ground-state electronic configurations (levels) of neutral and singly ionized berkelium were identified as $5f^97s^2$ ($^6H_{15/2}$) and $5f^97s^1$ ($^7H_8$), respectively (*82*). A nuclear magnetic dipole moment of 1.5 nuclear magnetons (*61*) and a quadrupole moment of 4.7 barns (*83*) were determined for ^{249}Bk, based on analysis of the hyperfine structure in the berkelium emission spectrum.

The wave numbers, wavelengths, and relative intensities of 1930 of the stronger emission lines from ^{249}Bk in the 254- to 980-nm wavelength region are available (*84*). The infrared emission spectrum of ^{249}Bk from 830 to 2700 nm has been recorded (*85*). An interpretation of 8 levels in the Bk(I) spectrum ($5f^97s^2$, neutral Bk) and 10 levels in the Bk(II) spectrum ($5f^97s$, Bk$^+$) has been made on the basis of a generalized parametric study of the $5f^N$ and $5f^N7s$ electron configurations of the actinides (*86*).

A preliminary report on the self-luminescence of ^{249}Bk(III) in a LaCl$_3$ host lattice was published in 1963 (*87*), and the self-luminescence spectra of ^{249}Bk-doped BaF_2 and $SrCl_2$ were reported in 1978 (*88*). The absence of ultraviolet-excited sharp-line sensitized luminescence of ^{249}Bk-doped gadolinium hexafluoroacetylacetonate has been noted (*89, 90*). Such luminescence was absent also in cesium berkelium hexafluoroacetylacetonate chelate in anhydrous ethanol (*89*). A study of Bk^{3+} fluorescence in H_2O and D_2O solutions has been reported, and a basis for assessing the use of fluorescence detection for transuranic ions has been established (*91*).

C. SOLUTION ABSORPTION SPECTRA

The first attempts to measure the absorption spectrum of Bk(III) involved the use of a single ion-exchange resin bead (*4*). Later the spectrum of a 3.6×10^{-3} *M* Bk(III) solution was recorded in a microcell (*92*). Sixteen absorption bands of Bk(III) were identified in the solution spectrum recorded in a "suspended drop" microcell over the wavelength range 320–680 nm (*93*). The results of additional observations identified a total of 23 absorption bands in the 280- to 1500-nm wavelength region (*94*).

The first attempts to record the Bk(IV) solution absorption spectrum were hindered by the presence of cerium impurities (*92*). The positions of the Bk(IV) absorption bands, superimposed on the strong Ce(IV) bands, suggested the assignment of $5f^7$ for the electronic configuration of Bk(IV), in agreement with the actinide hypothesis.

In later work, the absorption spectra of Bk(III) and Bk(IV) were recorded in various media (*95*). New absorption bands were reported as the result of using larger quantities of berkelium-249 of higher purity than had been previously available. Observations of the spectrum of Bk(III) were extended further into the ultraviolet wavelength region (to 200 nm), and nine new absorption bands were reported (*96*). Later the absorption spectra of Bk(III) and Bk(IV) in 2 *M* perchloric and 0.5 *M* nitric acid solutions have been obtained (*97*). An interpretation of the low-energy bands in the solution absorption spectra of Bk(III) and Bk(IV) has been published (*98*).

Solution absorption spectra of Bk(III) and Bk(IV) are shown in Figs. 2 and 3, respectively. The spectrum of Bk(III) is characterized by sharp absorption bands of low molar absorptivity attributed to "Laporte-forbidden" f–f transitions and by intense absorption bands in the ultraviolet region, which are attributed to f–d transitions (*96*). The spectrum of Bk(IV) is dominated by a strong absorption band at 250–290 nm, the peak position of which is strongly dependent on the degree of complexation of Bk(IV) by the solvent medium. This band is attributed to a charge-transfer mechanism (*96*).

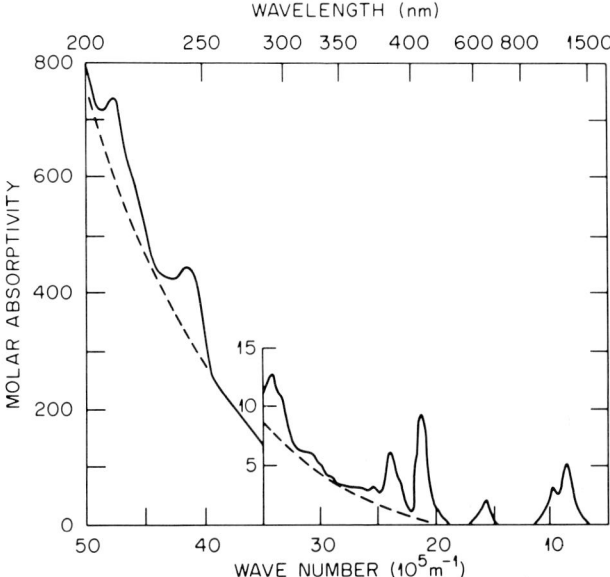

FIG. 2. Solution absorption spectrum of Bk(III) in 1 M DClO$_4$–99.9% D$_2$O. Reprinted with permission from the *J. Inorg. Nucl. Chem.* **34,** R. D. Baybarz, J. R. Stokely, and J. R. Peterson, 1972, Pergamon Press, Ltd. (*95*).

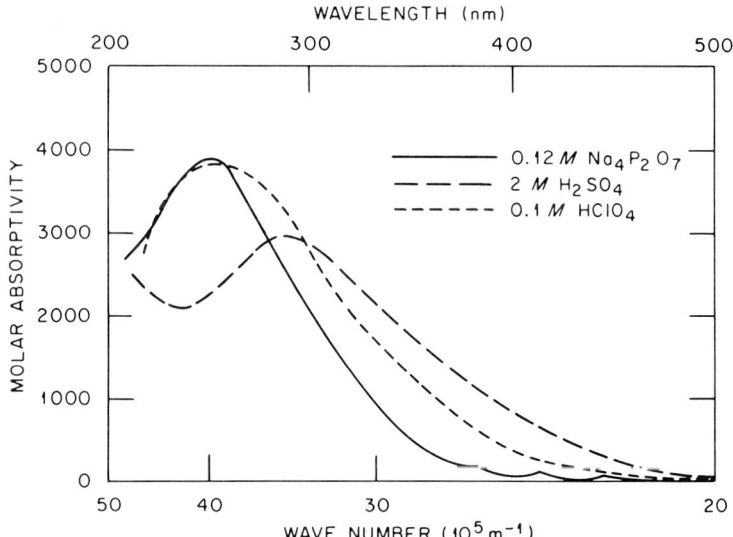

FIG. 3. Solution absorption spectra of Bk(IV) in various aqueous media. Reprinted with permission from the *J. Inorg. Nucl. Chem.* **35,** R. G. Gutmacher, D. D. Bodé, R. W. Lougheed, and E. K. Hulet, 1973, Pergamon Press, Ltd. (*96*).

Electronic spectra of Bk(III) (*99*) and Bk(IV) (*100*) and a prediction of the electronic spectrum of Bk(II) (*101*) have been published. Spin-orbit coupling diagrams for these berkelium ions, based on a free-ion interpretation of the f–f spectra, were proposed.

D. SOLID STATE ABSORPTION AND RAMAN SPECTRA

The absorption spectrum of Bk(III) in a lanthanum chloride host matrix at 77 K was first obtained in 1964 (*102*). A prediction of the energy level structure of Bk(III) was made by others the same year (*103*). Extensive, low-temperature spectroscopic studies of $BkCl_3$ showed the absence of transitions to excited $J = 0$ and $J = 1$ states (*104, 105*). This provided good evidence for a $\mu = 0$ ground level for Bk(III), consistent with that of Tb(III)–$LaCl_3$ (*106*). Experimental and theoretical studies of the crystal field parameters of Bk(III) in a $LaCl_3$ host lattice have also been reported (*107*).

Microscale spectrophotometric techniques, using 0.5–10 μg-sized berkelium samples, have been applied for identification and characterization of berkelium halides and oxyhalides (*108*). Spectra of orthorhombic and monoclinic $BkBr_3$, trigonal and orthorhombic BkF_3 (*13*), and monoclinic BkF_4 (*13*) are shown in Figs. 4–6, respectively. The coordination number of Bk(III) in the $PuBr_3$-type orthorhombic modification of $BkBr_3$ is eight, whereas it is six in the $AlCl_3$-type monoclinic form. The two spectra are readily distinguished by their differing fine structure and relative peak intensities. The more intense f–f transitions in orthorhombic $BkBr_3$ are expected because of the lower coordination site symmetry of Bk(III) in this form compared to that of Bk(III) in the monoclinic form of $BkBr_3$. This difference in absorption intensities was exploited in a study of the solid state transformation between the two forms of $BkBr_3$ (*109*).

The spectra of dimorphic BkF_3 (see Fig. 5) indicate the sensitivity of solid state absorption spectrophotometric analysis, since both crystallographic modifications exhibit a coordination number of nine for Bk(III). Small but reproducible changes are apparent in the relative absorbance and fine structure of the bands at 440, 475, 625, and 1050 nm. The less centrosymmetric, orthorhombic BkF_3 exhibits f–f transition intensities greater than those exhibited by the more centrosymmetric, trigonal BkF_3 (*13*). The absorption spectrum of BkF_4 (see Fig. 6) is characterized by four sharp peaks between 415 and 455 nm and major absorption bands at 562 and 592–617 nm (*13*). This solid-state spectrum is similar to, but exhibits better resolution than, the solution absorption spectrum of Bk(IV) in phosphate medium (*95*).

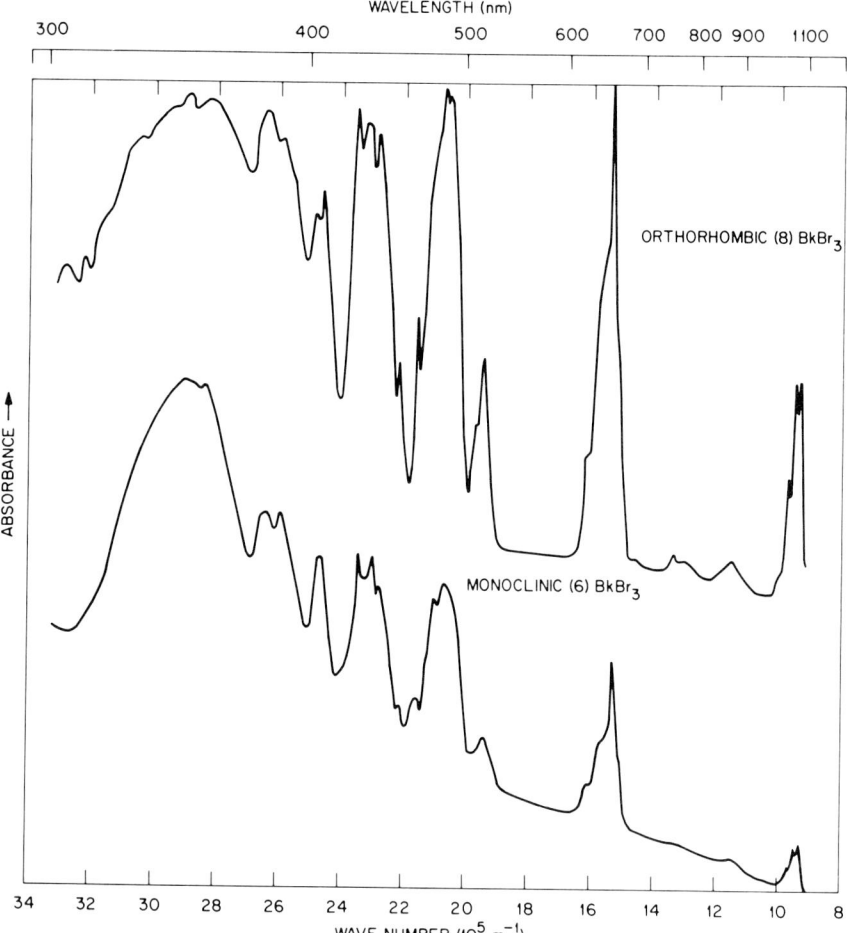

FIG. 4. Solid state absorption spectra of $BkBr_3$ in its orthorhombic and monoclinic modifications at room temperature.

This absorption spectrophotometric technique has also been applied to the study of the chemical consequences of radioactive decay in bulk-phase solid state samples (12, 110–112). It was found that the ^{249}Cf daughters growing into crystalline $^{249}BkBr_3$ and $^{249}BkCl_3$ exhibited the same oxidation state and crystal structure as their respective berkelium parents (12, 112).

The absorption spectra of Bk(III) and Bk(IV) hydroxides as suspensions in 1 M NaOH have been reported (113). The solid state absorp-

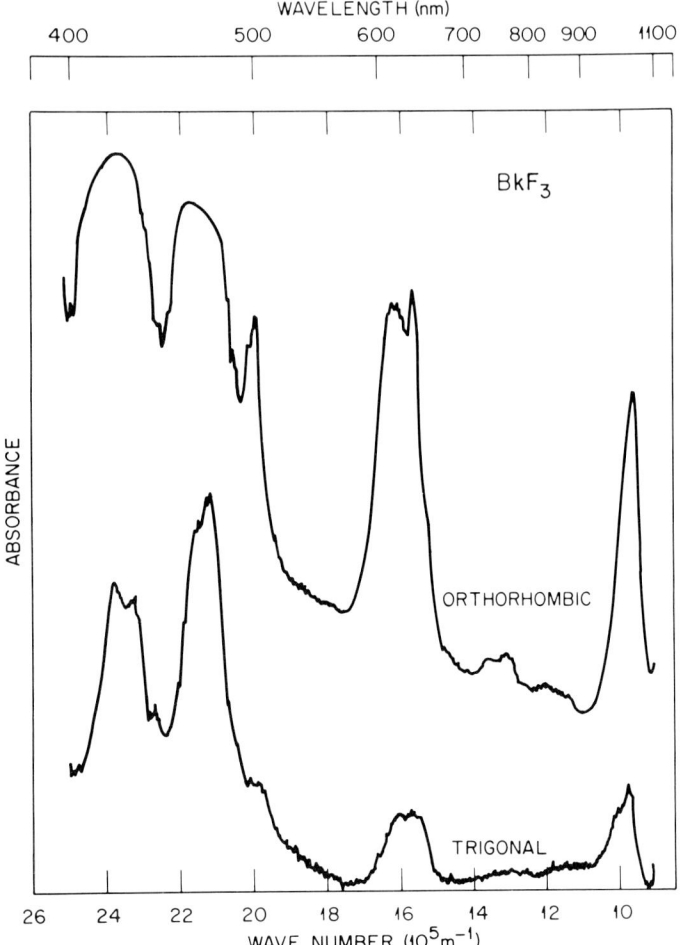

FIG. 5. Solid-state absorption spectra of BkF_3 in its orthorhombic and trigonal forms at room temperature. Reprinted with permission from the *J. Inorg. Nucl. Chem.* **43**, D. D. Ensor, J. R. Peterson, R. G. Haire, and J. P. Young, 1981, Pergamon Press, Ltd. (*13*).

tion (*111*) and Raman (*115*) spectra of berkelium(III) orthophosphate have been obtained. It was found that $BkPO_4$ fluoresced strongly when excited by the 514.5-nm emission of an argon-ion laser. By lowering the laser intensity below the fluorescence threshold, it was possible to obtain the Raman spectrum, the bands of which were assigned by direct analogy to those of the monazite-type monoclinic lanthanide orthophosphates (*115*).

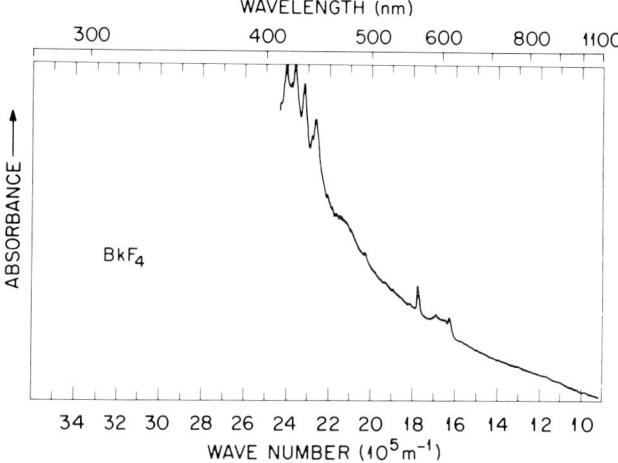

FIG. 6. Solid-state absorption spectrum of monoclinic BkF_4 at room temperature. Reprinted with permission from the *J. Inorg. Nucl. Chem.* **43**, D. D. Ensor, J. R. Peterson, R. G. Haire, and J. P. Young, 1981, Pergamon Press, Ltd. (*13*).

Line lists of the absorption bands of two organoberkelium compounds, $Bk(C_5H_5)_3$ (*116*) and $[Bk(C_5H_5)_2Cl]_2$ (*117*), have also been published. For additional information (*118*) and discussion of the development of the theoretical treatment of berkelium spectra, the reader is referred to other sources (*83, 106*).

V. Metallic State

A. PREPARATION

The first bulk samples of berkelium metal were prepared in early 1969 by the reduction at about 1300 K of BkF_3 with lithium metal vapor (*119*). The BkF_3 samples were suspended in a tungsten wire spiral above a charge of Li metal in a tantalum crucible. A photomicrograph of the first isolated bulk (1.7 μg) sample of berkelium metal is shown in Fig. 7.

Later berkelium metal samples of up to 0.5 mg each have been prepared via the same chemical procedure (*120*). Elemental berkelium can also be prepared by reduction of BkF_4 with lithium metal and by reduction of BkO_2 with either thorium or lanthanum metal. The latter reduction process is better suited to the preparation of thin metal foils unless multimilligram quantities of berkelium are available.

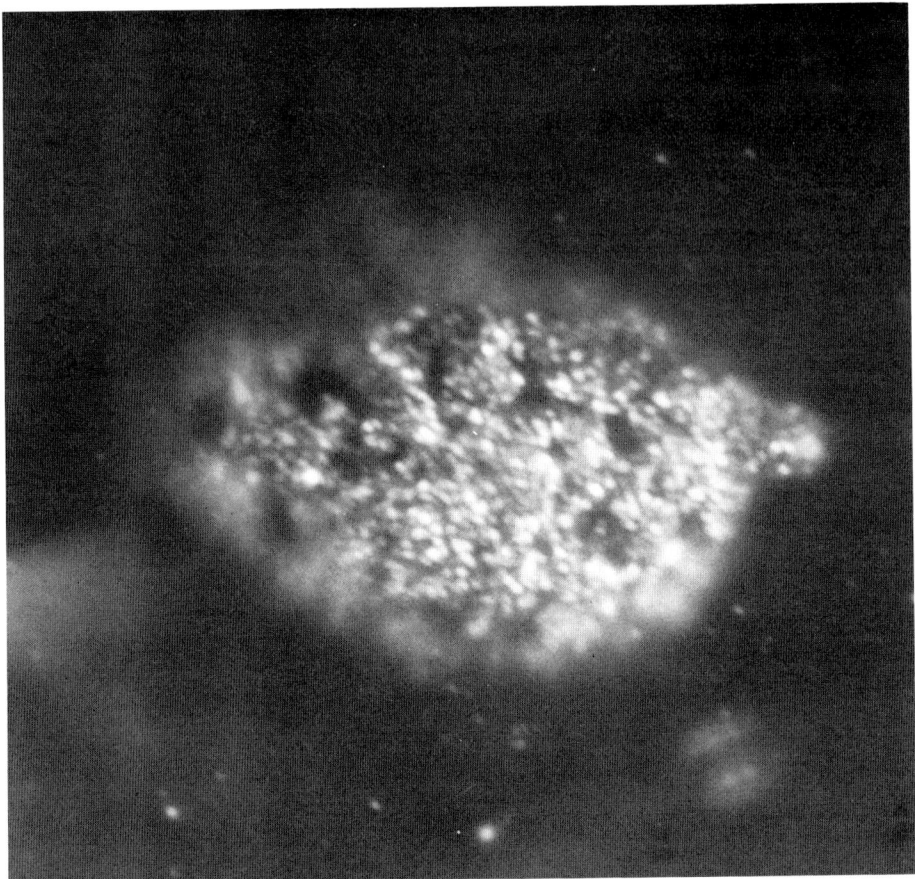

Fig. 7. A photomicrograph of the first isolated bulk (1.7 μg) sample of berkelium metal in a quartz X-ray capillary.

B. Physical Properties

Berkelium metal exhibits two stable crystallographic modifications, double hexagonal closest packed (dhcp) and face-centered cubic (fcc). Thus it is isostructural with the two preceding elements, all of which exhibit the fcc structure at high temperature. The room-temperature lattice constants of the dhcp form are $a_0 = 0.3416 \pm 0.0003$ nm and $c_0 = 1.1069 \pm 0.0007$ nm, yielding a calculated density of 1.478×10^4 kg/m^3 and a metallic radius (CN = 12) of 0.170 nm (119). The room-temperature fcc lattice parameter is $a_0 = 0.4997 \pm 0.0004$ nm from which the

X-ray density and the metallic radius (CN = 12) are calculated to be 1.325×10^4 kg/m^3 and 0.177 nm, respectively (119). The metallic radius of berkelium, assuming a metallic valence of 3- and 12-fold coordination, has been calculated to be 0.1739 nm (121). On the other hand, the radii (CN = 12) of berkelium were predicted to be 0.184 nm for trivalent metal and 0.1704 nm for tetravalent metal, so that the observed dhcp form would correspond to tetravalent metal, while the fcc form would represent a metallic valence of 3.5 (122).

Although berkelium metal is dimorphic, the transformation temperature is now known with certainty. A change in the appearance of Bk metal samples at 1203 ± 30 K during the course of two melting point determinations might correspond to the dhcp → fcc phase transformation, which should be accompanied by a 12% change in the volume of the sample (123). The melting point of berkelium metal has been determined to be 1259 ± 25 K from measurements on two samples (123). The melting and boiling points of elemental berkelium have been estimated to be 1323 ± 50 and 2900 ± 50 K, respectively (124).

The first data on berkelium metal under pressure were obtained with a diamond anvil pressure cell, using silicone oil as the pressure medium, ruby fluorescence to determine the applied pressure, and energy-dispersive X-ray powder diffraction analysis (125). Three metallic phases were observed with increasing pressure to 57 GPa: the normal-pressure dhcp modification was transformed to fcc at about 8 GPa; this fcc form was transformed to the α-uranium-type orthorhombic structure at ~22 GPa (125). Accompanying this latter transition was a shrinkage in volume of ~12% (see Fig. 8), which was associated with delocalization of the 5f electrons, i.e., with the onset of 5f-electron participation in the metallic bonding in berkelium (126). A bulk modulus of 30 ± 10 GPa was estimated for berkelium metal below 22 GPa (125). Thus berkelium metal under pressure behaves similarly to americium, californium, and some of the light lanthanide metals and does not appear to undergo an isostructural phase transition (corresponding to a change in metallic valence from three to four) before delocalization of its 5f electrons (127).

In the first experiments to measure the vapor pressure of metallic Bk, using Knudsen effusion target-collection techniques, the preliminary data were fitted with a least-squares line to give a provisional vaporization equation for the temperature range 1326-1582 K, and ΔH°_{298} was calculated to be 382 ± 18 kJ/mol (128). The crystal entropy of berkelium metal at 298 K (S°_{298}) had been estimated earlier to be 76.2 ± 1.3 J K^{-1} mol^{-1} (129), and later, to be 78.2 ± 1.3 J K^{-1} mol^{-1} (124).

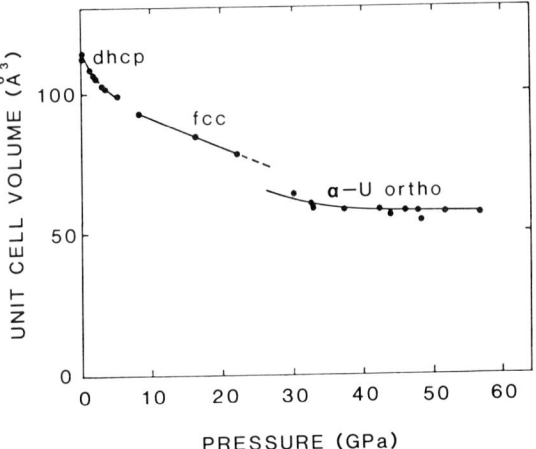

FIG. 8. Unit-cell volume of berkelium metal as a function of pressure. Adapted and reprinted with permission from the *J. Phys. F: Met. Phys.* **14**, U. Benedict, J. R. Peterson, R. G. Haire, and C. Dufour, 1984, The Institute of Physics, London (*126*).

Later data on the vapor pressure of berkelium metal over the temperature range 1100–1500 K, obtained by using combined Knudsen effusion mass spectrometric and target collection techniques, have been published in 1982 (*124*). The vaporization equations obtained were

$$\log P(\text{atm}) = (5.78 \pm 0.21) - (15{,}718 \pm 253)/T$$

for solid Bk between 1107 and 1319 K and

$$\log P(\text{atm}) = (5.14 \pm 0.17) - (14{,}902 \pm 244)/T$$

for liquid Bk between 1345 and 1528 K. The enthalpy of fusion was calculated to be 7.92 kJ/mol, and the enthalpy associated with the dhcp → fcc phase transition was calculated to be 3.66 kJ/mol (*124*). An average of the second and third law data yielded 310 ± 6 kJ/mol for its enthalpy of vaporization, $\Delta H°_{298}$ (*124*). Correlation systematics have suggested that the standard enthalpy of sublimation of berkelium metal [$\Delta H°_f \text{Bk(g)}$] is 280 kJ/mol and that the standard enthalpy of formation of aqueous Bk(III) [$\Delta H°_f \text{Bk}^{3+}(\text{aq})$] is −615 kJ/mol (*130, 131*). A later modification of the systematics (*130*) led to values of 320 ±

8 kJ/mol and -590 ± 21 kJ/mol for $\Delta H_f^\circ \text{Bk(g)}$ and $\Delta H_f^\circ \text{Bk}^{3+}\text{(aq)}$, respectively (132).

The enthalpy of solution of Bk metal (dhcp) to $\text{Bk}^{3+}\text{(aq)}$ in 1 M HCl at 298 K was determined from five measurements to be -576 ± 25 kJ/mol (120). The error limits reported did not reflect the precision of the calorimetric measurements but rather the uncertainties in the purity of the berkelium metal. A new determination of the enthalpy of solution of Bk metal (dhcp) in 1 M HCl at 298.15 ± 0.05 K has yielded a value of -600.2 ± 5.1 kJ/mol (133). From this value $\Delta H_f^\circ \text{Bk}^{3+}\text{(aq)}$ was derived to be -601 ± 5 kJ/mol, and using reasonable entropy estimates, the standard potential of the Bk(III)–Bk(0) couple was calculated to be -2.01 ± 0.03 V.

Studies of the magnetic susceptibility of berkelium metal have been hampered by the difficulty in obtaining well-characterized, single-phase, bulk samples containing minimal amounts of daughter californium. Recent results obtained from a 21-μg sample of dhcp Bk metal ($\sim$12 atom % Cf) indicated a transition to antiferromagnetic behavior at about 34 K and paramagnetic behavior between 70 and 250 K (134). Applying the Curie–Weiss susceptibility relationship to the berkelium data obtained at fields greater than 0.08 T (where the field dependency was saturated) yielded $\mu_{\text{eff}} = 9.69$ μ_B and $\theta = 101.6$ K. The agreement of this value with the theoretical free-ion effective moment (9.72 μ_B) calculated for trivalent berkelium with L–S coupling suggests that dhcp Bk metal exhibits high-temperature magnetic behavior like its lanthanide homolog terbium. The results of earlier magnetic measurements on smaller samples of berkelium metal exhibiting mixed phases were reported by others (94, 135).

C. CHEMICAL PROPERTIES

During the handling of microgram-sized samples of berkelium metal, it was observed that the rate of oxidation in air at room temperature is not extremely rapid, possibly because of the formation of a "protective" oxide film on the metal surface (135). Berkelium is a chemically reactive metal, and berkelium hydride (123), some chalcogenides (123, 136, 137) and pnictides (138, 139) have been prepared directly from the reaction of Bk metal with the appropriate nonmetallic element.

Berkelium metal dissolves rapidly in aqueous mineral acids, liberating hydrogen gas and forming Bk(III) in solution (120, 133). Undoubtedly it forms alloys and/or intermetallic compounds with a number of other metals.

D. THEORETICAL TREATMENT

In 1972 a hybridized nondegenerate 6d and 5f virtual-bound-states model was used to describe the properties of the actinide metals, including berkelium (140). It accounted for the occurrence of localized magnetism in Bk metal. In 1974 a review of the understanding of the electronic properties of berkelium metal, as derived from electronic band theory, was published (141). Included were the relativistic energy-band structure of face-centered cubic Bk metal ($5f^86d^17s^2$) and the conclusion that berkelium is a rare-earth-like metal with localized (ionic) 5f electrons resulting from less hybridization with the 6d and 7s itinerant bands than occurs in the lighter actinides.

A phenomenological model based on crystal structure, metallic radius, melting point, and enthalpy of sublimation has been used to arrive at the electronic configuration of berkelium metal (142). An energy difference of 0.92 eV was calculated between the $5f^97s^2$ ground state and the $5f^86d^17s^2$ first excited state. The enthalpy of sublimation of trivalent Bk metal was calculated to be 2.99 eV (288 kJ/mol), reflecting the fact that berkelium metal is more volatile than curium metal. It was also concluded that the metallic valence of the face-centered cubic form of berkelium metal is less than that of the double hexagonal closest packed modification (142).

A relativistic Hartree–Fock–Wigner–Seitz band calculation has been performed for Bk metal in order to estimate the Coulomb term U (the energy required for a 5f electron to hop from one atomic site to an adjacent one) and the 5f-electron excitation energies (143). The results for berkelium in comparison to those for the lighter actinides show increasing localization of the 5f states, i.e., the magnitude of the Coulomb term U increases through the first half of the actinide series with a concomitant decrease in the width of the 5f level.

Predictions based on electronic structure calculations have been made for the pressures at which the 5f electrons in the heavier actinide metals will delocalize. For berkelium metal, such calculations indicated a delocalization pressure higher than 45 GPa (127). In addition, it was suggested that a change in metallic valence (corresponding to the change from a localized $5f^8$ to a localized $5f^7$ electron configuration) might occur at a pressure below 45 GPa. Such a change in metallic valence would be accompanied by a sudden and significant volume decrease and could possibly appear as an isostructural phase transition (127).

VI. Compounds

A. GENERAL SUMMARY

The trivalent oxidation state of berkelium prevails in the known berkelium compounds, although the tetravalent state is exhibited in BkO_2, BkF_4, and Cs_2BkCl_6. Selected crystallographic data for Bk metal and a number of berkelium compounds are collected in Table II. In cases where there have been multiple reports of lattice parameters for a particular compound, the later values or the ones considered more reliable by the present authors are given in Table II. The interested reader is encouraged to refer to the citations given in the table and text for complete details. An inherent difficulty, not addressed here, in the determination of lattice constants of "pure" berkelium-249 compounds concerns the ingrowth of daughter californium-249 at the rate of about 0.22% per day. Two experimental methods to address this problem are: (1) the determination of the lattice parameters of berkelium compounds as a function of californium content and then extrapolation to zero californium content; and (2) the utilization of Vegard's Law to correct measured berkelium lattice parameters for the presence of a known amount of californium (this assumes, of course, that the lattice parameters of the isostructural "pure" californium compound are known).

An empirical set of "effective" ionic radii in oxides and fluorides, taking into account the electronic spin state and coordination of both the cation and anion, have been calculated (*114*). For six-coordinate Bk(III), the radii values are 0.096 nm, based on a six-coordinate oxide ion radius of 0.140 nm, and 0.110 nm, based on a six-coordinate fluoride ion radius of 0.119 nm. For eight-coordinate Bk(IV), the corresponding values are 0.093 and 0.107 nm, based on the same anion radii (*114*). Other self-consistent sets of trivalent and tetravalent lanthanide and actinide ionic radii, based on isomorphous series of oxides (*145, 157*) and fluorides (*148, 157*), have been published. Based on a crystal radius for Cf(III), the ionic radius of isoelectronic Bk(II) was calculated to be 0.114 nm (*158*). It is important to note, however, that meaningful comparisons of ionic radii can be made only if the values compared are calculated in like fashion from the same type of compound, both with respect to composition and crystal structure.

The thermal decomposition of $Bk(NO_3)_3 \cdot 4H_2O$, $BkCl_3 \cdot 6H_2O$, $Bk_2(SO_4)_3 \cdot 12H_2O$, and $Bk_2(C_2O_4)_3 \cdot 4H_2O$ has been studied in air,

TABLE II

SELECTED CRYSTALLOGRAPHIC DATA FOR BERKELIUM METAL AND COMPOUNDS[a]

Substance	Structure type	Crystal system	Lattice parameters[a]				Other[b]	Reference
			a_0 (nm)	b_0 (nm)	c_0 (nm)	β (deg)		
Metal								
Bk	α-La	Hexagonal (dhcp)	0.3416		1.1069		ρ14.78 V27.96	119
Bk	Cu(A1)	Cubic (fcc)	0.4997				ρ13.25 V31.19	119
Oxides								
BkO_2	CaF_2	Cubic (fcc)	0.53315					144
Bk_2O_3	$(Fe, Mn)_2O_3$	Cubic (bcc)	1.0887					145
Bk_2O_3	Sm_2O_3	Monoclinic	1.4197	0.3606	0.8846	100.23		146
Bk_2O_3	La_2O_3	Hexagonal	0.3754		0.5958		V72.7	146
Halides								
BkF_4	UF_4	Monoclinic	1.2396	1.0466	0.8118	126.33	V70.7	147
BkF_3	LaF_3	Trigonal	0.697		0.714		ρ10.15	148
BkF_3	YF_3	Orthorhombic	0.670	0.709	0.441		ρ9.70	148
$BkCl_3$	UCl_3	Hexagonal	0.7382		0.4127			149
Cs_2BkCl_6	Rb_2MnF_6	Hexagonal	0.7451		1.2097		ρ4.155	150
$Cs_2NaBkCl_6$	$(NH_4)_3AlF_6$	Cubic (fcc)	1.0805				ρ3.952	150
$BkCl_3 \cdot 6H_2O$	$GdCl_3 \cdot 6H_2O$	Monoclinic	0.966	0.654	0.797	93.77	ρ3.06	151
$BkBr_3$	$PuBr_3$	Orthorhombic	0.403	1.271	0.912		V116.8	152
$BkBr_3$	$AlCl_3$	Monoclinic	0.723	1.253	0.683	110.6	V144.8	152
BkI_3	BiI_3	Hexagonal	0.7584		2.087			153

Oxyhalides					
BkOCl	PbFCl	Tetragonal	0.3966		0.6710	154
BkOBr	PbFCl	Tetragonal	0.395		0.81	155
BkOI	PbFCl	Tetragonal	0.3986		0.9149	153
Pnictides					
BkN	NaCl	Cubic (fcc)	0.4951			138
BkP	NaCl	Cubic (fcc)	0.5669			138
BkAs	NaCl	Cubic (fcc)	0.5829			138
BkSb	NaCl	Cubic (fcc)	0.6191			138
Chalcogenides					
BkS_{2-x}	Anti-Fe_2As	Tetragonal	0.3902		0.792	137
Bk_2S_3	Deficit Th_3P_4	Cubic (bcc)	0.8358			137
$BkSe_{2-x}$	Anti-Fe_2As	Tetragonal	0.404		0.828	137
Bk_2Se_3	Deficit Th_3P_4	Cubic (bcc)	0.8712			137
$BkTe_3$	$NdTe_3$	Orthorhombic	0.4318	0.4319	2.5467	136
						137
$BkTe_{2-x}$	Anti-Fe_2As	Tetragonal	0.4314		0.8945	137
Bk_2Te_3	Sc_2S_3	Orthorhombic	1.226	0.8685	2.605	137
Miscellaneous					
BkH_{2+x}	CaF_2	Cubic (fcc)	0.523			$\rho9.45$	123
$Bk_2O_2SO_4$	$La_2O_2SO_4$	Orthorhombic	0.4195	0.4083	1.3110	156
Bk_2O_2S	Pu_2O_2S	Trigonal	0.3861		0.6686	156
$Bk(C_5H_5)_3$	$Pr(C_5H_5)_3$	Orthorhombic	1.411	1.755	0.963	$\rho2.47$	116
					V298	

[a] See original source for precision claimed on these room-temperature values and for information regarding sample purity.
[b] ρ = density in 10^3 kg/m^3(g/cm^3); V = formula volume in 10^6 pm^3(Å^3).

argon, and H_2–Ar atmospheres and compared to that of the corresponding hydrates of cerium, gadolinium, and terbium (159). In air or Ar, the final berkelium product was BkO_2; in H_2–Ar, it was Bk_2O_3.

B. OXIDES

The first compound of berkelium identified on the basis of its characteristic X-ray powder diffraction pattern was BkO_2 (5). Other workers have since confirmed its CaF_2-type face-centered cubic structure with $a_0 = 0.533$ nm (144–146, 160, 161). The thermal expansion of BkO_2 in 1 atm oxygen was determined and shown to be reversible with temperature (144). The data were fitted by the expression $a_0(T) = 5.3304 + (4.32 \times 10^{-5})T + (15.00 \times 10^{-9})T^2$, where $a_0(T)$ is the unit cell edge in Å at temperature T in °C. In addition, the instantaneous expansion coefficients at 25°C (298 K) and 900°C (1173 K) were calculated to be 8.25×10^{-6}/°C and 13.2×10^{-6}/°C, respectively (144).

The results of a preliminary study of a sample of berkelium oxide (BkO_2, Bk_2O_3, or a mixture of the two) via X-ray photoelectron spectroscopy (XPS) included measured core- and valence-electron binding energies (162). The valence-band XPS spectrum, which was limited in resolution by photon broadening, was dominated by 5f-electron emission.

A capacitance manometer system was used to measure the equilibrium oxygen decomposition pressures over nonstoichiometric BkO_x ($1.5 < x < 2.0$) (163). Three broad nonstoichiometric phases were defined: $BkO_{1.50-1.77}$, $BkO_{1.81-1.91}$, and $BkO_{1.93-2.00}$. Later an X-ray diffraction investigation of this BkO_x system under equilibrium conditions was undertaken to correlate the above data with structural behavior (164). A phase diagram was suggested, showing above 673 K two widely nonstoichiometric phases: body-centered cubic for $1.5 < O/Bk < \sim 1.70$ and face-centered cubic for $\sim 1.78 < O/Bk < 2.00$. Interestingly, no evidence was found for the formation of Bk_7O_{12}, expected to exhibit a rhombohedral structure based on its common presence in other MO_x ($1.5 < x < 2.0$) systems.

This same BkO_2 sample (164) was used in a later study to investigate, under controlled conditions of temperature and oxygen activity, the redox behavior of the daughter californium-249 at compositions of $Bk_{0.76}Cf_{0.24}O_x$ and $Cf_{0.64}Bk_{0.36}O_x$ (165). Based on measured lattice parameters, the former ternary compound cooled in oxygen produced MO_2 and the latter produced $MO_{1.8}$, in contrast to the behavior of pure CfO_x, which produces $CfO_{1.71}$ (Cf_7O_{12}). Another observation was that oxidation of the californium to Cf(IV) or reduction of the californium to Cf(III) occurred at much lower temperatures in these ternary oxides than those in pure CfO_x (*165*).

The stable room-temperature form of berkelium sesquioxide exhibits the bixbyite-type body-centered cubic structure with $a_0 = 1.0887$ nm (*145*). This has been corroborated by an independent worker (*146, 160*). The cubic sesquioxide has also been analyzed by electron diffraction (*166*). The high-temperature behavior of Bk_2O_3 has been studied with the finding that the cubic-to-monoclinic transition at 1473 ± 50 K is irreversible, while the monoclinic-to-hexagonal transition at about 2025 K is reversible (*146*). In addition, the melting point of Bk_2O_3 was determined to be 2193 ± 25 K. Thus berkelium continues the trend of actinide sesquioxides exhibiting trimorphism: with increasing temperature, the structure of Bk_2O_3 changes from body-centered cubic (C form) to monoclinic (B form) to hexagonal (A form).

The possibility of the existence of BkO has been raised (*123*). The true identity of the brittle, gray material exhibiting a face-centered cubic structure with $a_0 = 0.4964$ nm is still in doubt. Not to be excluded from consideration is that this phase represents a nitride or an oxide nitride.

C. HALIDES AND OXYHALIDES

The only reported binary Bk(IV) halide is BkF_4 (*147, 167, 168*), prepared by fluorination of BkO_2 or BkF_3. Although these workers agree that it exhibits the UF_4-type monoclinic structure, there is some variance in the reported lattice parameters. This stems from the complexity of the X-ray powder diffraction pattern of BkF_4. A molecular volume of 7.07×10^7 pm^3 is calculated from the lattice constants given in Table II in contrast to those of 7.148×10^7 pm^3 (*167*) and 7.28×10^7 pm^3 (*168*) derived from the other reported lattice parameters. The solid state absorption spectrum of BkF_4 was reported in 1981 (*13*). Mixed alkali metal (M)—Bk(IV) fluoride compounds of the types $MBkF_5$, M_2BkF_6, M_3BkF_7, and $M_7Bk_6F_{31}$, although unreported, should be readily prepared. The structural systematics of such actinide fluoride complexes have been discussed elsewhere (*169, 170*).

One other Bk(IV) halide compound, Cs_2BkCl_6, has been characterized by its crystallographic properties (*150*). This orange compound precipitated upon dissolution of Bk(IV) hydroxide in chilled, concentrated HCl solution containing CsCl and was found to crystallize in the Rb_2MnF_6-type hexagonal structure with $a_0 = 0.7451$ and

$c_0 = 1.2097$ nm. Using a separated halogen atom model, the lattice energy of this compound has been calculated to be 1295 kJ/mol and the average radius of the $BkCl_6^{2-}$ ion to be 0.270 nm (171).

The trihalides of berkelium can be prepared by hydrohalogenation of BkO_2, Bk_2O_3, or a lighter halide of berkelium. BkF_3 (13, 148), $BkCl_3$ (108, 112, 149), and $BkBr_3$ (152, 172) have been shown by X-ray powder diffraction and absorption spectrophotometric studies to be dimorphic. Berkelium is the first actinide whose trifluoride exhibits the YF_3-type orthorhombic structure as the room-temperature, α phase and the LaF_3-type trigonal structure as the high-temperature phase (148).

In the case of dimorphic $BkCl_3$, the UCl_3-type hexagonal structure (149) represents the low-temperature form, while the $PuBr_3$-type orthorhombic structure is exhibited by the high-temperature modification (112). The phase transition temperature appears to be close (112) to the $BkCl_3$ melting point, 876 K (173). The volatilization behavior of many of the binary actinide chlorides, including $BkCl_3$, has been studied and correlated with the oxidation state and atomic number (<92 or ≥ 92) of the actinide (174). White $Cs_2NaBkCl_6$ was crystallized from aqueous CsCl–HCl solution by increasing the HCl concentration and cooling and found to exhibit a face-centered cubic structure in which the Bk(III) ions (O_h site symmetry) are octahedrally coordinated by chloride ions (150). The unique properties of such compounds stimulated the synthesis and study of an isostructural set of Cs_2NaMCl_6 compounds containing trivalent cations (M) whose ionic radii ranged from 0.065 to 0.106 nm (175).

X-Ray diffraction by a powder sample of $BkCl_3 \cdot 6H_2O$ showed that it is isostructural with $AmCl_3 \cdot 6H_2O$, whose structure was refined by single-crystal diffraction methods (151). By analogy, the basic units of the berkelium structure are $BkCl_2(OH_2)_6^+$ cations and Cl^- anions, the latter ones being octahedrally coordinated by water molecules.

From X-ray powder diffraction patterns of $BkBr_3$ obtained as a function of the sample's thermal treatment, it was concluded that the $PuBr_3$-type orthorhombic structure is the low-temperature form of $BkBr_3$ and the $AlCl_3$-type monoclinic structure is the high-temperature form (152). Since these two crystallographic modifications differ by two in the Bk(III) coordination number, absorption spectrophotometric analysis easily distinguishes between them (172). The possibility of a third polymorph of $BkBr_3$ has been suggested on the basis of eight lines of low intensity in one powder pattern (152). If it does exist, it would be the form intermediate between the $PuBr_3$- and $AlCl_3$-type structures and exhibit the $FeCl_3$-type rhombohedral structure with $a_0 = 0.766$ nm and $\alpha = 56.6°$ (152). There is one additional report (155) with lattice parameters for the orthorhombic form of BkBr$_3$ and for BiI$_3$-type hexagonal BkI$_3$.

BkOCl (*154*), BkOBr (*155*), and BkOI (*153, 155*) have been synthesized and found to exhibit the PbFCl-type tetragonal structure. Although presently unreported, BkOF certainly can be prepared and probably exhibits polymorphism.

D. PNICTIDES, CHALCOGENIDES, AND OTHER COMPOUNDS

The berkelium monopnictides have been prepared on the multimicrogram scale by direct combination of the elements (*138*). In all cases, the lattice constants of the NaCl-type cubic structures were smaller than those of the corresponding curium monopnictides but comparable to those of the corresponding terbium compounds. This supports the semimetallic classification for these compounds. One additional report of BkN has appeared (*139*). The lattice parameter derived from the sample exhibiting a single phase was 0.5010 ± 0.0004 nm, whereas that extracted from the mixed-phase sample of BkN resulting from incomplete conversion of a hydride was 0.4948 ± 0.0003 nm. Clearly, additional samples of BkN should be prepared to establish more firmly its lattice constant.

The only other crystallographic result reported for a berkelium chalcogenide besides those summarized in Table II is a cubic lattice parameter of 0.844 nm for Bk$_2$S$_3$ (*155*). The microscale synthesis of the brownish-black sesquisulfide was carried out by treatment of berkelium oxide at 1400 K with a mixture of H$_2$S and CS$_2$ vapors. In later work (*136, 137*), the higher chalcogenides were prepared on the 20- to 30-μg scale in quartz capillaries by direct combination of the elements. These were then thermally decomposed *in situ* to yield the lower chalcogenides. The stoichiometries of these compounds have not been determined directly.

The preparation of only one sample of berkelium hydride has been reported and that was by treatment of berkelium metal at 500 K with H$_2$ gas derived from thermal decomposition of UH$_3$ (*123*). The product exhibited a face-centered cubic structure with a_0 = 0.523 ± 0.0001 nm, determined from nine observed X-ray powder diffraction lines. By analogy with the behavior of the lanthanide hydrides (*176*), the superdihydride stoichiometry, BkH$_{2+x}$ (0 < x < 1), was assigned. The stoichiometric dihydride should exhibit the largest cubic unit cell. One would anticipate also the existence of BkH$_3$, which should exhibit a hexagonal unit cell. Additional work is required to characterize fully the berkelium–hydrogen system.

Both the oxysulfate (body-centered orthorhombic) and oxysulfide (trigonal) of Bk(III) have been studied by X-ray powder diffraction (*156*). $Bk_2O_2SO_4$ resulted from the decomposition of $Bk_2(SO_4)_3 \cdot nH_2O$ in an argon atmosphere (to prevent oxidation to BkO_2) at about 875 K, whereas Bk_2O_2S formed upon thermal decomposition of the sulfate hydrate in a 4% H_2–96% Ar atmosphere. No decomposition of the oxysulfide was observed up to 1300 K in the H_2–Ar gas mixture (*156*). Both $Bk_2O_2SO_4$ and Bk_2O_2S are isostructural with the corresponding lanthanide and actinide compounds.

Berkelium(III) orthophosphate has been characterized both by X-ray powder diffraction and by solid state absorption and Raman spectroscopies (*111, 115*). The X-ray data confirm the hypothesis that $BkPO_4$ is isostructural (monazite-type monoclinic) with the lighter lanthanide orthophosphates (*177*) and exhibits unit-cell dimensions similar to those of $SmPO_4$ and $EuPO_4$ (*111*).

Two organoberkelium compounds have been reported, but only one of them, $Bk(C_5H_5)_3$, has been characterized crystallographically (*116*). In addition to the data given in Table II, the formula volume of this compound is 2.98×10^8 pm^3. The amber-colored tricyclopentadienylberkelium(III) was isolated from a reaction mixture of $BkCl_3$ and molten $Be(C_5H_5)_2$ by sublimation in vacuum at 475–495 K. It decomposes to an orange melt at 610 K. By vacuum sublimation at temperatures above 500 K (up to 600 K), a second berkelium fraction was obtained (*117*). Its identity was established to be dicyclopentadienylberkelium(III) chloride dimer, $[Bk(C_5H_5)_2Cl]_2$, based on the similarities of its X-ray powder diffraction pattern and sublimation behavior to those of known $[Sm(C_5H_5)_2Cl]_2$. The solid state absorption spectrum of $[Bk(C_5H_5)_2Cl]_2$ was obtained and noted to be very similar to that of $Bk(C_5H_5)_3$ (*117*).

E. MAGNETIC BEHAVIOR OF BERKELIUM IONS

In order to improve upon the precision ($\pm 10\%$) of the initial measurements of the magnetic susceptibility of Bk(III) ions (*4*) and to extend the range of measurements to lower temperatures, single beads of cation-exchange resin were saturated with Bk(III) and subjected to susceptibility measurements over the temperature range 9–298 K (*94*). The magnetic behavior of Bk(III) over the entire temperature range was described well by the Curie–Weiss relationship with $\mu_{\text{eff}} = 9.40 \pm 0.06 \, \mu_B$ and $\theta = 11.0 \pm 1.9$ K. The magnetic susceptibility of Bk(III) in an octahedral environment of host matrix $Cs_2NaLuCl_6$ was measured, and temperature-independent paramagnetism was observed over the temperature range 10–40 K; the lowest level of Bk(III) was determined to be Γ_1, and a $\Gamma_1 - \Gamma_4$ separation of 8.5×10^3 m^{-1} was derived (178).

Results of electron paramagnetic resonance (179) and magnetic susceptibility (180) studies of Bk(IV) in ThO$_2$ have been reported. The eight-line hyperfine pattern confirmed the hypothesis that the nuclear spin of ^{249}Bk is 7/2; the estimated nuclear moment was 2.2 ± 0.4 μ_N (179). Two regions of temperature-dependent paramagnetism of BkO$_2$ in ThO$_2$ were observed over the temperature range 10–220 K; the possibility of an antiferromagnetic transition at 3 K was noted (180).

The first measurements of the magnetic susceptibilities of bulk-phase samples of some berkelium compounds (BkO$_2$, BkF$_3$, BkF$_4$, and BkN) have been made (181). The effective moments were found to agree with the calculated free-ion values, assuming Bk(III) or Bk(IV) cores and L–S coupling. Of particular note was the apparent agreement between the measured moment (7.8 μ_B) for BkN and that calculated (7.9 μ_B) for Bk(IV). However, the interpretation of this result is complicated by the unknown magnitudes of the covalency and conduction electron polarization effects (181).

VII. Solution Chemistry

A. Ionic Species

Berkelium exhibits both the III and IV oxidation states, as would be expected from the oxidation states displayed by its lanthanide counterpart, terbium. Bk(III) is the most stable oxidation state in noncomplexing aqueous solution. Bk(IV) is reasonably stable in solution, undoubtedly because of the stabilizing influence of the half-filled 5f^7 electronic configuration. Bk(III) and Bk(IV) exist in aqueous solution as the simple hydrated ions Bk^{3+}(aq) and Bk^{4+}(aq), respectively, unless complexed by ligands. Bk(III) is green in most mineral acid solutions. Bk(IV) is yellow in HCl solution and is orange-yellow in H$_2$SO$_4$ solution. A discussion of the absorption spectra of berkelium ions in solution can be found in Section IV,C.

The possible existence of divalent berkelium was studied by polarography in acetonitrile solution. Because of high background currents (caused by radiolysis products) obscuring the polarographic wave, evidence for Bk(II) was not obtained (182). Divalent berkelium has been reported to exist in mixed lanthanide chloride–strontium chloride melts. The claim is based on the results of the distribution of trace amounts of berkelium between the melt and a solid crystalline phase (cocrystallization technique) (*183, 184*).

B. THERMODYNAMIC QUANTITIES

Values of thermodynamic quantities for the formation of berkelium ions in solution, according to the reactions

$$Bk(c,\alpha) + 3H^+(aq) \longrightarrow Bk^{3+}(aq) + \tfrac{3}{2}H_2(g)$$

and

$$Bk(c,\alpha) + 4H^+(aq) \longrightarrow Bk^{4+}(aq) + 2H_2(g)$$

are summarized in Table III.

An electrostatic hydration model has been applied to the trivalent lanthanide and actinide ions in order to predict the standard free energy (ΔG_t°) and enthalpy (ΔH_t°) of hydration for these series. Assuming crystallographic and gas-phase radii for Bk(III) to be 0.096 and 0.1534 nm, respectively, and using 6.1 as the primary hydration number, ΔG_{298}° was calculated to be -3357 kJ/mol, and ΔH_{298}° was calculated to be -3503 kJ/mol (*187*).

Activity coefficients for Bk(III) in aqueous $NaNO_3$ solutions have been calculated from distribution data for the ion between the aqueous phase and a tertiary alkylamine organic phase (*188*). The activity coefficient values were reported as a function of the $NaNO_3$ concentration.

C. STABILITY CONSTANTS AND OTHER PROPERTIES

Although Bk(IV) is well known in solution, only stability constants of complexes with Bk(III) have been reported, most of which were determined during investigations of separation procedures. A compilation of the stability constants of Bk(III) complexes with various anions is given in Table IV. In most cases, the lack of replicate results precludes an assessment here of the accuracy of the reported values. The reader should consult the original sources for any information regarding the precision of the stability constant values. Although the number of directly measured stability constants for complexes of Bk(III) is rather small, a number of additional, reasonably accurate values for other complexes of Bk(III) can be obtained by interpolation of the stability constant data for the corresponding complexes of Am(III), Cm(III), and Cf(III).

TABLE III

Thermodynamic Quantities for Aqueous Berkelium Ions at 298 K

	ΔH_f° (kJ/mol)	ΔG_f° (kJ/mol)	S° (J K^{-1} mol^{-1})	Reference
Bk^{3+}(aq)	-601 ± 5	-581 ± 7	-188 ± 17	*133, 185*
Bk^{4+}(aq)	-483 ± 5	-417 ± 13	-406 ± 42	*185, 186*

TABLE IV

Stability Constants of Bk(III) Complexes with Various Anions

Ligand	Conditions[a]	Stability constants[b]	Reference
Fluoride ion, F$^-$	Solv. extrn., 298 K, $\mu = 1.0$, pH = 2.72	$\beta_1 = 7.8 \times 10^2$	*189*
Chloride ion, Cl$^-$	Solv. extrn., 298 K, $\mu = 1.0$, pH = 2	$\beta_1 = 0.96$	*190*
	Solv. extrn., ca. 293 K, $\mu = 3.0$, pH = 0.82	$\beta_1 = 0.59$, $\beta_2 = 0.25$	*239*
Bromide ion, Br$^-$	Solv. extrn., ca. 293 K, $\mu = 3.0$, pH = 0.82	$\beta_1 = 0.15$, $\beta_2 = 0.29$	*239*
Hydroxide ion, OH$^-$	$\mu = 0.1$	$\beta_1 = 7.9 \times 10^8$	*191*
Sulfate ion, SO$_4^{2-}$	Solv. extrn., 298 K, Calc. values for $\mu = 0$ (meas. $\mu \leq 0.5$)	$\beta_1 = 5.1 \times 10^3$, $\beta_2 = 3.9 \times 10^5$, $\beta_3 = 1.1 \times 10^5$	*192*
Thiocyanate ion, SCN$^-$	Solv. extrn., 298 K, $\mu = 5.0$	$\beta_1 = 7.21$	*193*
	$\mu = 1.0$, pH = 2	$\beta_1 = 3.11$, $\beta_2 = 0.31$, $\beta_3 = 2.34$	*194*
Oxalate ion, C$_2$O$_4^{2-}$	Electromigrn. rates, 298 K, $\mu = 0.1$, pH $\simeq$ 1.8	$\beta_1 = 2.8 \times 10^5$, $\beta_2 = 1.4 \times 10^9$	*195*
Acetate ion, CH$_3$COO$^-$	Solv. extrn., 298 K, 2.0 M NaClO$_4$	$\beta_1 = 1.11 \times 10^2$	*196*
Glycolate ion, CH$_2$(OH)COO$^-$	Solv. extrn., 298 K, 2.0 M NaClO$_4$	$\beta_1 = 4.4 \times 10^2$, $\beta_2 = 5.0 \times 10^4$	*197*
Lactate ion, CH$_3$CH(OH)COO$^-$	Ion exch., $\mu = 0.5$	$\beta_3 = 7.9 \times 10^5$ (est.)	*198*
2-Methyllactate ion, (CH$_3$)$_2$C(OH)COO$^-$	Solv. extrn., 10^{-2}–1 M	$\beta_1 = 6.39 \times 10^3$	*47*
α-Hydroxyisobutyrate ion, (CH$_3$)$_2$C(OH)COO$^-$	Ion exch., $\mu = 0.5$	$\beta_3 = 4.0 \times 10^6$ (est.)	*198*
Malate ion, CH(OH)(COO)CH$_2$COO^{2-}	Solv. extrn., 10^{-2}–1 M	$\beta_1 = 1.07 \times 10^7$	*47*
Tartrate ion, [CH(OH)COO]$_2^{2-}$	Solv. extrn., 10^{-2}–1 M	$\beta_1 = 6.80 \times 10^5$	*47*

(continued)

TABLE IV (continued)

Ligand	Conditions[a]	Stability constants[b]	Reference
Citrate ion, $C(OH)(COO)(CH_2COO)_2^{3-}$	Electromigrn. rates, 298 K, $\mu = 0.1$	$\beta_1 = 7.8 \times 10^7$ $\beta_2 = 1.5 \times 10^{11}$	195
	Solv. extrn., 10^{-2}–1 M	$\beta_1 = 3.00 \times 10^{11}$	47
Ethylenediamine-tetraacetate ion (EDTA), $C_2H_4N_2(CH_2COO)_4^{4-}$	Ion exch., 298 K, $\mu = 0.1$	$\beta_1 = 7.59 \times 10^{18}$	199
1,2-Diaminecyclohexane-tetraacetate ion (DACTA), $C_6H_{10}N_2(CH_2COO)_4^{4-}$	Ion exch., 298 K, $\mu = 0.1$	$\beta_1 = 1.44 \times 10^{19}$	200
Diethylenetriamine-pentaacetate ion (DTPA), $C_4H_8N_3(CH_2COO)_5^{5-}$	Ion exch., 298 K, $\mu = 0.1$	$\beta_1 = 6.2 \times 10^{22}$	201

[a] Solv. extrn., solvent extraction; calc., calculated; electromigrn., electromigration; exch., exchange; meas., measured.

[b] Overall stability constants, e.g., $\beta_1 = \frac{[BkL^{(3-n)+}]}{[Bk^{3+}][L^{n-}]}$, $\beta_2 = \frac{[BkL_2^{(3-2n)+}]}{[Bk^{3+}][L^{n-}]^2}$, and $\beta_3 = \frac{[BkL_3^{(3-3n)+}]}{[Bk^{3+}][L^{n-}]^3}$.

Attempts to obtain thermodynamic data for solvent extraction of Bk(III) by thenoyltrifluoroacetone in benzene and for complexation of Bk(III) by hydroxide and citrate ions were unsuccessful (202). The high extractability and complexability of the easily accessible tetravalent state of berkelium probably accounts for the difficulty encountered in this work.

Aside from the fact that no quantitative information was reported for a Bk(IV) complex with nitrate ions, a 1979 report is worthy of note (203). The hexanitrato complexes of Bk(IV) were studied in nitric acid solution by electromigration. An ionic mobility corresponding to a negatively charged Bk(IV) ion was evident only at HNO_3 concentrations higher than 10 M. The data indicated that at concentrations between 3 and 6 M HNO_3, Bk(IV) exists mainly as $[Bk(H_2O)_x(NO_3)_3]^+$. This study provides an explanation for the differences observed in the ion-exchange and solvent extraction behavior of Bk(IV) as compared to that of Ce(IV), Th(IV), and Np(IV) (203).

The extraction of Bk(IV) from sulfuric acid solutions by decylamine in chloroform has been studied. The extraction mechanism, the optimized extraction parameters, and the identity of the extracted species have been reported (31).

The extraction of Bk(III) by phosphoorganic acids has been studied, and activity coefficients for Bk(III) in nitric acid solution were estimated (204). In addition, a mechanism for the extraction of Bk(IV) by hydrogen di(2-ethylhexyl)orthophosphoric acid (HDEHP) from nitric acid solutions was proposed.

Mixtures of the linear polyether 1,13-bis[8-quinolyl]-1,4,7,10,13-pentaoxatridecane (K-5) and thenoyltrifluoroacetone (HTTA) were found to exhibit synergism in Bk(III) extraction from 0.5 M NaNO$_3$ solution. The stability constant for the Bk(TTA)$_3 \cdot$ K5 species in HTTA–chloroform mixtures was calculated (205). Berkelium(III) hexafluoroacetylacetonate, Bk(III) and Bk(IV) thenoyltrifluoroacetonates, and Bk(IV) pivaloyltrifluoroacetonate have been obtained in tracer quantities by extraction of Bk(III) or Bk(IV) from aqueous solutions with xylene solutions of the respective fluorine-containing β-diketones (206). Thermoradiometric studies carried out in a stream of carrier gas saturated with hexafluoroacetylacetone indicated substantial volatilities for these berkelium species in the temperature range 413–483 K. Although the identities of the gas-phase berkelium species are unknown, it was suggested that they consist mainly of Bk(III) hexafluoroacetylacetonate, the first reported volatile compound of berkelium (206).

The kinetics of exchange of Bk(III) with EuEDTA$^-$ in aqueous acetate solutions of 0.1 M ionic strength has been studied (207). The exchange was found to be first order with both acid-dependent and acid-independent rate terms. Rate values were calculated and compared to other actinide reaction rates (207).

The solubilities of Bk(III) oxalate and Bk(IV) iodate have been reported to be 1.5 and 10 mg/liter, respectively (208).

D. OXIDATION–REDUCTION BEHAVIOR AND POTENTIALS

Berkelium(III) in solution can be oxidized by strong oxidizing agents such as BrO$_3^-$ (15–19, 209), AgO (210), Ag(I)/S$_2$O$_8^{2-}$ (20, 23, 211), perxenate (212), and ozone (24, 213, 214).

Oxidation of green Bk(III) hydroxide as a suspension in 1 M NaOH to yellow Bk(IV) hydroxide was performed by bubbling ozone through the slurry (113). In basic solution, Bk(III) is unstable toward oxidation by radiolytically produced peroxide. This "auto-oxidation" had been previously observed in carbonate solution (95). Bk(III) can be stabilized in alkaline media by the presence of a reducing agent such as hydrazine hydrate (113).

$BkCl_3$ is reported to be soluble in acetonitrile saturated with tetraethylammonium chloride (215). The colorless, 7.6×10^{-4} M $BkCl_6^{3-}$ solution formed could be completely oxidized to red-orange $BkCl_6^{2-}$ by treatment with chlorine gas. The color of this Bk(IV) solution was quite similar to that observed for crystalline Cs_2BkCl_6 (150).

Investigation of the amalgamation behavior of trivalent actinides in acetate and citrate solutions by treatment with sodium amalgam showed that Bk(III) does not readily form an amalgam. This behavior is in contrast to that of the heavier actinides californium, einsteinium, and fermium, which readily amalgamate (216, 217).

Bk(IV) is a strong oxidizing agent, comparable to Ce(IV) (218). It can be coprecipitated with cerium iodate (17) or zirconium phosphate (4). The stability of Bk(IV) solutions is a function of the degree of complexation of Bk(IV) by the solvent medium (95). Bk(IV) is reduced by radiolytically generated peroxide in acidic and neutral solutions (14, 97, 219). The rate of reduction of Bk(IV) can be accelerated electrochemically (220) or by the introduction of a reducing agent such as hydrogen peroxide (14, 219), hydroxylamine hydrochloride (219), or ascorbic acid (219).

The first estimate of the Bk(IV)–Bk(III) potential was made in 1950, only a short time after the discovery of the element. A value of 1.6 V was reported, based on tracer experiments (3). Later, in 1959, a refined value of 1.62 ± 0.01 V was reported for the couple, based on the results of experiments with microgram quantities of berkelium (4). The potential of the Bk(IV)–Bk(III) couple has subsequently been determined by several workers using direct potentiometry (220–224) or indirect methods (218, 225, 226). All of the above-mentioned determinations were performed in media of relatively low complexing capability. The formal potential of the Bk(IV)–Bk(III) couple is significantly shifted to less positive values in media containing anions that strongly complex Bk(IV), such as PO_4^{3-} and CO_3^{2-} ions (227). This behavior closely parallels that of the Ce(IV)–Ce(III) couple (228). In fact, the Bk(IV)–Bk(III) couple markedly resembles the Ce(IV)–Ce(III) couple in its oxidation-reduction chemistry.

The potential of the Bk(III)–Bk(0) couple has been investigated, using radiopolarography (229, 230) and theoretical calculations (231), as well as by correlation with enthalpy of formation data (132, 133). Estimates of the potentials of berkelium redox couples have also been made from correlation plots of electron-transfer and f–d absorption band energies versus redox potential and by theoretical calculations (215, 221, 232, 233).

TABLE V

POTENTIALS OF BERKELIUM REDOX COUPLES

Redox couple	Potential (V versus NHE)	Conditions[a]	Reference
Bk(IV)–Bk(III)	1.6 ± 0.2	Calc.	221, 232
	1.664	Calc.	234
	1.54 ± 0.1	1 M HClO$_4$, dir. pot.	227
	1.597 ± 0.005	1 M HClO$_4$, dir. pot.	224
	1.735 ± 0.005	9 M HClO$_4$, dir. pot.	224
	1.54 ± 0.1	1 M HNO$_3$, dir. pot.	227
	1.562 ± 0.005	1 M HNO$_3$, dir. pot.	224
	1.56	6 M HNO$_3$, solv. extrn.	226
	1.6	3–8 M HNO$_3$, coprecip.	3
	1.543 ± 0.005	8 M HNO$_3$, dir. pot.	224
	1.43	0.1 M H$_2$SO$_4$, dir. pot.	222
	1.44	0.25 M H$_2$SO$_4$, solv. extrn.	226
	1.38	0.5 M H$_2$SO$_4$, dir. pot.	223
	1.42	0.5 M H$_2$SO$_4$, solv. extrn.	226
	1.37	1 M H$_2$SO$_4$, dir. pot.	220, 223
	1.36	2 M H$_2$SO$_4$, dir. pot.	220, 223
	1.12 ± 0.1	7.5 M H$_3$PO$_4$, dir. pot.	227
	0.85	0.006 M K$_{10}$P$_2$W$_{17}$O$_{61}$, pH = 0, dir. pot.	235
	0.65	0.006 M K$_{10}$P$_2$W$_{17}$O$_{61}$, pH > 4, dir. pot.	235
	0.26 ± 0.1	2 M K$_2$CO$_3$, dir. pot.	227
Bk(III)–Bk(II)	−2.8 ± 0.2	Calc.	232
	−2.75	Calc.	236
Bk(III)–Bk(0)	−2.03 ± 0.05	Calc.	233
	−2.4	Calc.	231
	−1.99 ± 0.09	Calc.	132
	−2.18 ± 0.09	0.1 M LiCl, radiopol.	229, 230
	−2.01 ± 0.03	Calc.	133

[a] Calc., calculated value; dir. pot., direct potentiometry; solv. extrn., solvent extraction; coprecip., coprecipitation; radiopol., radiopolarography.

Potentials of berkelium redox couples are summarized in Table V. Replicate values for the Bk(IV)–Bk(III) couple are in reasonable agreement with one another. The effect of anions that strongly complex Bk(IV) is clearly reflected in the values of the formal potential for the Bk(IV)–Bk(III) couple and can be seen in the Nernst equation plots for the couple in various media given in Fig. 9 (227). Values of 1.36 (220, 223) and 1.12 V (227) have been reported for the couple in sulfuric and phosphoric acid solutions, respectively. Carbonate ions, apparently

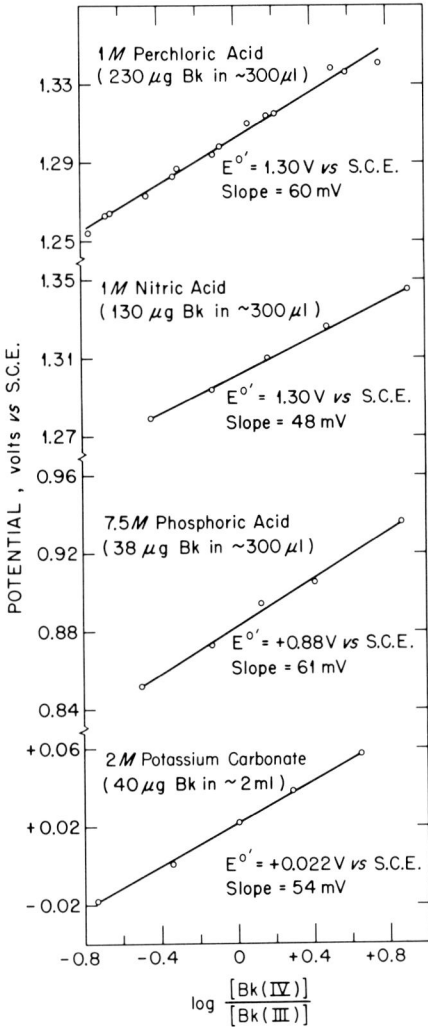

FIG. 9. Nernst equation plots for the Bk(IV)–Bk(III) couple in various aqueous media. Reprinted with permission from the *J. Inorg. Nucl. Chem.* **34**, J. R. Stokely, R. D. Baybarz, and J. R. Peterson, 1972, Pergamon Press, Ltd. (*227*).

forming the strongest complex with Bk(IV) of the anions listed in Table V, provide conditions for the least positive potential, 0.26 V (*227*), as compared to the potential of 1.6 V for the couple in noncomplexing perchlorate solution (*224, 227*). The overall thermodynamic and electrochemical data support a value of 1.67 ± 0.07 V for the standard potential ($E°$) for the Bk(IV)–Bk(III) couple, which is 0.07 V less positive than the accepted value of 1.74 V for the corresponding cerium couple (*185*). The scatter in the potential values for the Bk(III)–Bk(II) and Bk(III)–Bk(0) couples reflects the necessary requirement of making estimates of thermodynamic quantities that have not been directly determined. The potential for the Bk(III)–Bk(II) couple in mixed lanthanide chloride–strontium chloride melts has been reported to be 0.12 ± 0.02 V more positive than that for the Nd(III)–Nd(II) couple in the same media (*237*). As is the case with most nonaqueous solvent systems, it is difficult to relate this measured difference in potentials to the standard potentials in aqueous solutions that are referenced to the normal hydrogen electrode (NHE).

Theoretical estimates of the potentials for the Bk(V)–Bk(0) and Bk(V)–Bk(IV) couples have been reported as 0.2 and 3.5 V, respectively (*132*). These estimates suggest that Bk(V) is very unstable in aqueous solution.

Additional information on the oxidation-reduction behavior of berkelium can be found in a comprehensive review (*238*).

VIII. Concluding Remarks

Knowledge of both the accessibility and stability of the various oxidation states of an element is fundamental to the understanding and predictability of that element's physicochemical behavior. For element 97, oxidation states 0, III, and IV are presently known and reasonably well characterized.

The possibility of studying the physical and chemical properties of Bk(II) in bulk is intriguing. It has been suggested that this new oxidation state of berkelium is produced by nature via the α decay of einsteinium-253 dihalides (*110, 240*); however, in these absorption spectrophotometric studies, only the divalent parent species and the divalent californium-249 granddaughter species have been directly identified. Other work has shown that the β decay of berkelium-249 in the bulk-phase solid state results in the production of daughter californium-249 species exhibiting the same oxidation state and structural environment as those of the berkelium parent (*12, 13*). Direct synthesis of Bk(II) via the reaction Bk + 2BkBr$_3$ → 3BkBr$_2$ should be attempted to establish with certainty the existence of Bk(II) in the bulk-phase solid state and to characterize it via absorption spectrophotometry and X-ray powder diffraction.

Studies of berkelium metal under pressure should be continued to determine more precisely its bulk modulus and to search for the existence of a distorted face-centered cubic (fcc) modification between the known fcc and α-uranium-type orthorhombic phases. An interesting extension of this research would be to investigate the behavior of BkN under pressure to see if it might undergo a sudden volume collapse corresponding to a change in metallic valence from three to four.

The preparation and characterization of intermetallic compounds and alloys of berkelium should be pursued, as well as the determination of the stability constants of Bk(IV) complexes. The range of oxidation states accessible to berkelium might be expanded by stabilizing Bk(II) and/or Bk(V) in highly complexing aqueous, nonaqueous, or even molten salt media and/or in appropriate solid-state matrices.

Further work is necessary to elucidate more completely both the solid state and solution chemistries of berkelium. Complete knowledge of its physicochemical behavior is important for more accurate extrapolations to the behavior of the still heavier elements, for which experimental studies are often precluded by intense self-irradiation and/or by lack of material.

ACKNOWLEDGMENTS

Research sponsored by the Division of Chemical Sciences, U.S. Department of Energy, under contracts DE-AS05-76ER04447 with the University of Tennessee (Knoxville), W-7405-eng-26 with Union Carbide Corporation, and W-7405-eng-36 with the University of California.

The authors gratefully acknowledge the early assistance of J. N. Stevenson and D. L. Raschella in gathering some of the references included here. The more recent cooperation of numerous colleagues in supplying and/or verifying information, identifying oversights, and in critiquing the assembled manuscript was greatly appreciated. One of the present authors (JRP) wishes to thank posthumously Professor B. B. Cunningham, whose guidance, inspiration, and personal interest during the first systematic study of the physicochemical properties of berkelium are still fondly remembered.

REFERENCES[1]

1. Thompson, S. G., Ghiorso, A., and Seaborg, G. T., *Phys. Rev.* **77**, 838 (1950).
2. Thompson, S. G., Ghiorso, A., and Seaborg, G. T., *Phys. Rev.* **80**, 781 (1950).
3. Thompson, S. G., Cunningham, B. B., and Seaborg, G. T., *J. Am. Chem. Soc.* **72**, 2798 (1950).

[1] The authors have focused this review of the chemistry of berkelium on open-literature references in English or English translation, except where it was deemed necessary to cite a research institution report or technical memorandum. Thus also minimized are

4. Cunningham, B. B., *J. Chem. Educ.* **36,** 32 (1959).
5. Cunningham, B. B., *Proc. Robert A. Welch Found. Conf. Chem. Res. VI. Top. Mod. Inorg. Chem., Houston, Nov. 1962,* 237 (1963).
6. Browne, E., Dairiki, J. M., and Doebler, R. E., "Table of Isotopes" (C. M. Lederer and V. S. Shirley, eds.), p. 1479. Wiley, New York, 1978.
7. Bigelow, J. E., Corbett, B. L., King, L. J., McGuire, S. C., and Sims, T. M., *Am. Chem. Soc. Symp. Ser.* **161,** 3 (1981).
8. Ferguson, D. E., and Bigelow, J. E., *Actinides Rev.* **1,** 213 (1969).
9. King, L. J., Bigelow, J. E., and Collins, E. D., *Am. Chem. Soc. Symp. Ser.* **161,** 133 (1981).
10. Bigelow, J. E., personal communication (1984).
11. IUPAC Inorganic Chemistry Division Commission on Atomic Weights, *Pure Appl. Chem.* **47,** 75 (1976).
12. Young, J. P., Haire, R. G., Peterson, J. R., Ensor, D. D., and Fellows, R. L., *Inorg. Chem.* **19,** 2209 (1980).
13. Ensor, D. D., Peterson, J. R., Haire, R. G., and Young, J. P., *J. Inorg. Nucl. Chem.* **43,** 1001 (1981).
14. Peppard, D. F., Moline, S. W., and Mason, G. W., *J. Inorg. Nucl. Chem.* **4,** 344 (1957).
15. Fardy, J. J., and Weaver, B., *Anal. Chem.* **41,** 1299 (1969).
16. Knauer, J. B., and Weaver, B., U.S. Atomic Energy Commission Document ORNL-TM-2428, Oak Ridge National Laboratory, Oak Ridge, 1968.
17. Weaver, B., *Anal. Chem.* **40,** 1894 (1968).
18. Overman, R. F., *Anal. Chem.* **43,** 600 (1971).
19. Erin, E. A., Vityutnev, V. M., Kopytov, V. V., and Vasil'ev, V. Ya., *Sov. Radiochem.* **21,** 487 (1979).
20. Milyukova, M. S., Malikov, D. A., and Myasoedov, B. F., *Sov. Radiochem.* **22,** 267 (1980).
21. Shafiev, A. I., Efremov, Yu. V., and Yakovlev, G. N., *Sov. Radiochem.* **16,** 31 (1974).
22. Moore, F. L., *Anal. Chem.* **38,** 1872 (1966).
23. Milyukova, M. S., Malikov, D. A., and Myasoedov, B. F., *Sov. Radiochem.* **20,** 762 (1978).
24. Myasoedov, B. F., *Sov. Radiochem.* **16,** 716 (1974).
25. Myasoedov, B. F., Barsukova, K. V., and Radionova, G. N., *Radiochem. Radioanal. Lett.* **7,** 269 (1971).

---

references to theses, dissertations, and patents. The biologic and metabolic effects on man and animals of exposure to and/or ingestion of berkelium have not been reviewed here. Also excluded are references dealing with the determination and/or use of the nuclear properties of the various isotopes of berkelium. The references cited are not necessarily inclusive or the original ones, yet they should be more than adequate to permit the interested reader to access easily the broader areas beyond. Where necessary the present authors converted published data to SI or other commonly used units. Errors are solely the responsibility of the authors, who welcome notification of same. Other pertinent work on the chemistry of berkelium not included herein (or excluded above) should be made known to the authors for inclusion in any revisions of this text.

Earlier reviews of the physicochemical properties of berkelium are available in several new supplement series volumes of *Gmelin Handbuch der Anorganischen Chemie* (G. Koch, editor, Springer-Verlag, New York) and in *The Chemistry of the Transuranium Elements* (C. Keller, 1971, Verlag Chemie, Weinheim).

26. Kosyakov, V. N., Yakovlev, N. G., Kazakova, G. M., Erin, E. A., and Kopytov, V. V., *Sov. Radiochem.* **19**, 397 (1977).
27. Kosyakov, V. N., and Yakovlev, N. G., *Sov. Radiochem.* **25**, 172 (1983).
28. Karalova, Z. K., Myasoedov, B. F., Rodionova, L. M., and Kuznetsova, V. S., *Sov. Radiochem.* **25**, 175 (1983).
29. Moore, F. L., *Anal. Chem.* **41**, 1658 (1969).
30. Milyukova, M. S., and Myasoedov, B. F., *Sov. Radiochem.* **20**, 324 (1978).
31. Malikov, D. A., Milyukova, M. S., Kuzovkina, E. V., and Myasoedov, B. F., *Sov. Radiochem.* **25**, 293 (1983).
32. Milyukova, M. S., Malikov, D. A., Kuzovkina, E. V., and Myasoedov, B. F., *Radiochem. Radioanal. Lett.* **48**, 355 (1981).
33. Yakovlev, N. G., Kosyakov, V. N., and Kazakova, G. M., *J. Radioanal. Chem.* **75**, 113 (1982).
34. Kooi, J., Boden, R., and Wijkstra, J., *J. Inorg. Nucl. Chem.* **26**, 2300 (1964).
35. Kooi, J., and Boden, R., *Radiochim. Acta* **3**, 226 (1964).
36. Erin, E. A., Vityutnev, V. M., Kopytov, V. V., and Vasil'ev, V. Ya., *Sov. Radiochem.* **21**, 85 (1979).
37. Shafiev, A. I., and Efremov, Yu. V., *Sov. Radiochem.* **14**, 754 (1972).
38. Moore, F. L., *Anal. Chem.* **39**, 1874 (1967).
39. Chudinov, E. G., and Pirozhkov, S. V., *J. Radioanal. Chem.* **10**, 41 (1972).
40. Guseva, L. I., Gregor'eva, S. I., and Tikhomirova, G. S., *Sov. Radiochem.* **13**, 802 (1971).
41. Horwitz, E. P., Bloomquist, C. A. A., Henderson, D. J., and Nelson, D. E., *J. Inorg. Nucl. Chem.* **31**, 3255 (1969).
42. Harbour, R. M., *J. Inorg. Nucl. Chem.* **34**, 2680 (1972).
43. Farrar, L. G., Cooper, J. H., and Moore, F. L., *Anal. Chem.* **40**, 1602 (1968).
44. Moore, F. L., and Jurriaanse, A., *Anal. Chem.* **39**, 733 (1967).
45. Horwitz, E. P., and Bloomquist, C. A. A., *J. Inorg. Nucl. Chem.* **35**, 271 (1973).
46. Korpusov, G. V., Patrusheva, E. N., and Dolidze, M. S., *Sov. Radiochem.* **17**, 230 (1975).
47. Aly, H. F., and Latimer, R. M., *Radiochim. Acta* **14**, 27 (1970).
48. Baybarz, R. D., Knauer, J. B., and Orr, P. B., U.S. Atomic Energy Commission Document ORNL-4672, Oak Ridge National Laboratory, Oak Ridge, 1973.
49. Liu, Y.-F., Luo, C., von Gunten, H. R., and Seaborg, G. T., *Inorg. Nucl. Chem. Lett.* **17**, 257 (1981).
50. Liu, Y.-F., Luo, C., Moody, K. J., Lee, D., Seaborg, G. T., and von Gunten, H. R., *J. Radioanal. Chem.* **76**, 119 (1983).
51. Korkisch, J., *Oesterr. Chem. Ztg.* **67**, 273 (1966).
52. Ulstrup, J., *At. Energy Rev.* **4**, 3 (1966).
53. Myasoedov, B. F., Guseva, L. I., Lebedev, I. A., Milyukova, M. S., and Chmutova, M. K., "Analytical Chemistry of Transplutonium Elements." (D. Slutzkin, ed.), p. 122. Wiley, New York, 1974.
54. Bigelow, J. E., "Gmelin Handbuch der Anorganischen Chemie, System No. 71 (Transurane), New Supplement Series" (G. Koch, ed.), Vol. 7b, Part A1, II: The Elements. p. 326. Springer-Verlag, Berlin and New York, 1974.
55. Müller, W., *Actinides Rev.* **1**, 71 (1967).
56. Collins, E. D., Benker, D. E., Chattin, F. R., Orr, P. B., and Ross, R. G., *Am. Chem. Soc. Symp. Ser.* **161**, 147 (1981).
57. Campbell, D. O., *Am. Chem. Soc. Symp. Ser.* **161**, 189 (1981).
58. Lebedev, I. A., and Myasoedov, B. F., *Sov. Radiochem.* **24**, 583 (1982).

59. Sugar, J., *J. Chem. Phys.* **60,** 4103 (1974).
60. Krestov, G. A., *Sov. Radiochem.* **8,** 200 (1966).
61. Worden, E. F., Gutmacher, R. G., Conway, J. G., and Mehlhorn, R. J., *J. Opt. Soc. Am.* **59,** 1526 (1969).
62. Brewer, L., *J. Opt. Soc. Am.* **61,** 1101 (1971).
63. Nugent, L. J., and Vander Sluis, K. L., *J. Opt. Soc. Am.* **61,** 1112 (1971).
64. Vander Sluis, K. L., and Nugent, L. J., *Phys. Rev. A* **6,** 86 (1972).
65. Vander Sluis, K. L., and Nugent, L. J., *J. Opt. Soc. Am.* **64,** 687 (1974).
66. Brewer, L., *J. Opt. Soc. Am.* **61,** 1666 (1971).
67. Hollander, J. M., Holtz, M. D., Novakov, T., and Graham, R. L., *Ark. Fys.* **28,** 375 (1965).
68. Dittner, P. F., and Bemis, C. E., Jr., *Phys. Rev. A* **5,** 481 (1972).
69. Carlson, T. A., Nestor, C. W., Jr., Malik, F. B., and Tucker, T. C., *Nucl. Phys.* **A135,** 57 (1969).
70. Lu, C. C., Malik, F. B., and Carlson, T. A., *Nucl. Phys.* **A175,** 289 (1971).
71. Barreau, G., Börner, H. G., von Egidy, T., and Hoff, R. W., *Z. Phys. A* **308,** 209 (1982).
72. Chen, M. H., Crasemann, B., Huang, K., Aoyagi, M., and Mark, H., *At. Data Nucl. Data Tables* **19,** 97 (1977).
73. Huang, K., Aoyagi, M., Chen, M. H., Crasemann, B., and Mark, H., *At. Data Nucl. Data Tables* **18,** 243 (1976).
74. Chen, M. H., Crasemann, B., Aoyagi, M., Huang, K.-N., and Mark, H., *At. Data Nucl. Data Tables* **26,** 561 (1981).
75. Carlson, T. A., and Nestor, C. W., Jr., *At. Data Nucl. Data Tables* **19,** 153 (1977).
76. Sevier, K. D., *At. Data Nucl. Data Tables* **24,** 323 (1979).
77. Snijders, J. G., Vernooijs, P., and Baerends, E. J., *At. Data Nucl. Data Tables* **26,** 483 (1981).
78. Scofield, J. H., *At. Data Nucl. Data Tables* **14,** 121 (1974).
79. Manson, S. T., and Kennedy, D. J., *At. Data Nucl. Data Tables* **14,** 111 (1974).
80. Gutmacher, R. G., Hulet, E. K., and Lougheed, R., *J. Opt. Soc. Am.* **55,** 1029 (1965).
81. Worden, E. F., Hulet, E. K., Lougheed, R., and Conway, J. G., *J. Opt. Soc. Am.* **57,** 550 (1967).
82. Worden, E. F., Gutmacher, R. G., Lougheed, R. W., and Conway, J. G., *J. Opt. Soc. Am.* **60,** 1555 (1970).
83. Conway, J. G., U.S. Energy Research and Development Administration Document LBL-4366, p. 70 (1976).
84. Worden, E. F., and Conway, J. G., *At. Data Nucl. Data Tables* **22,** 329 (1978).
85. Conway, J. G., Worden, E. F., Blaise, J., Camus, P., and Vergès, J., *Spectrochim. Acta* **32B,** 101 (1977).
86. Blaise, J., Wyart, J.-F., Conway, J. G., and Worden, E. F., *Phys. Scripta* **22,** 224 (1980).
87. Gutmacher, R. G., Hulet, E. K., Worden, E. F., and Conway, J. G., *J. Opt. Soc. Am.* **53,** 506 (1963).
88. Finch, C. B., Fellows, R. L., and Young, J. P., *J. Lumin.* **16,** 109 (1978).
89. Nugent, L. J., Burnett, J. L., Baybarz, R. D., Werner, G. K., Tanner, S. P., Tarrant, J. R., and Keller, O. L., Jr., *J. Phys. Chem.* **73,** 1540 (1969).
90. Nugent, L. J., Baybarz, R. D., Werner, G. K., and Friedman, H. A., *Chem. Phys. Lett.* **7,** 179 (1970).
91. Beitz, J. V., Carnall, W. T., Wester, D. W., and Williams, C. W., U.S. Dept. of Energy Document LBL-12441, p. 108 (1981).

92. Gutmacher, R. G., Hulet, E. K., Lougheed, R., Conway, J. G., Carnall, W. T., Cohen, D., Keenan, T. K., and Baybarz, R. D., *J. Inorg. Nucl. Chem.* **29**, 2341 (1967).
93. Peterson, J. R., *Am. Chem. Soc. Symp. Ser.* **131**, 221 (1980); Peterson, J. R., U.S. Atomic Energy Commission Document UCRL-17875, 1967.
94. Fujita, D. K., U.S. Atomic Energy Commission Document UCRL-19507, 1969.
95. Baybarz, R. D., Stokely, J. R., and Peterson, J. R., *J. Inorg. Nucl. Chem.* **34**, 739 (1972).
96. Gutmacher, R. G., Bodé, D. D., Lougheed, R. W., and Hulet, E. K., *J. Inorg. Nucl. Chem.* **35**, 979 (1973).
97. Erin, E. A., Vityutnev, V. M., Kopytov, V. V., and Vasil'ev, V. Ya., *Sov. Radiochem.* **24**, 148 (1982).
98. Carnall, W. T., Sjoblom, R. K., Barnes, R. F., and Fields, P. R., *Inorg. Nucl. Chem. Lett.* **7**, 651 (1971).
99. Varga, L. P., Baybarz, R. D., Reisfeld, M. J., and Volz, W. B., *J. Inorg. Nucl. Chem.* **35**, 2787 (1973).
100. Varga, L. P., Baybarz, R. D., and Reisfeld, M. J., *J. Inorg. Nucl. Chem.* **35**, 4313 (1973).
101. Varga, L. P., Baybarz, R. D., Reisfeld, M. J., and Asprey, L. B., *J. Inorg. Nucl. Chem.* **35**, 2775 (1973).
102. Gutmacher, R. G., U.S. Atomic Energy Comission Document UCRL-12275-T, 1964.
103. Fields, P. R., Wybourne, B. G., and Carnall, W. T., U.S. Atomic Energy Commission Document ANL-6911, 1964.
104. Carnall, W. T., Fried, S., and Wagner, F., Jr., *J. Chem. Phys.* **58**, 3614 (1973).
105. Carnall, W. T., Fried, S. Wagner, F., Jr., Barnes, R. F., Sjoblom, R. K., and Fields, P. R., *Inorg. Nucl. Chem. Lett.* **8**, 773 (1972).
106. Carnall, W. T., and Fried, S., U.S. Energy Research and Development Administration Document LBL-4366. p. 61 (1976).
107. Carnall, W. T., Crosswhite, H. M., Crosswhite, H., Hessler, J. P., Aderhold, C., Caird, J. A., Paszek, A., and Wagner, F. W., *Proc. Int. Conf. the Electron. Struct. Actinides, 2nd, Wrocław, Sept. 1976*, 105 (1977).
108. Young, J. P., Haire, R. G., Fellows, R. L., and Peterson, J. R., *J. Radioanal. Chem.* **43**, 479 (1978).
109. Peterson, J. R., Fellows, R. L., Young, J. P., and Haire, R. G., *Proc. Int. Conf. Electron. Struct. Actinides, 2nd, Wrocław, Sept. 1976* 111–116; Peterson, J. R., Fellows, R. L., Young, J. P., and Haire, R. G., *Rev. Chim. Min.* **14**, 172 (1977).
110. Young, J. P., Haire, R. G., Peterson, J. R., Ensor, D. D., and Fellows, R. L., *Inorg. Chem.* **20**, 3979 (1981).
111. Haire, R. G., Hellwege, H. E., Hobart, D. E., and Young, J. P., *J. Less-Common Metals* **93**, 358 (1983).
112. Peterson, J. R., Young, J. P., Ensor, D. D., and Haire, R. G., U.S. Dept. of Energy Document LBL-12441, p. 118 (1981).
113. Cohen, D., *J. Inorg. Nucl. Chem. Suppl.* 41 (1976).
114. Shannon, R. D., and Prewitt, C. T., *Acta Crystallogr.* **B25**, 925 (1969).
115. Hobart, D. E., Begun, G. M., Haire, R. G., and Hellwege, H. E., *J. Raman Spectrosc.* **14**, 59 (1983).
116. Laubereau, P. G., and Burns, J. H., *Inorg. Chem.* **9**, 1091 (1970).
117. Laubereau, P. G., *Inorg. Nucl. Chem. Lett.* **6**, 611 (1970).
118. Carnall, W. T., "Gmelin Handbuch der Anorganischen Chemie, System No. 71

(Transurane), New Supplement Series" (G. Koch, ed.), Vol. 8, Part A2, p. 35. Springer-Verlag, Berlin and New York, 1973.
119. Peterson, J. R., Fahey, J. A., and Baybarz, R. D., *J. Inorg. Nucl. Chem.* **33,** 3345 (1971).
120. Fuger, J., Peterson, J. R., Stevenson, J. N., Noé, M., and Haire, R. G., *J. Inorg. Nucl. Chem.* **37,** 1725 (1975).
121. Sarkisov, E. S., *Dokl. Akad. Nauk SSSR* **166,** 627 (1966).
122. Zachariasen, W. H., *J. Inorg. Nucl. Chem.* **35,** 3487 (1973).
123. Fahey, J. A., Peterson, J. R., and Baybarz, R. D., *Inorg. Nucl. Chem. Lett.* **8,** 101 (1972).
124. Ward, J. W., Kleinschmidt, P. D., and Haire, R. G., *J. Chem. Phys.* **77,** 1464 (1982).
125. Haire, R. G., Peterson, J. R., Benedict, U., and Dufour, C., *J. Less-Common Metals*, **102,** 119 (1984).
126. Benedict, U., Peterson, J. R., Haire, R. G., and Dufour, C., *J. Phys. F: Met. Phys.* **14,** L43 (1984).
127. Johansson, B., Skriver, H. L., and Andersen, O. K., "Physics of Solids Under High Pressure" (J. S. Schilling and R. N. Shelton, eds.), p. 245. North-Holland Publ., Amsterdam (1981).
128. Ward, J. W., Kleinschmidt, P. D., Haire, R. G., and Brown, D., *Am. Chem. Soc. Symp. Ser.* **131,** 199 (1980).
129. Ward, J. W., and Hill, H. H., "Heavy Element Properties" (W. Müller and H. Blank, eds.), p. 65. North-Holland Publ., Amsterdam, 1976.
130. Nugent, L. J., Burnett, J. L., and Morss, L. R., *J. Chem. Thermodyn.* **5,** 665 (1973).
131. Johansson, B., and Rosengren, A., *Phys. Rev. B* **11,** 1367 (1975).
132. David, F., Samhoun, K., Guillaumont, R., and Nugent, L. J., "Heavy Element Properties" (W. Müller and H. Blank, eds.), p. 97. North-Holland Publ., Amsterdam, 1976.
133. Fuger, J., Haire, R. G., and Peterson, J. R., *J. Inorg. Nucl. Chem.* **43,** 3209 (1981).
134. Nave, S. E., Huray, P. G., and Haire, R. G., "Crystalline Electric Field and Structural Effects in f-Electron Systems" (J. E. Crow, R. P. Guertin, and T. W. Mihalisin, eds.), p. 269. Plenum, New York, 1980.
135. Peterson, J. R., Fahey, J. A., and Baybarz, R. D., *Nucl. Metall.* **17,** 20 (1970).
136. Damien, D. A., Haire, R. G., and Peterson, J. R., *J. Phys.* **40,** C4:95 (1979).
137. Damien, D., Haire, R. G., and Peterson, J. R., Oak Ridge National Laboratory, preliminary values of results to be published, 1981.
138. Damien, D., Haire, R. G., and Peterson, J. R., *J. Inorg. Nucl. Chem.* **42,** 995 (1980).
139. Stevenson, J. N., and Peterson, J. R., *J. Less-Common Metals* **66,** 201 (1979).
140. Jullien, R., Galleani d'Agliano, E., and Coqblin, B., *Phys. Rev. B* **6,** 2139 (1972).
141. Freeman, A. J., and Koelling, D. D., "The Actinides: Electronic Structure and Related Properties" (A. J. Freeman and J. B. Darby, Jr., eds.), Vol. 1, p. 51. Academic Press, New York, 1974.
142. Fournier, J. M., *J. Phys. Chem. Solids* **37,** 235 (1976).
143. Herbst, J. F., Watson, R. E., and Lindgren, I., *Phys. Rev. B* **14,** 3265 (1976).
144. Fahey, J. A., Turcotte, R. P., and Chikalla, T. D., *Inorg. Nucl. Chem. Lett.* **10,** 459 (1974).
145. Peterson, J. R., and Cunningham, B. B., *Inorg. Nucl. Chem. Lett.* **3,** 327 (1967).
146. Baybarz, R. D., *J. Inorg. Nucl. Chem.* **35,** 4149 (1973).
147. Haug, H. O., and Baybarz, R. D., *Inorg. Nucl. Chem. Lett.* **11,** 847 (1975).
148. Peterson, J. R., and Cunningham, B. B., *J. Inorg. Nucl. Chem.* **30,** 1775 (1968).
149. Peterson, J. R., and Cunningham, B. B., *J. Inorg. Nucl. Chem.* **30,** 823 (1968).

150. Morss, L. R., and Fuger, J., *Inorg. Chem.* **8**, 1433 (1969).
151. Burns, J. H., and Peterson, J. R., *Inorg. Chem.* **10**, 147 (1971).
152. Burns, J. H., Peterson, J. R., and Stevenson, J. N., *J. Inorg. Nucl. Chem.* **37**, 743 (1975).
153. Fellows, R. L., Young, J. P., and Haire, R. G., U.S. Energy Research and Development Administration Document ORO-4447-048, p. 5 (1977).
154. Peterson, J. R., and Cunningham, B. B., *Inorg. Nucl. Chem. Lett.* **3**, 579 (1967).
155. Cohen, D., Fried, S., Siegel, S., and Tani, B., *Inorg. Nucl. Chem. Lett.* **4**, 257 (1968).
156. Haire, R. G., and Fahey, J. A., *J. Inorg. Nucl. Chem.* **39**, 837 (1977).
157. Shannon, R. D., *Acta Crystallogr.* **A32**, 751 (1976).
158. Ionova, G. V., Spitsyn, V. I., and Mikheev, N. B., *Proc. Int. Conf. Electron. Struct. Actinides, 2nd, Wrocław, Sept. 1976*, p. 39 (1977).
159. Haire, R. G., *Proc. Rare Earth Res. Conf., 10th, Carefree, Arizona, April–May,* p. 882 (1973).
160. Baybarz, R. D., *J. Inorg. Nucl. Chem.* **30**, 1769 (1968).
161. Sudakov, L. V., Erin, E. A., Kopytov, V. V., Baranov, A. Yu., Shimbarev, E. V., Vasil'ev, V. Ya., and Kapshukov, I. I., *Sov. Radiochem.* **19**, 394 (1977).
162. Veal, B. W., Lam, D. J., Diamond, H., and Hoekstra, H. R., *Phys. Rev. B* **15**, 2929 (1977).
163. Turcotte, R. P., Chikalla, T. D., and Eyring, L., *J. Inorg. Nucl. Chem.* **33**, 3749 (1971).
164. Turcotte, R. P., Chikalla, T. D., Haire, R. G., and Fahey, J. A., *J. Inorg. Nucl. Chem.* **42**, 1729 (1980).
165. Turcotte, R. P., *J. Inorg. Nucl. Chem.* **42**, 1735 (1980).
166. Haire, R. G., and Baybarz, R. D., *J. Inorg. Nucl. Chem.* **35**, 489 (1973).
167. Asprey, L. B., and Haire, R. G., *Inorg. Nucl. Chem. Lett.* **9**, 1121 (1973).
168. Asprey, L. B., and Kennan, T. K., *Inorg. Nucl. Chem. Lett.* **4**, 537 (1968); Kennan, T. K., and Asprey, L. B., *Inorg. Chem.* **8**, 235 (1969).
169. Penneman, R. A., Ryan, R. R., and Rosenzweig, A., *Struct. Bond.* **13**, 1 (1973).
170. Thoma, R. E., *Inorg. Chem.* **1**, 220 (1962).
171. Jenkins, H. D. B., and Pratt, K. F., *Prog. Solid State Chem.* **12**, 125 (1979).
172. Peterson, J. R., Fellows, R. L., Young, J. P., and Haire, R. G., *Rev. Chim. Min.* **14**, 172 (1977).
173. Peterson, J. R., and Burns, J. H., *J. Inorg. Nucl. Chem.* **35**, 1525 (1973).
174. Merinis, J., Legoux, Y., and Bouissières, G., *Radiochem. Radioanal. Lett.* **3**, 255 (1970).
175. Morss, L. R., Siegal, M., Stenger, L., and Edelstein, N., *Inorg. Chem.*, **9**, 1771 (1970).
176. Holley, C. E., Jr., Mulford, R. N. R., Ellinger, F. H., Koehler, W. C., and Zachariasen, W. H., *J. Phys. Chem.* **59**, 1226 (1955).
177. Pepin, J. G., and Vance, E. R., *J. Inorg. Nucl. Chem.* **43**, 2807 (1981).
178. Hendricks, M. E., Jones, E. R., Jr., Stone, J. A., and Karraker, D. G., *J. Chem. Phys.* **60**, 2095 (1974).
179. Boatner, L. A., Reynolds, R. W., Finch, C. B., and Abraham, M. M., *Phys. Lett.* **42A**, 93 (1972).
180. Karraker, D. G., *J. Chem. Phys.* **62**, 1444 (1975).
181. Nave, S. E., Haire, R. G., and Huray, P. G., U.S. Dept. of Energy Document LBL-12441, p. 144 (1981).
182. Friedman, H. A., and Stokely, J. R., *Inorg. Nucl. Chem. Lett.* **12**, 505 (1976).

183. Mikheev, N. B., D'yachkova, R. A., and Spitsyn, V. I., *Dokl. Chem. (Chem. Sect.)* **244,** 18 (1979).
184. D'yachkova, R. A., Auérman, L. N., Mikheev, N. B., and Spitsyn, V. I., *Sov. Radiochem.* **22,** 234 (1980).
185. Fuger, J., and Oetting, F. L., "The Chemical Thermodynamics of Actinide Elements and Compounds Part 2: The Actinide Aqueous Ions" (V. Medvedev, M. H. Rand, E. F. Westrum, Jr., and F. L. Oetting, eds.), p. 50. IAEA, Vienna, 1976.
186. Fuger, J., "Actinides in Perspective" (N. M. Edelstein, ed.), p. 409. Pergamon, Oxford, 1982.
187. Goldman, S., and Morss, L. R., *Can. J. Chem.* **53,** 2695 (1975).
188. Chudinov, E. G., and Pirozhkov, S. V., *Sov. Radiochem.* **15,** 195 (1973).
189. Choppin, G. R., and Unrein, P. J., "Transplutonium Elements 1975" (W. Müller and R. Lindner, eds.), p. 97. North-Holland Publ., Amsterdam, 1976.
190. Harmon, H. D., Peterson, J. R., and McDowell, W. J., *Inorg. Nucl. Chem. Lett.* **8,** 57 (1972).
191. Hussonnois, M., Hubert, S., Brillard, L., and Guillaumont, R., *Radiochem. Radioanal. Lett.* **15,** 47 (1973); Désiré, B., Hussonnois, M., and Guillaumont, R., *C.R. Acad. Sci. Paris C* **269,** 448 (1969).
192. McDowell, W. J., and Coleman, C. F., *J. Inorg. Nucl. Chem.* **34,** 2837 (1972).
193. Kinard, W. F., and Choppin, G. R., *J. Inorg. Nucl. Chem.* **36,** 1131 (1974).
194. Harmon, H. D., Peterson, J. R., McDowell, W. J., and Coleman, C. F., *J. Inorg. Nucl. Chem.* **34,** 1381 (1972).
195. Stepanov, A. V., *Russ. J. Inorg. Chem.* **16,** 1583 (1971).
196. Choppin, G. R. and Schneider, J. K., *J. Inorg. Nucl. Chem.* **32,** 3283 (1970).
197. Choppin, G. R., and Degischer, G., *J. Inorg. Nucl. Chem.* **34,** 3473 (1972).
198. Stary, J., *Talanta* **13,** 421 (1966).
199. Fuger, J., *J. Inorg. Nucl. Chem.* **18,** 263 (1961).
200. Baybarz, R. D., *J. Inorg. Nucl. Chem.* **28,** 1055 (1966).
201. Baybarz, R. D., *J. Inorg. Nucl. Chem.* **27,** 1831 (1965).
202. Hubert, S., Hussonnois, M., Brillard, L., and Guillaumont, R., "Transplutonium, 1975" (W. Müller and R. Lindner, eds.) p. 109. North-Holland Publ., Amsterdam, 1976.
203. Makarova, T. P., Fridkin, A. M., Kosyakov, V. N., and Yerin, E. A., *J. Radioanal. Chem.* **53,** 17 (1979).
204. Chudinov, E. G., Kosyakov, V. N., Shvetsov, I. K., and Vereshchaguin, Yu. I., "Transplutonium 1975" (W. Müller and R. Lindner, eds.) p. 49. North-Holland Publ., Amsterdam, 1976.
205. Ensor, D. D., and Shah, A. H., *J. Less-Common Metals* **93,** 358 (1983).
206. Fedoseev, E. V., Ivanova, L. A., Travnikov, S. S., Davydov, A. V., and Myasoedov, B. F., *Sov. Radiochem.* **25,** 343 (1983).
207. Williams, K. R., and Choppin, G. R., *J. Inorg. Nucl. Chem.* **36,** 1849 (1974).
208. Erin, E. A., Kopytov, V. V., Vasil'ev, V. Ya., and Vityutnev, V. M., *Sov. Radiochem.* **19,** 380 (1977).
209. Malikov, D. A., Almasova, E. V., Milyukova, M. S., and Myasoedov, B. F., *Radiochem. Radioanal. Lett.* **44,** 297 (1980).
210. Erin, E. A., Kopytov, V. V., and Vityutnev, V. M., *Sov. Radiochem.* **18,** 446 (1976).
211. Milyukova, M. S., Malikov, D. A., and Myasoedov, B. F., *Radiochem. Radioanal. Lett.* **29,** 93 (1977).
212. Lebedev, I. A., Chepovoy, V. I., and Myasoedov, B. F., *Radiochem. Radioanal. Lett.* **22,** 239 (1975).

213. Myasoedov, B. F., Chepovoy, V. I., and Lebedev, I. A., *Radiochem. Radioanal. Lett.* **15,** 39 (1973).
214. Myasoedov, B. F., Chepovoy, V. I., and Lebedev, I. A., *Radiochem. Radioanal. Lett.* **22,** 233 (1975).
215. Nugent, L. J., Baybarz, R. D., Burnett, J. L., and Ryan, J. L., *J. Inorg. Nucl. Chem.* **33,** 2503 (1971).
216. Malý, J., *Inorg. Nucl. Chem. Lett.* **3,** 373 (1967).
217. Malý, J., *J. Inorg. Nucl. Chem.* **31,** 1007 (1968).
218. Weaver, B., and Stevenson, J. N., *J. Inorg. Nucl. Chem.* **33,** 1877 (1971).
219. Kazakova, G. M., Kosyakov, V. N., and Erin, E. A., *Sov. Radiochem.* **17,** 315 (1975).
220. Kulyako, Yu. M., Frenkel, V. Ya., Lebedev, I. A., Trofimov, T. I., Myasoedov, B. F., and Mogilevskii, A. N., *Radiochim. Acta* **28,** 119 (1981).
221. Nugent, L. J., Baybarz, R. D., Burnett, J. L., and Ryan, J. L., *J. Phys. Chem.* **77,** 1528 (1973).
222. Propst, R. C., and Hyder, M. L., *J. Inorg. Nucl. Chem.* **32,** 2205 (1970).
223. Stokely, J. R., Barybarz, R. D., and Shults, W. D., *Inorg. Nucl. Chem. Lett.* **5,** 877 (1969).
224. Simakin, G. A., Kosyakov, V. N., Baranov, A. A., Erin, E. A., Kopytov, V. V., and Timofeev, G. A., *Sov. Radiochem.* **19,** 302 (1977).
225. Weaver, B., and Fardy, J. J., *Inorg. Nucl. Chem. Lett.* **5,** 145 (1969).
226. Musikas, C., and Berger, R., *Am. Chem. Soc. Adv. Chem. Ser.* **71,** 296 (1967).
227. Stokely, J. R., Baybarz, R. D., and Peterson, J. R., *J. Inorg. Nucl. Chem.* **34,** 392 (1972).
228. Hobart, D. E., Samhoun, K., Young, J. P., Norvell, V. E., Mamantov, G., and Peterson, J. R., *Inorg. Nucl. Chem. Lett.* **16,** 321 (1980).
229. Samhoun, K., and David, F., "Transplutonium Elements 1975" (W. Müller and R. Lindner, eds.), p. 297. North-Holland Publ., Amsterdam 1976.
230. Samhoun, K., and David, F., *J. Inorg. Nucl. Chem.* **41,** 357 (1979).
231. Krestov, G. A., *Sov. Radiochem.* **7,** 69 (1965).
232. Nugent, L. J., Baybarz, R. D., Burnett, J. L., and Ryan, J. L., *J. Inorg. Nucl. Chem. Suppl.* 37 (1976).
233. Nugent, L. J., *J. Inorg. Nucl. Chem.* **37,** 1767 (1975).
234. Simakin, G. A., Baranov, A. A., Kosyakov, V. N., Timofeev, G. A., Erin, E. A., and Lebedev, I. A., *Sov. Radiochem.* **19,** 307 (1977).
235. Baranov, A. A., Simakin, G. A., Kosyakov, V. N., Erin, E. A., Kopytov, V. V., Timofeev, G. A., and Rykov, A. G., *Sov. Radiochem.* **23,** 104 (1981).
236. Lebedev, I. A., *Sov. Radiochem.* **20,** 556 (1978).
237. Mikheev, N. B., D'yachkova, R. A., and Auérman, L. N., *Acta Chim. Acad. Sci. Hung.* **108,** 249 (1981).
238. Martinot, L., "Encyclopedia of Electrochemistry of the Elements" (A. J. Bard, ed.), Chap. VIII-2, p. 196. Dekker, New York, 1978.
239. Fukasawa, T., Kawasuji, I., Mitsugashira, T., Satô, A., and Suzuki, S., *Bull. Chem. Soc. Jpn.* **55,** 726 (1982).
240. Peterson, J. R., Ensor, D. D., Fellows, R. L., Haire, R. G., and Young, J. P., *J. Phys.* **40,** C4:111 (1979).

PREPARATIONS AND REACTIONS OF OXIDE FLUORIDES OF THE TRANSITION METALS, THE LANTHANIDES, AND THE ACTINIDES

JOHN H. HOLLOWAY and DAVID LAYCOCK[1]

Department of Chemistry, The University, Leicester, England

I. Introduction 73
II. The Oxide Fluorides of the Transition Metals 74
   A. The Oxide Fluorides of Scandium and Yttrium 74
   B. The Oxide Fluorides of Titanium, Zirconium, and Hafnium 74
   C. The Oxide Fluorides of Vanadium, Niobium, and Tantalum 75
   D. The Oxide Fluorides of Chromium, Molybdenum, and Tungsten . . . 77
   E. The Oxide Fluorides of Manganese, Technetium, and Rhenium . . . 82
   F. The Oxide Fluorides of Iron, Ruthenium, and Osmium 84
   G. The Oxide Fluorides of Iridium and Platinum 85
III. The Oxide Fluorides of the Lanthanide Elements 85
IV. The Oxide Fluorides of the Actinide Elements 86
   A. The Trivalent Actinide Oxide Fluorides 86
   B. The Tetravalent Actinide Oxide Fluorides 86
   C. The Pentavalent Actinide Oxide Fluorides 86
   D. The Hexavalent Actinide Oxide Fluorides 88
   E. Other Actinide Oxide Fluorides. 91
   References 91

I. Introduction

The aim of this article is to summarize the preparations and chemical properties of the oxide fluorides of the transition metals, the lanthanides, and the actinides.

In general, the replacement of fluorine atoms with oxygen atoms tends to stabilize higher oxidation states. This stabilization is exempli-

---

[1] Present address: Mobil Oil Company, Ltd., Research and Technical Service Laboratory, The Manorway, Coryton, Stanford-le-Hope, Essex SS17 9LN, England.

fied among the transition metals by chromium, manganese, technetium, and osmium, which form Cr(VI), Mn(VII), Tc(VII), and Os(VIII) oxide fluorides, yet their highest oxidation-state binary fluorides are Cr(V), Mn(IV), Tc(VI), and Os(VI). The fluorides CrF_6 (1, 2) and OsF_7 (3) have been claimed but are not well authenticated. In addition, with respect particularly to transition-metal compounds, the substitution of fluorine atoms by oxygen atoms tends to destabilize lower oxidation states. For example, the lowest oxidation-state oxide fluorides appear for Cr(III), Mn(VII), Tc(VI), and Os(VI), whereas the lowest binary fluorides are CrF_2, MnF_2, TcF_5, and OsF_4.

II. The Oxide Fluorides of the Transition Metals

The known oxide fluorides of the transition metals and of scandium and yttrium are shown in Table I.

A. The Oxide Fluorides of Scandium and Yttrium

The oxide fluoride ScOF can be prepared from ScF_3 and Sc_2O_3 at 800–1000°C and 100 kbar pressure (4), and by the hydrolysis of ScF_3 in a stream of moist nitrogen gas at 800°C (5). In addition, the system $HF-H_2O-Sc_2O_3$ at equilibrium shows the solid phases ScOF and ScOF · $2H_2O$ (6). A crystalline solid, Sc_3OF_7, has been detected in the ScOF–ScF_3 system at 1200°C and 100 kbar pressure (7).

Yttrium monoxide monofluoride (YOF) is formed as one of the products when YF_3 is thermally decomposed in air between 450 and 650°C (8, 9) or when YF_3 is heated with Y_2O_3 at high temperature (4, 10, 11). The low-temperature annealing of a $YF_3-Y_2O_3$ mixture yields several stoichiometric oxide fluoride phases, including $Y_7O_6F_9$ (12). The ^{19}F-NMR spectrum of YOF suggests the presence of F–F interactions similar to those exhibited by Hg_2F_2 (13).

B. The Oxide Fluorides of Titanium, Zirconium, and Hafnium

The reaction of Ti_2O_3 with TiF_3 at 100°C and 60 kbar pressure yields the black crystalline solid TiOF (14).

Early claims of the preparation of $TiOF_2$ from the hydrolysis of TiF_4 and $TiClF_3$ (15, 16) were incorrect, and TiO(OH)F was the true product (17). The oxide difluoride $TiOF_2$ has since been prepared as a yellow solid from the reaction of $TiCl_2F_2$ and liquid Cl_2O at 4°C (17) and also by the reaction of TiO_2 with CHF_3 (18). In the latter preparation, the solid formed has a large surface area and has proved to be a useful catalyst carrier, especially for hydrocarbon reactions.

TABLE I

The Oxide Fluorides of the Transition Metals[a]

Oxidation state	Group IIIa	Group IVa	Group Va	Group VIa	Group VIIa	Group VIII
+3	ScOF, YOF	TiOF	VOF	CrOF, MoOF		FeOF
+4		TiOF$_2$, ZrOF$_2$	VOF$_2$	CrOF$_2$		
+5			VOF$_3$, VO$_2$F, NbOF$_3$, NbO$_2$F, Nb$_3$O$_7$F, TaOF$_3$, TaO$_2$F, Ta$_3$O$_7$F	CrOF$_3$, MoOF$_3$	ReOF$_3$	PtOF$_3$
+6				CrOF$_4$, CrO$_2$F$_2$, MoOF$_4$, MoO$_2$F$_2$, WOF$_4$, WO$_2$F$_2$	TcOF$_4$, ReOF$_4$	RuOF$_4$, OsOF$_4$, IrOF$_4$
+7					MnO$_3$F, TcO$_3$F, TcO$_2$F$_3$, (TcO$_2$F$_2$)$_2$O, ReOF$_5$, ReO$_2$F$_3$, ReO$_3$F	OsOF$_5$, OsO$_2$F$_3$
+8						OsO$_3$F$_2$

[a] For cobalt, rhodium, nickel, palladium, silver, and gold no oxide fluorides have been reported.

Zirconium oxide fluoride, ZrOF$_2$, is produced by the dehydration of ZrF$_4 \cdot$ 3H$_2$O (19) or ZrF$_2$(OH)$_2$ (20) at temperatures higher than 140°C. It is also formed as an intermediate during the hydrofluorination of ZrO$_2$ (21). Other more complex oxide fluorides of zirconium have been reported; for example, Zr$_2$OF$_6 \cdot$ 2H$_2$O, Zr$_3$O$_2$F$_8$, Zr$_4$O$_3$F$_{10}$, Zr$_4$O$_5$F$_6$, Zr$_4$OF$_{14}$, Zr$_7$O$_9$F$_{10}$, and Zr$_{10}$O$_{13}$F$_{14}$ (19–25).

C. The Oxide Fluorides of Vanadium, Niobium, and Tantalum

Black, crystalline VOF was claimed to be the product when VF$_5$ and V$_2$O$_3$ were heated to 1000°C at 60 kbar (14). However, there were indications that the product may have been nonstoichiometric.

Yellow VOF_2 is produced when $VOBr_2$ reacts with anhydrous HF (26) and when VO_2 reacts with aqueous HF (27). The $[VO]^{2+}$ unit is rather stable and capable of persisting in reactions with ligands in the first coordination sphere of the vanadium, and so considerable interest has been shown in oxovanadium(IV) complexes (28). Fluorooxovanadates(IV), mostly prepared with alkali metal and ammonium ions (29–35), exhibit a range of stoichiometries. The basic structural units of the fluoro anions are $[VOF_5]^{3-}$ octahedra, which are found isolated in $M_3^I[VOF_5]$ (29, 30) and $[M^{III}(NH_3)_6][VOF_5]$ (36), connected to cis chains in $K_2[VOF_4]$ (37) and $[NH_4]_2[VOF_4]$ (38) or to complex chains in Cs-$[VOF_3] \cdot 0.5H_2O$ (39). In the case of $Cs_2[VOF_4(OH_2)]$ the structure contains $[VOF_4(OH_2)]^{2-}$ octahedra (40), while in $Cs_2[V_2O_2F_7]$ two $[VOF_5]^{3-}$ octahedra sharing a common face are found (41). The complex $[NMe_4]_2[V_2O_2F_6(OH_2)_2]$ consists of dimeric anionic units $[\{VOF_3(OH_2)\}_2]^{2-}$ linked in chains by short hydrogen bonds (42).

The most studied oxide fluoride of vanadium is VOF_3 (43), which has been prepared by many methods such as the fluorination of V_2O_5, using elemental fluorine (44–46), ClF_3, or BrF_3 (47, 48), the reaction of $VOCl_3$ with BrF_3 (49), the reaction of oxygen with VF_3 at red heat (26), the reaction of V_2O_5 with NF_3 (50) or NOF (51), and the thermal decomposition of VO_2F (52). It is a yellow solid, monomeric in the vapor phase, which hydrolyzes to V_2O_5. The adducts $2VOF_3 \cdot 3IOF_3$ (53) and $VOF_3 \cdot 2SeF_4$ (54, 55) are formed from the reaction of VOF_3 with IF_5 and of V_2O_5 with SeF_4, respectively. The latter adduct probably is best formulated as a salt, $[SeF_3]_2[VOF_5]$. Vanadium oxide trifluoride reacts with Me_3SiNEt_2 to give $VOF_{3-n}(NEt_2)_n$ and reacts similarly with Me_3SiOMe. However, with $(Me_3Si)_2O$, VO_2F is the only product (56).

The action of a fluorine–nitrogen mixture on VO_2Cl at 75–80°C yields the compound VO_2F. It is stable up to 300°C and is insoluble in nonpolar solvents (52). Its infrared spectrum suggests that it may be formulated as $[VO_2]^+F^-$, and this is consistent with its reaction with the Lewis acid SbF_5; with an excess of antimony pentafluoride at 250°C, VO_2F forms a 1:1 adduct, $[VO_2]^+[SbF_6]^-$; at 165°C, $[VO_2]^+[Sb_2F_{11}]^-$ is formed; and at 100–110°C, $[VO_2]^+[Sb_3F_{16}]^-$ is the product (52).

The oxide fluoride NbO_2F is prepared by treating Nb_2O_5 with aqueous HF (57) or by the reaction of NbF_5 with ground glass above 400°C (58). Thermal decomposition of NbO_2F (59, 60) yields $NbOF_3$ at 700°C and Nb_3O_7F above 840°C. When Nb_2O_5 reacts with potassium fluoride, NbO_2F and Nb_3O_7F are the products (61).

A number of complex niobium(V) oxide fluoride derivatives, such as

$Nb_{59}O_{147}F$ and $Nb_{65}O_{161}F_3$ (*65*), have been reported to be formed in the Nb_2O_5–NbO_2F system at 1250°C (*62–66*).

Reaction of tantalum metal or tantalum(V) oxide with aqueous HF yields TaO_2F (*57, 67*), which on strong heating yields Ta_3O_7F and TaF_5 (*68*). The reaction of TaO_2F with an NH_4F–HF mixture in a 1:3 ratio at 130–190°C yields $[NH_4]_2[TaF_7]$, which decomposes at 200–300°C to give $[NH_4][TaF_6]$ (*69*).

$TaOF_3$ has been reported as the product of the reaction of TaF_5 vapor with silica at high temperatures (*70*).

D. THE OXIDE FLUORIDES OF CHROMIUM, MOLYBDENUM, AND TUNGSTEN

1. Chromium Compounds and the Lower Oxide Fluorides of Molybdenum and Tungsten

The compounds $CrOF_4$ and CrO_2F_2 are well characterized. The former can be prepared as a dark red solid (mp 55°C) by fluorinating CrO_3 (*71*) or chromium metal (*72*) in a flow system. Chromyl fluoride (CrO_2F_2) has been prepared by numerous methods (*51, 55, 73–79*). This reasonably reactive oxide fluoride is also dark red (mp 31.6°C) but polymerizes to yield a white solid on exposure to sunlight. The gas phase Raman and matrix infrared spectra of CrO_2F_2 have been measured, and clear assignments of the fundamental frequencies have been made (*80*). With graphite, CrO_2F_2 forms a lamellar compound (*81*). Thermal decomposition of CrO_2F_2 at 500°C results in the production of the lower chromium oxide fluoride $CrOF_2$ (*82*), which is stable *in vacuo* up to 1600°C. This reacts with NO_2F and NOF to give $[NO_2][CrOF_3]$ and $[NO][CrOF_3]$, respectively (*83*).

Although magnetic measurements indicated the occurrence of $CrOF_3$ (*84, 85*), early attempts to isolate it failed (*84–86*), but in 1982 its existence was fully confirmed (*87*). The red solid $CrOF_3 \cdot x$ClF (x = 0.10–0.21), prepared by reaction between CrO_3 or CrO_2F_2 and ClF, was converted to $CrOF_3$ by multiple treatments with fluorine at 393 K. The compound decomposes to CrF_3 at 773 K and is mildly hygroscopic, giving Cr(VI) and Cr(III). A fluorine-bridged polymeric structure has been suggested for the compound. With potassium fluoride in HF, it yields $K[CrOF_4]$ (*87*).

The reaction of Cr_2O_3 with HF gives a dark olive-green compound at high temperatures, and this has been identified as CrOF (*88*). Further evidence for the existence of this chromium(III) oxide fluoride was provided during the study of the Cr_2O_3–CrF_3 system (*89*). The related molybdenum(III) oxide fluoride, MoOF, has been prepared by the fusion of MoOCl with ammonium fluoride (90), and MoOF$_3$ has been similarly prepared from MoF$_4$ and MoOF$_4$ (91).

A gray, involatile solid product of the reaction of WO$_2$ with anhydrous HF at 500–800°C was initially reported to be WOF$_2$ (92). It has since been recognized that the product of the reaction is actually tungsten metal.

The oxide tetrafluorides and dioxide difluorides of molybdenum and tungsten are the most studied transition-metal oxide fluorides, and their preparation and properties are discussed separately below.

2. Molybdenum and Tungsten Oxide Tetrafluorides

Both molybdenum and tungsten oxide tetrafluorides were first reported by Ruff and his co-workers who obtained them from the appropriate oxide tetrachloride with anhydrous HF (93, 94). Reaction of the metals Mo and W with potassium nitrate in anhydrous HF produced MoOF$_4$ and WOF$_4$, respectively, together with by-products such as WF$_6$ (76). These oxide fluorides can also be prepared by heating the metals in an oxygen–fluorine mixture (95), by treating the trioxides with lithium fluoride (96, 97), and by the reaction of the hexafluorides with boron oxide, B$_2$O$_3$ (98). The molybdenum compound can also be obtained from the hydrolysis of MoF$_6$ with water in hydrogen fluoride solvent (99). Tungsten oxide tetrafluoride can be obtained by fluorination of WO$_3$, using CrF$_2$ (100) or WF$_6$ (101), fluorination of WO$_2$ by difluorodichloromethane at 500°C (102), and from thermal decomposition of EtWOF$_5$ (103). Undoubtedly, the most convenient method for the preparation of large quantities of materials is the thermal reaction of the metals with an oxygen–fluorine mixture in a dynamic system (95).

MoOF$_4$ and WOF$_4$ are colorless, crystalline solids melting at 95 and 104°C, respectively, and their volatility is such that sublimation *in vacuo* is easily accomplished. Vapor-density measurements (104) and gas-phase infrared data (96, 105) have shown that both are predominantly monomeric in the vapor phase. Electron diffraction (106) and photoelectron (107) spectra have been interpreted in terms of the C_{4v} symmetry for the gaseous molecules first demonstrated by infrared work (108) and later confirmed by matrix-isolation studies in nitrogen (109). In the solid state, MoOF$_4$ has been found as a chain structure, in which the MoOF$_4$ molecules are linked by cis fluorine bridges (110) and as a trimeric, fluorine-bridged ring structure (111). The existence of these and possibly other polymeric forms is supported by Raman spectra for sublimed $MoOF_4$ and for $MoOF_4$ obtained by crystallization from a melt, each indicating a different structure (112). The crystal structure of WOF_4 (113) was interpreted at first as an oxygen-bridged tetrameric arrangement, but studies of the solid and liquid phases by vibrational spectroscopy (104, 112, 114) gave results inconsistent with this model. Further infrared studies, using labeled $W^{18}OF_4$, clearly showed the presence of terminal oxygen atoms, and ^{19}F-NMR evidence showed the presence of both terminal and bridging fluorine atoms (115).

Chemically, the oxide tetrafluorides of molybdenum and tungsten are weak fluoride-ion acceptors (i.e., weak Lewis acids) with WOF_4 the stronger (116, 117). Their hydrolysis in liquid hydrogen fluoride has resulted in the isolation of $[H_3O]^+[WOF_5]^-$ and observation of Raman evidence for $[MOF_5]^-$ (99). In anhydrous HF, the dimeric ions $[M_2O_2F_9]^-$ (M = Mo or W) have been identified (116). The oxide fluorides react with the strong bases NOF and $ClOF_3$ to give the ionic adducts $[NO][M_2O_2F_9]$, $[NO]_2[MOF_6]$ (M = Mo or W), $[ClOF_2]$-$[Mo_2O_2F_9]$, and $[ClOF_2][MoOF_5]$ (116), and with CsF, WOF_4 yields $Cs[WOF_5]$ (118). The synthesis and some of the chemistry of $Cs[WOF_5]$, $Cs[W_2O_2F_9]$, and their relatives have been checked and discussed, and the preparation of $[NH_4]^+$ salts have been reported (119). Ionic complexes containing the $[MOF_5]^-$ (M = Mo or W) anion have been reported also from reactions of the hexafluorides with alkali metal fluorides in the presence of moisture (120, 121), with $WF_6 \cdot P(CH_3)_3$ in SO_2 (101), MF_6 with $(RO)_2SO$, $(MeO)_3P$ (103), or EtOH (122), or with SO_2 (120, 121). The reaction of WO_3 with NF_3 (30) and with KF in SeF_4 (55) yield $[NO][WOF_5]$ and $K[WOF_5]$, respectively, and WF_6 reacts with moist sodium iodide in IF_5 to give $Na[WOF_5]$ (121). The dimeric, fluorine-bridged anions $[M_2O_2F_9]^-$, referred to earlier, have also been identified by ^{19}F-NMR studies of MOF_4 (M = Mo or W) and acetylacetone in MeCN (123) and the tungsten anion in the reaction of $WOCl_4$ with anhydrous HF in MeCN (124). The mixed anion $[F_4OMoFWOF_4]^-$ has also been observed in the reaction of $MoOF_4$ with $[WOF_5]^-$ in MeCN (125).

In liquid ammonia, $MoOF_4$ forms the adducts $MoOF_4 \cdot 5NH_3$ and $MoOF_4 \cdot 2NH_3$, which decompose thermally to give $MoOF_3 \cdot NH_3$ and $MoOF_2 \cdot NH_3$ (126). The reactions of MOF_4 (M = Mo or W) with alcohols and other donor ligands have been studied, and a range of addition and substitution-with-addition products (127–129) have been produced. For example, $MoOF_4 \cdot MeCN$, $MoOF_3 \cdot acac$, $MoF_3(OEt) \cdot MeCN$, $MOF_3(OEt) \cdot EtOH$, $WOF_2(OEt)_2 \cdot EtOH$, $WOF_3(OMe) \cdot MeCN$, and $WOF_{4-n}(OMe)_n \cdot MeOH$ have been identified by ^{19}F-NMR spectroscopy. The addition of WF_6 to EtOH results in the formation of $WOF_4 \cdot$ EtOH (*122*), and furthermore, the reaction of WOF_4 with hydrogen peroxide produces 10 hydroxo- and peroxofluorotungsten complexes (*130*).

Oxygen substitution reactions have also been reported to result from the reaction of WOF_4 with primary amines and diamines (*131*), for example, reaction with $BuNH_2$ in MeCN gives $W(NBu)F_4 \cdot$ MeCN, $[W(NBu)F_5]^-$, $[W_2(NBu)_2F_9]^-$, and $W(NBu)F_4 \cdot BuNH_2$.

Both $MoOF_4$ and WOF_4 react with xenon difluoride to yield the adducts $XeF_2 \cdot nMOF_4$ (M = Mo or W; n = 1 or 2), which are, on the basis of their vibrational and NMR spectroscopy, essentially covalent and contain Xe-F-M bridges both in the solid state and in solution (*132–135*), and the covalent nature of the $XeF_2 \cdot WOF_4$ adduct in the solid state has been confirmed by a single crystal structure determination (*133*). Longer chain species (n = 1–4) have been characterized at low temperatures by ^{19}F-NMR studies in SO_2ClF, and in solutions containing $XeF_2 \cdot nWOF_4$ (n = 2, 3) there is evidence for isomerization between oxygen- and fluorine-bridged Xe–F groups (*134*). The first adducts of krypton difluoride with an oxide fluoride have recently been prepared by reaction of KrF_2 with MOF_4 (M = Mo or W) in SO_2ClF at low temperature, and their ^{19}F-NMR and Raman spectra have been recorded (*136*). The ^{19}F-NMR spectra of $KrF_2 \cdot nMoOF_4$ (n = 1–3) and $KrF_2 \cdot WOF_4$ in solution show that they are essentially covalent, with Kr–F–M bridging to the metal oxide fluoride moieties. Studies of equilibria involving Kr–O–W- and Kr–F–W-bridged species suggest that stable Kr—O bonds are unlikely. Raman spectra of solid 1:1 adducts suggest that, as in solution, the compounds have covalent, bridged structures (*136*).

The adducts $WOF_4 \cdot IF_5$, $WOF_4 \cdot SeF_4$, and $WOF_4 \cdot SeOF_2$ have been produced indirectly by reaction of WO_3 with IF_5 and SeF_4, respectively (*55*).

No adducts are formed when WOF_4 is allowed to interact with AsF_5, PF_5, or BF_3, but with SbF_5, both $MoOF_4$ and WOF_4 form colorless fluorine-bridged polymeric adducts. There is also some evidence for $MoOF_4 \cdot 2SbF_5$, but as expected, no evidence for $WOF_4 \cdot 2SbF_5$ was found, the WOF_4 adducts being less stable than those of $MoOF_4$. The complexes possess little ionic character, this having been amply demonstrated by a single-crystal X-ray structure determination on $MoOF_4 \cdot SbF_5$, which has shown that it is best regarded as a polymeric zigzag chain with alternating antimony and molybdenum atoms linked through fluorine bridges. Only minor contributions to the bonding from the ionic formulation $[MoOF_3][SbF_6]$ are evident (*137*).

3. Molybdenum and Tungsten Dioxide Difluorides

Molybdenum dioxide difluoride, a moisture-sensitive, white solid, which sublimes at 273°C, was first reported in 1907 from the reaction of MoO_2Cl_2 with anhydrous HF (93). This high-temperature, high-pressure reaction has been confirmed (135). A more convenient method of preparation, however, is the fluorination of MoO_2F_2 with XeF_2 at low temperature (135). The hydrolysis of MoF_6 in hydrogen fluoride solution at $-5°C$ (138), also produces MoO_2Cl_2, but this reaction is difficult to control. The isolation of MoO_2F_2 from the reaction of MoO_3 with IF_5 has been claimed (112), and it has been suggested that it can be obtained from the reaction of $MoOF_4$ with MoO_3 (91). Subsequent work suggests that these claims are in error; however, it appears that MoO_2F_2 is present in reaction mixtures of MoO_2 with CrF_2 at 700°C, with MnF_2 at above 900°C (100), and in MoO_3–LiF mixtures under oxygen at 500°C (96).

Early attempts to isolate WO_2F_2 from the hydrolysis products of WOF_4 were unsuccessful (94); however, the compound was finally obtained by controlled hydrolysis in 1978 (135). Claims that it can be formed by the hydrolysis of WF_6 dissolved in anhydrous HF have been made (138), and it has been identified as one of the products of the reaction of WO_3 with CrF_2 at 600°C (100). Several attempts to prepare it by the interaction of WO_3 and WOF_4 have failed (139).

Reports of the chemical behavior of dioxide difluorides are many fewer than for the corresponding oxide tetrafluorides. Like $MoOF_4$, MoO_2F_2 yields molecular adducts with SeF_4 and IF_5 (55); however, the reactions of MO_2F_2 (M = Mo and W) with XeF_2 do not yield $XeF_2 \cdot MO_2F_2$ adducts, but $XeF_2 \cdot MOF_4$ complexes are formed (135). With oxalic acid, MO_2F_2 (M = Mo or W) forms the dimeric anions $[M_2O_4F_4(C_2O_4)]^{2-}$, which contain the nonlinear $[MO_2]^{2+}$ groups (140).

Many anionic complexes derived from the dioxide difluorides have been reported, but these have not been prepared from the oxide fluorides directly. Complexes of the general formulas $MO_2F_2L_2$ and $MO_2F_2L(H_2O)$ [M = W; L = MeCN, DMSO, acac, or $(CH_2OH)_2$] [M = Mo; L = H_2O, MeCN, DMSO, $(CH_3)_2CO$, and a series of alcohols] have been prepared (141, 142) as have the related adducts $WO_2F_2 \cdot$ bipy, $WOF_2 \cdot$ phen, and $WO_2F_2 \cdot 2(SCN)$ (143). The salts $Cs[WO_2F_3]$ (143) and $NO[MoO_2F_3]$ (50) have been reported, and the anions $[MoO_2F_4]^{2-}$ and $[WO_2F_4]^{2-}$ have been identified in the HF–MoO_3–H_2O system (144, 145), and their reactions with H_2O_2 have produced oxoperoxometallates such as $K_2[MoO(O_2)F_4] \cdot H_2O$, which has been studied by X-ray diffraction (146). The anions $[WO_2F_4]^{2-}$ and

[WO$_2$F$_3$L]$^-$ have been identified by ^{19}F-NMR when aqueous solutions of methanol or ethanol were added to a solution of WO$_3$ in 40% HF (*141*). Other reported anions include [WO$_2$F$_5$]$^{3-}$ and [W$_2$O$_4$F$_7$]$^{3-}$ (*147*).

4. Anionic Clusters Incorporating Molybdenum and Tungsten Oxide Fluoride Species

Trinuclear clusters play an important role in the chemistry of molybdenum and tungsten. Deep red, isostructural clusters containing both [Mo$_3$O$_4$F$_9$]$^{5-}$ (*148, 149*) and [W$_3$O$_4$F$_9$]$^{5-}$ (*149–151*) are later additions to this family of simplest types of cluster species. Their basic structures conform to the B_1-type of trinuclear electron-poor transition-metal clusters where the metals are in a distorted octahedral environment.

E. THE OXIDE FLUORIDES OF MANGANESE, TECHNETIUM, AND RHENIUM

The only known manganese oxide fluoride is MnO$_3$F. It is a dark-green solid, which melts at $-78°$C to give a dark green liquid but decomposes explosively at room temperature. It is prepared from potassium permanganate by reaction with anhydrous HF (*76, 152*), IF$_5$ (*153*), or HSO$_3$F (*152*). It reacts with gaseous hydrogen chloride to give MnO$_3$Cl, and there is also evidence for the existence of MnO$_3$F · HF complexes (*154*).

Fluorination of technetium metal produces blue and green polymorphs of TcOF$_4$ (*155, 156*), the oxygen apparently originating from the surface of the metal or the reactor walls. The monoclinic blue form has been shown to be isostructural with the corresponding rhenium compound, which has a chain structure (*105, 157*). The green hexagonal polymorph contains octahedrally coordinated units linked into trimers by cis-bridged fluorine atoms (*158*). The first evidence for TcO$_3$F was provided by a mass-spectrometric study of the reaction between Tc$_2$O$_7$ and UF$_4$ (*159*), and the compound was later isolated as a yellow solid by direct fluorination of TcO$_2$ (*160*). Franklin *et al.* (1982) have employed ^{99}Tc, ^{17}O, and ^{19}F-NMR spectroscopy in the characterization of two new technetium oxide fluorides, F$_2$O$_2$TcOTcO$_2$F$_2$ and TcO$_2$F$_3$ (*161*).

There are five oxide fluorides of rhenium, ReOF$_4$, ReOF$_3$, ReOF$_5$, ReO$_2$F$_3$, and ReO$_3$F. Early reports of ReO$_2$F$_2$ and ReOF$_2$ (*162, 163*) have not been substantiated and must be considered doubtful.

ReOF$_4$ can be prepared from ReF$_6$ by reaction with metal carbonyls

(*164*), with B_2O_3 (*98*), with ReO_3 at 300°C (*165*), or by hydrolysis (*166*). It can also be obtained by the reaction of $ReOF_5$ with Re (*167*). It is a blue, crystalline solid, isostructural with the chain structure forms of $TcOF_4$ and $MoOF_4$ (*105*). Electron-diffraction (*106*) and photoelectron spectroscopic (*106*) studies have confirmed the expected C_{4v} symmetry for $ReOF_4$ in the gaseous state. The oxide tetrafluoride reacts slowly with Pyrex glass at 250°C to give a second oxide fluoride, $ReOF_3$ (*163*), which is formed as a black, nonvolatile solid.

$ReOF_5$ is prepared by treating rhenium metal with a fluorine–oxygen mixture in a flow system (*95*) or by the reaction of anhydrous rhenium dioxide with fluorine at about 250°C in a static reactor (*168*). Both $ReOF_5$ and ReO_2F_3 are obtained from the direct fluorination of ReO_2 and from the reaction of potassium perrhenate, $KReO_4$, with elemental fluorine (*165, 169*). The other rhenium(VII) oxide fluoride, ReO_3F, is prepared by refluxing $KReO_4$ with IF_5 (*165*), by treating ReO_3Cl (*152*), Re_2O_7, or $KReO_4$ with anhydrous HF (*170*), or by the reaction of $ReOF_5$ with Re_2O_7 (*167*).

Electron diffraction studies on gaseous $ReOF_5$ (*171*) have confirmed that the molecule has C_{4v} symmetry as indicated by earlier vibrational work (*168*).

There are rather few reported reactions of rhenium oxide fluorides. However, fluoride ion acceptor properties have been demonstrated for $ReOF_4$, $ReOF_5$, ReO_2F_3, and ReO_3F by the formation of the salts $MReOF_5$, $MReOF_{5.5(6.0)}$, $MReO_2F_4$, $MReO_3F_2$, and $M_2ReO_3F_3$ by reaction with the fluorides MF (M = Na, K, Rb, and Cs) (*167, 172*) and of $NO[ReOF_6]$ and $NO_2[ReOF_6]$ by reaction with NOF and NO_2F, respectively (*173*). Conductivity measurements on solutions of the oxide tetrafluorides of molybdenum, tungsten, and rhenium have shown $ReOF_4$ to be a weak fluoride ion acceptor, though stronger than $MoOF_4$ and WOF_4 (*117*). Raman and ESR measurements have shown that partial hydrolysis of ReF_6 in hydrogen fluoride gives green $[ReOF_5]^-$, and transient, blue colors in the solution have been attributed to polymeric intermediates (*174*). Recent work on the reaction of $ReOF_4$ with SbF_5 (*137*) has shown that the 1:1 adduct formed, $ReOF_4 \cdot SbF_5$, is also a fluorine-bridged structure. Unlike $MoOF_4 \cdot SbF_5$ and $ReOF_4$ itself, which have chain structures, the structure of $ReOF_4 \cdot SbF_5$ contains pairs of the adduct unit linked via fluorine bridges into distorted ruthenium pentafluoride like tetramers. Calculated ionicity values for the structure show that the molecule is essentially covalent and suggest that ionic contributions are not a requirement for the formation of a stable and ordered adduct (*174*).

Rhenium oxide fluorides have not escaped activity concerned with intercalation of fluorides into graphite and other polymers. Intercalation of $ReOF_5$ into graphite is accompanied by the liberation of large amounts of ReF_6. Oxidation of the host graphite is evident from the formation of CO_2 and COF_2, and the intercalated species are believed to be predominantly ReF_4 and ReF_6 (175).

F. THE OXIDE FLUORIDES OF IRON, RUTHENIUM, AND OSMIUM

Strong heating of a mixture of FeF_3 and Fe_2O_3 under an atmosphere of oxygen yields the only known iron oxide fluoride as the dimer $Fe_2O_2F_2$ (14, 88, 176).

The only ruthenium oxide fluoride, $RuOF_4$, was first reported in 1963. The stable, pale green solid, which was the product of the reaction of ruthenium metal with a bromine trifluoride–bromine mixture in Pyrex glass at 20°C, had a vapor pressure and magnetic susceptibility in accord with those expected for an oxide tetrafluoride (177). Later work (178) suggested that $RuOF_4$ is the unstable product formed by the fluorination of RuO_2 at temperatures in excess of 400°C. Although chemical analysis in this second case was not successful, a F:Ru mole ratio of 4.0 was established, an excellent mass spectrum was obtained, and mass spectrometry also confirmed the release of oxygen from the product. This suggests that further characterization of the earlier product may be necessary.

Similar difficulties surround reports of the preparation of $OsOF_4$. Early reports claiming its successful synthesis (179, 180) are unreliable. However, the compound has been definitively characterized as a golden-yellow solid, resulting from the reduction of $OsOF_5$, using a hot tungsten filament (181), by the reaction of OsF_6 with B_2O_3 (98), and by the hydrolysis of OsF_6 with an equimolar quantity of water in anhydrous HF at room temperature (182). Electron-diffraction studies have confirmed the expected C_{4v} symmetry for the molecule in the gas phase (106), and the photoelectron spectrum has been interpreted on a similar basis (107).

Two osmium(VII) oxide fluorides can be obtained. Passage of a fluorine–oxygen mixture (in a 2:1 ratio) over osmium metal or fluorination of OsO_2 at 250°C yields green crystals of $OsOF_5$ (183). The dioxide trifluoride (OsO_2F_3) has been synthesized thermally by the reaction of OsO_4 with OsF_6 or of $OsOF_4$ with OsO_3F_2. This yellow-green compound is quite stable at room temperature, but at 60°C, at which it has a significant vapor pressure, it disproportionates to form an equilibrium mixture with OsO_3F_2 and $OsOF_4$, and at higher temperatures (~110°C), OsF_6 and OsO_4 are also involved in the equilibrium. Crystals of OsO_2F_3 are apparently isomorphous with the monoclinic α

phase of OsO_3F_2, and although the single-crystal structure is unknown, Raman data suggest a fluorine-bridged polymeric arrangement (184).

The possibility that a third osmium(VII) oxide fluoride exists remains, since a dimorphic crystalline material of unknown composition, but thought to be either OsO_2F_2 or OsO_3F, was produced during the reaction of OsF_6 with Pyrex glass (185).

OsO_3F_2 is formed as an orange solid when OsO_4 is treated with BrF_3 at 50°C or when a 1 : 2 fluorine–oxygen mixture is passed over osmium metal (179). One monoclinic (α) and two orthorhombic (β and γ) forms are known (186). Efforts to prepare other osmium(VIII) oxide fluorides, such as $OsOF_6$ and OsO_2F_4, have not only failed, but the results of this (186) and later (184) work suggest that such compounds will disproportionate readily.

G. THE OXIDE FLUORIDES OF IRIDIUM AND PLATINUM

The early report by Ruff that $IrOF_4$ is formed when IrF_6 is contacted with glass or moisture (187) has not been confirmed.

The platinum oxide fluoride, originally assigned the formula $PtOF_4$, which was reported to be one of the products of fluorination of platinum sponge or anhydrous platinum compounds, including the tetraiodide or dichloride (188), is now well known to be $[O_2]^+[PtF_6]^-$. A light-brown solid, prepared by treating PtO_2 with fluorine at 200°C or by passing a fluorine–nitrogen stream over a hot mixture of platinum and powdered glass, is reported to be $PtOF_3$ (189). A further, as yet unidentified, nonstoichiometric oxide fluoride was also observed in this latter reaction.

III. The Oxide Fluorides of the Lanthanide Elements

The lanthanides, unlike the transition metals and the actinides, tend not to form compounds over a range of oxidation states. The +3 oxidation state is characteristic of all of the lanthanides, and the oxide fluorides of formula LnOF (Ln = lanthanide metal) are well known. The less stable oxidation states of +2 and +4 are known, but the latter is represented only by the dioxides and tetrafluorides of cerium, praseodymium, and terbium, and no tetravalent oxide fluorides have been reported.

The trivalent oxide fluorides LnOF can generally be prepared by heating the oxide Ln_2O_3 with the corresponding fluoride LnF_3 in air at

1000–1000°C (*190–196*) and by the thermal decomposition of the trifluoride at 800°C in air (*191, 197, 198*). The lanthanum compound itself may also be prepared by hydrolysis of the trifluoride (*199*) and by the reaction of the oxide with molten sodium fluoride (*200*). On treatment with $CFCl_3$ (*201*), it is converted back to the trifluoride. The cerium analog has been prepared from CeO_2 by reaction with CeF_3 at 2750°C (*202*) or with CeF_3 and cerium metal at 900°C in a nickel tube (*203*). The infrared spectra of these solids have been reported (*204*).

IV. The Oxide Fluorides of the Actinide Elements

The known actinide oxide fluorides are shown in Table II.

A. THE TRIVALENT ACTINIDE OXIDE FLUORIDES

The oxide fluorides AcOF, PuOF, and CfOF have been prepared by the hydrolysis of the corresponding trifluoride (*197, 205, 206*). Thorium oxide fluoride is prepared from a stoichiometric Th–ThF_4–ThO_2 mixture kept at 1200°C in a sealed tube for 4 days (*207*). All of the compounds are high-melting-point, nonvolatile solids of which AcOF is white, ThOF is gray-white, and CfOF is light green. Their crystal structures have been extensively studied (*197, 206, 208, 209*).

B. THE TETRAVALENT ACTINIDE OXIDE FLUORIDES

Thorium oxide fluoride ($ThOF_2$) is produced when ThF_4 is heated in air above 300°C (*210*), when ThF_4 reacts with ThO_2 at 900°C in an inert atmosphere (*211*), and when $ThF_4 \cdot 2H_2O$ is heated to red heat *in vacuo* (*212*). The crystal structure of the compound was reported by Zachariasen (*208*), who pioneered much of the early solid-state work on the actinide oxide fluorides. Green precipitates of $UOF_2 \cdot 2H_2O$ and $UOF_2 \cdot H_2O$ are reported to form in aqueous solutions of U(IV) (*213*), and heating of the latter hydrate to 100–280°C yields the anhydrous UOF_2 (*214*). This is reported to be stable up to 900°C; however, this claim seems to be inconsistent with the results of studies on the UF_4–UO_2 and UF_4–ThO_2 systems at 400–1100°C, which failed to produce UOF_2 (*215*).

C. THE PENTAVALENT ACTINIDE OXIDE FLUORIDES

The oxide fluoride U_2OF_8 (*216*) was reported as one of the intermediate products of the reaction of UO_2 with UF_6, and evidence for UOF_3

TABLE II
ACTINIDE OXIDE FLUORIDES

Oxidation state	Ac	Th	Pa	U	Np	Pu	Am	Cm	Bk	Cf	Es
+3	AcOF									CfOF	
+4		ThOF$_2$									
+5			Pa$_2$OF$_8$	UOF$_2$ U$_2$OF$_8$	NpOF$_3$ NpO$_2$F	PuOF (PuOF$_3$)					
			PaO$_2$F Pa$_3$O$_7$F	UO$_2$F							
+6				UOF$_4$ UO$_2$F$_2$ U$_2$O$_3$F$_6$ U$_2$O$_5$F$_2$ U$_3$O$_5$F$_8$	NpOF$_4$ NpO$_2$F$_2$	PuOF$_4$ PuO$_2$F$_2$	AmO$_2$F$_2$				

was found in the same reaction. The white solid Pa_2OF_8 is produced when hydrated Pa_2O_5 reacts with gaseous HF at 140°C (217) or when hydrates of PaF_5 are thermally decomposed (218).

Thermal decomposition of Pa_2OF_8 (217) at 220–290°C yields a white solid, PaO_2F, which also decomposes to Pa_3O_7F, which is also white, at 560–580°C. Thermal decomposition of U_2OF_8 does not follow the same path as that of Pa_2OF_8, and UO_2F is not formed (215). However, uranium oxide phases close to this composition have been reported, and their structures have been shown to contain O–U–O chain structures (219).

The only actinide oxide trifluoride isolated so far is $NpOF_3$, which has been prepared as a green solid from the reaction of Np_2O_5 with gaseous HF at 140°C (220). A solid that gave an analysis corresponding closely to that for the compound NpO_2F has been observed in the reduction of NpO_2F_2 by hydrogen (220). Finally, although $PuOF_3$ has never been isolated, it has been detected in the vapor phase by thermochromatography (221); the existence of its relative, $PaOF_3$, has been postulated (222), but as yet it has not been successfully synthesized.

D. The Hexavalent Actinide Oxide Fluorides

There are three known oxide tetrafluorides: UOF_4, $NpOF_4$, and $PuOF_4$. The first, uranium oxide tetrafluoride, was prepared by Wilson by the reaction of UF_6 with traces of water in anhydrous hydrogen fluoride (223, 224) and has since been prepared from the hexafluoride by reaction with quartz wool (SiO_2) in anhydrous HF (225) or boron oxide (B_2O_3) (98). It is an orange, hygroscopic powder, which yields UO_2F_2 with moist air, is almost insoluble in anhydrous HF and many organic solvents, and reacts with Nujol (224). It is thermally unstable, decomposing above 230°C to give UO_2F_2 and UF_6 via the intermediate $U_2O_3F_6$ (224, 226). Unlike the transition metal oxide tetrafluorides $MoOF_4$ and WOF_4, it is nonvolatile (224).

Uranium oxide tetrafluoride exists in two structural modifications. In the trigonal α form, there is a pentagonal bipyramidal arrangement of light atoms about the uranium with the two axial positions occupied by nonbridging oxygen and fluorine atoms. Of the five equatorial fluorine atoms, four are bridging and one is terminal (225, 227). The β form is tetragonal. This also has a pentagonal array of light atoms but with two axial terminal fluorines, a terminal equatorial oxygen, and four equatorial bridging fluorines (228). The interpretations of the infrared and Raman spectra (225, 229) are in accord with the crystallographic data.

The acid character of the compound has been demonstrated by its reaction with monovalent alkali metal and ammonium fluorides, which give rise to adducts of formulas $MUOF_5$ and M_3UOF_7 (M = alkali metal or ammonium) (116, 230–233), and by its reaction with nitrosyl fluoride to give the unstable $[NO][UOF_5]$ (116, 231). Weak donor properties in UOF_4 have also been revealed in the formation of the adducts $UOF_4 \cdot nSbF_5$ (n = 1–3), prepared by the reaction of UOF_4 with SbF_5 in SbF_5 or anhydrous HF solvents (234, 235). The structure of the 1:2 adduct $UOF_4 \cdot 2SbF_5$ has been shown to consist of a fluorine-bridged network of UOF_4 and SbF_5 molecules in which there is a slight tendency toward the ionic formulation $[U^{VI}OF_2][Sb^VF_6]_2$, and it has been suggested that the 1:1 and 1:3 adducts are related (235). The extent of this donor behavior has been investigated by examination of the reactions of a range of other Lewis-acid pentafluorides in addition to SbF_5 (236). Arsenic pentafluoride was found to be unreactive, but the adducts $UOF_4 \cdot 3MF_5$ (M = Nb, Ta) and $UOF_4 \cdot 2BiF_5$ were readily prepared, and like $UOF_4 \cdot 2SbF_5$, these are essentially fluorine bridged. Thermal decomposition of these does not yield lower UOF_4 adducts, but uranyl fluoride adducts and UF_6 are formed instead (236). This is particularly interesting since it has been shown that, from a thermochemical point of view, UOF_4 behaves as a loosely bound complex of UO_2F_2 and UF_6 (237).

The oxide fluorides $NpOF_4$ and $PuOF_4$, like UOF_4 were prepared by hydrolysis of the corresponding hexafluoride (238, 239), usually in anhydrous HF, and $NpOF_4$ has also been prepared by the reaction of NpO_2 with krypton difluoride (240, 241). The plutonium compound $PuOF_4$ is a dark-brown solid. It is isostructural with the trigonal forms of UOF_4 and $NpOF_4$, is stable at room temperature, and yet is unstable in anhydrous HF in which it readily dismutates to PuO_2F_2 and PuF_6 (239).

The dioxide difluorides of U, Np, Pu, and Am have all been isolated. Uranyl fluoride, being an important intermediate in the conversion of enriched UF_6 to UO_2 for the production of fuel rods for Advanced Gas-Cooled Nuclear Reactors, is undoubtedly the most studied. The majority of papers on UO_2F_2, therefore, are concerned with its formation from the reaction of UF_6 with steam or its conversion to UO_2 by reduction with hydrogen.

In the laboratory, anhydrous uranyl fluoride may be prepared by a variety of methods, the most favored being by the reaction of uranium trioxide with anhydrous hydrogen fluoride at 300°C or fluorine at 270–350°C (*215, 242*) and the thermal decomposition of H[UO$_2$F$_3$] · H$_2$O, prepared from UO$_3$ and aqueous HF (*243*).

Uranyl fluoride is a pale-yellow solid, which, on exposure to moist air, readily forms the dihydrate. It is stable in air up to 400°C and up to 700°C in a closed system. Above 760°C it dissociates to give U$_3$O$_8$, UF$_6$, and O$_2$ (*244, 245*). It is very soluble in water and UF$_6$ and is usually present in the latter.

Structural work on UO$_2$F$_2$ (*24, 246, 247*) has gradually been refined over the years, and a neutron diffraction study has shown that UO$_2$F$_2$ has a trigonal structure in which the uranium is eight-coordinate (*247*).

The infrared and Raman spectra have also been closely studied (*220, 242, 248*). A curious feature of the infrared spectra is that the position of the UO$_2$ asymmetric stretch is dependent on the method by which the sample is prepared (*220, 242*). When anhydrous conditions are employed, ν_3 (UO$_2^{2+}$) appears at 990 cm^{-1}, but if "wet" methods of preparation involving aqueous HF are used, the band appears at 1000 cm^{-1} (*248*). A recent spectroscopic study of UO$_2$F$_2$ in anhydrous HF–AsF$_5$ solutions has demonstrated the stability of the dioxouranium(VI) ion in this medium (*249*).

The majority of the reactions of uranyl fluoride fall into three categories. These are reactions resulting in the conversion of UO$_2$F$_2$ to the hexafluoride, reactions in which adducts are formed with neutral donor molecules, and reactions in which UO$_2$F$_2$ acts as fluoride ion acceptor. Evidence of weak basic character in UO$_2$F$_2$ has also been observed. The dioxide difluoride is converted to UF$_6$ by reaction with fluorine at temperatures above 300°C (*250*), with BrF$_3$ (*251*) or XeF$_6$ (*252*) at room temperature, with crystalline XeF$_2$ at 140°C and 10^{-1} torr (*253*), with ClF (*254*), ClF$_3$ (*254–256*), ClF$_5$ (*229*), and ClO$_2$F (*255*) at temperatures ranging from 50 to 160°C, with SF$_4$ above 250°C (*257*), and with VF$_5$ at 100°C (*258*). The hydrates UO$_2$F$_2$ · nH$_2$O (n = 1–4) have been identified (*259, 260*), and their crystal structures have been reported (*260*). Indeed, a wide variety of oxygen and nitrogen donor ligands such as NH$_3$, DMSO, DMF, 2,2'-dipyridyl, phosphine oxides, and urea form adducts of the type UO$_2$F$_2$ · L · (H$_2$O) and UO$_2$F$_2$ · nL (n = 1–4) (*215, 261–263*). Acid properties of uranyl fluoride have been observed several times in its reactions with alkali metals (*264–267*) and ammonium fluorides (*264, 265, 268, 269*), and salts incorporating the anions [UO$_2$F$_3$]$^-$, [UO$_2$F$_4$]$^{2-}$, [UO$_2$F$_5$]$^{3-}$, [UO$_2$F$_6$]$^{4-}$, [(UO$_2$)$_2$F$_5$]$^-$, and [(UO$_2$)$_2$F$_7$]$^{3-}$ have been observed. The isolation of the adducts UO$_2$F$_2$ · 2SbF$_5$ and UO$_2$F$_2$ · 3SbF$_5$ from the combination of UO$_2$F$_2$ with SbF$_5$

has also provided evidence of weak basic character (*270*). The room-temperature decomposition of UOF_4–SbF_5–HF mixtures, of $UOF_4 \cdot 2SbF_5$ (*271*), and of the related adducts with NbF_5, TaF_5, or BiF_5 (*236*) in anhydrous HF produces uranyl fluoride adducts also. The X-ray single-crystal structure of $UO_2F_2 \cdot 3SbF_5$ has shown that the solid consists of a three-dimensional network of UO_2F_2 and SbF_5 molecules linked by fluorine bridges (*270*). These observations give added weight to the suggestion that the product formed on heating a 1:1 mixture of UO_2F_2 and thorium tetrafluoride to 775°C might be $UO_2[ThF_6]$ (*245*).

Neptunyl fluoride (NpO_2F_2) is a pink solid prepared by the reaction of the hydrated trioxide with gaseous HF and by direct fluorination of Np_2O_5 at 350°C or $NpO_3 \cdot H_2O$ at 225°C (*208, 220*). Plutonyl fluoride (PuO_2F_2) is prepared by the hydrolysis of PuF_6 in aqueous solution (*272*) or in moist air (*273*) and by the reaction of PuO_2Cl_2 with anhydrous HF (*274*). It is a white solid, isomorphous with UO_2F_2 and forms the dihydrate readily.

Other actinide(VI) oxide fluorides exist. The thermal decomposition of UOF_4 at 290°C (*226*) and the reaction of uranyl fluoride with SeF_4 (*275*) yield a yellow solid, which is $U_2O_3F_6$. The compound $U_2O_5F_2$ is formed as the dihydrate in the UO_3–HF–H_2O system (*276*); and when UF_6 reacts with a small quantity of water, $U_3O_5F_8$ is the product (*275*).

E. OTHER ACTINIDE OXIDE FLUORIDES

Thorium oxide fluoride (Th_2OF_5) (*277*) has been prepared from ThF_4 and ThOF, and the reaction of uranium oxides with UF_4 at 400–500°C is said to produce U_2O_5F as one of the products (*278*).

Attempts to prepare $NpOF_5$ by the reaction of $NpOF_4$ with KrF_2 have failed (*238*).

REFERENCES

1. Glemser, O., Roesky, H. W., and Hellberg, K. H., *Angew Chem., Int. Ed. Engl.* **2**, 266 (1963).
2. Bethuel, L., *Commis. Energ. At. Fr. Serv. Doc., Ser. Bibliogr.* CEA-BIB-134, 17 (1969).
3. Glemser, O., Roesky, H. W., Hellberg, K.-H., and Werther, H.-U., *Chem. Ber.* **99**, 2652 (1966).
4. Bendeliani, N. A., *Dokl. Akad. Nauk SSSR* **223**, 1112 (1975).
5. Kutek, F., *Russ. J. Inorg. Chem.* **9**, 1499 (1964).
6. Ikrami, D. D., Dzhuraev, K. S., and Pirmatova, A. N., *Dokl. Akad. Nauk Tadzh. SSR* **12**, 34 (1969).
7. Atabaeva, E. Y., and Bendeliani, N. A., *Dokl. Akad. Nauk. SSSR* **233**, 359 (1977).

8. Fridman, Ya. D., Moshkina, V. A., Gorokhov, S. D., and Nitsevich, E. A., *Russ. J. Inorg. Chem.* **10,** 1347 (1965).
9. Orlovskii, V. P., *Russ. J. Inorg. Chem.* **12,** 10 (1967).
10. Samsonov, G. V., Paderno, Yu. B., and Fomenko, V. S., *Poroshk. Metall., Akad. Nauk. Ukr. SSR* **3,** 24 (1963).
11. Markovskii, L. Ya., Pesina, E. Ya., Loev, L. M., and Omel'chenko, Yu. A., *Russ. J. Inorg. Chem.* **15,** 2 (1970).
12. Bevan, D. J. M., and Mann, A. W., *Proc. Rare Earth Res. Conf., 7th.* **1,** 149 (1969).
13. Moskovich, Yu. N., Buznik, V. M., Fedorov, P. P., and Sobolev, B. P., *Krystallografiya* **23,** 416 (1978).
14. Chamberland, B. L., and Sleight, A. W., *Solid State Commun.* **5,** 765 (1967).
15. Vorres, K. S., and Dutton, F. B., *J. Am. Chem. Soc.* **77,** 2019 (1955).
16. Vorres, K. S., and Donohue, J., *Acta Crystallogr.* **8,** 25 (1955).
17. Dehnicke, K., *Naturwissenschaften* **52,** 660 (1965).
18. McVicker, G. B., and Eggert, J. J., *Ger. Offen.* **2,** 900, 854, 19 Jul. 37 (1979).
19. Waters, T. N., *J. Inorg. Nucl. Chem.* **15,** 320 (1960).
20. Kolditz, L., and Feltz, A., *Z. Anorg. Allg. Chem.* **310,** 204 (1961).
21. Kunaev, A. M., Galkin, N. P., Koslov, A. M., Mashirev, V. P., and Koslov, V. A., *Vestn. Akad. Nauk Kaz SSSR* **6,** 33 (1977).
22. Holmberg, B., *Acta Crystallogr., Sect. B* **26,** 830 (1970).
23. Kolditz, L., and Feltz, A., *Z. Anorg. Allg. Chem.* **310,** 217 (1961).
24. Nikolaev, N. S., Buslaev, Y. A., and Gustyakova, M. P., *Russ. J. Inorg. Chem.* **7,** 870 (1962).
25. Joubert, P., and Gaudreau, B., *Rev. Chim. Mineral.* **12,** 289 (1975).
26. Ruff, O., and Lickfett, R., *Ber. Dtsch. Chem. Gres.* **44,** 2539 (1911).
27. Buslaev, Y. A., and Gustyakova, M. P., *Izv. Akad. Nauk SSSR, Ser. Khim.* 1533 (1963).
28. Selbin, J., *Chem. Rev.* **65,** 153 (1963).
29. Pausewang, G., and Rüdorff, W., *Z. Anorg. Allg. Chem.* **364,** 69 (1969).
30. Pausewang, G., *Z. Anorg. Allg. Chem.* **381,** 189 (1971).
31. Baker, A. E., and Haendler, H. M., *Inorg. Chem.* **1,** 127 (1962).
32. Selbin, J., and Holmes, L. H., Jr., *J. Inorg. Nucl. Chem.* **24,** 1111 (1962).
33. Selbin, J., *J. Inorg. Nucl. Chem.* **31,** 433 (1969).
34. Chakraworti, M. C., and Sarkar, A. R., *J. Fluorine Chem.* **9,** 315 (1977).
35. Davidovich, P. L., Kharlamova, L. G., and Samarec, L. V., *Koord. Khim.* **3,** 850 (1977).
36. Demšar, A., and Bukovec, P., *Inorg. Chim. Acta* **25,** L121 (1977).
37. Waltersson, K., and Karlsson, *Cryst. Struct. Commun.* **7,** 459 (1978).
38. Bukovec, P., and Golič, L., *Acta Crystallogr., Sect. B* **36,** 1925 (1980).
39. Waltersson, K., *J. Solid State Chem.* **28,** 121 (1979).
40. Waltersson, K., *J. Solid State Chem.* **29,** 195 (1979).
41. Waltersson, K., *Cryst. Struct. Commun.* **7,** 507 (1978).
42. Bukovec, P., Milocev, S., Demšar, A., and Golič, L., *J. Chem. Soc. Dalton Trans.* 1802 (1981).
43. Selig, H., and Claassen, H. H., *J. Chem. Phys.* **44,** 1404 (1966).
44. Smalc, A., *Monatsh. Chem.* **98,** 163 (1967).
45. Haendler, H. M., Bartram, S. F., Becker, R. S., Bernard, W. J., and Bukata, S. W., *J. Am. Chem. Soc.* **76,** 2177 (1954).
46. Trevorrow, L. E., *J. Phys. Chem.* **62,** 362 (1958).
47. Eméleus, H. J., and Gutmann, V., *J. Chem. Soc.* 2979 (1949).

48. Emeléus, H. J., and Woolf, A. A., *J. Chem. Soc.* 164 (1950).
49. Blanchard, S., *J. Chem. Phys.* **61,** 749 (1964).
50. Glemser, O., Wegener, J., and Mews, R., *Chem. Ber.* **100,** 2474 (1967).
51. Aynsley, E. E., Heatherington, G., and Robinson, P. L., *J. Chem. Soc.* 1119 (1954).
52. Weidlein, J., and Dehnicke, K., *Z. Anorg. Allg. Chem.* **348,** 278 (1966).
53. Aynsley, E. E., Nichols, R., and Robinson, P. L., *J. Chem. Soc.* 623 (1953).
54. Peacock, R. D., *J. Chem. Soc.* 3617 (1953).
55. Bartlett, N., and Robinson, P. L., *J. Chem. Soc.* 3549 (1961).
56. Kolta, G. A., Sharp, D. W. A., and Winfield, J. M., *J. Fluorine Chem.* **14,** 153 (1979).
57. Frevel, L. K., and Rinn, H. W., *Acta Crystallogr.* **9,** 626 (1956).
58. Schafer, H., Schnering, H. G., Niehues, K. J., and Nieder-Vahrenholz, H. G., *J. Less-Common Metals* **9,** 95 (1965).
59. Andersson, S., and Astrom, A., *Acta Chem. Scand.* **19,** 2136 (1965).
60. Astrom, A., *Acta Chem. Scand.* **21,** 915 (1967).
61. Vaskresenskaya, N. K., and Budova, G. P., *Dokl. Akad. Nauk. SSSR* **170,** 329 (1966).
62. Andersson, S., and Astrom, A., *Acta Chem. Scand.* **18,** 2233 (1964).
63. Andersson, S., *Acta Chem. Scand.* **18,** 2339 (1964).
64. Andersson, S., *Acta Chem. Scand.* **19,** 1401 (1965).
65. Norin, R., *Acta Chem. Scand.* **25,** 347 (1971).
66. Gruehn, R., *Naturwissenschaften* **54,** 645 (1967).
67. Jahnberg, L., and Andersson, S., *Acta Chem. Scand.* **21,** 615 (1967).
68. Buslaev, Y. A., Kharitinov, Y. Y., and Sinitsyna, S. M., *Russ. J. Inorg. Chem.* **10,** 287 (1965).
69. Bratishko, V. D., Rakov, E. G., Sudarikov, B. N., Gromov, B. V., and Schlyakhter, *Zh. Neorg. Khim.* **18,** 712 (1973).
70. Schafer, H., Bauer, D., Beckmann, W., Gerken, R., Nieder-Vahrenholz, H. G., Niehues, K. J., and Schloz, H., *Naturwissenschaften* **51,** 241 (1964).
71. Edwards, A. J., Falconer, W. E., and Sunder, W. A., *J. Chem. Soc. Dalton Trans.* 541 (1974).
72. Edwards, A. J., *Proc. Chem. Soc.* 205 (1963).
73. Von Wartenburg, H., *Z. Anorg. Allg. Chem.* **247,** 135 (1941).
74. Engelbrecht, A., and Gross, A. V., *J. Am. Chem. Soc.* **74,** 5262 (1952).
75. Flesch, G. D., and Svec, H. J., *J. Am. Chem. Soc.* **80,** 3189 (1958).
76. Wiechert, K., *Z. Anorg. Allg. Chem.* **261,** 310 (1950).
77. Krauss, H. L., and Schwartzbach, F., *Chem. Ber.* **94,** 1205 (1961).
78. Smith, W. T., *Angew. Chem., Int. Ed. Engl.* **1,** 467 (1962).
79. Green, P. J., and Gard, G. L., *Inorg. Chem.* **16,** 1243 (1977).
80. Beattie, I. R., Marsden, C. J., and Ogden, J. S., *J. Chem. Soc. Dalton Trans.* 535 (1980).
81. Croft, R. C., and Thomas, R. C., *Nature (London)* **168,** 32 (1951).
82. Rochat, W. V., Gerlach, J. N., and Gard, G. L., *Inorg. Chem.* **9,** 998 (1970).
83. Green, P. J., and Gard, G. L., *Inorg. Nucl. Chem. Lett.* **14,** 179 (1978).
84. Clark, H. C., and Sadana, Y. N., *Can. J. Chem.* **42,** 50 (1964).
85. Clark, H. C., and Sadana, Y. N., *Can. J. Chem.* **42,** 702 (1964).
86. Sharpe, A. G., and Woolf, A. A., *J. Chem. Soc.* 798 (1951).
87. Green, P. J., Johnson, B. M., Loehr, T. M., and Gard, G. L., *Inorg. Chem.* **21,** 3562 (1982).
88. Von Wartenberg, H., *Z. Anorg. Allg. Chem.* **249,** 100 (1942).
89. Petit, G., and Bourlange, C., *C.R. Acad. Sci. Paris, Ser C* **273,** 1065 (1971).

90. Wardlaw, W., and Wormwell, R. L., *J. Chem. Soc.* 2370 (1924).
91. Opalovskii, A. A., Anufrienko, V. F., and Khaldoyanidi, K. A., *Dokl. Akad. Nauk SSSR* **184**, 860 (1969).
92. Priest, H. F., and Schumb, W. C., *J. Am. Chem. Soc.* **70**, 3348 (1948).
93. Ruff, O., and Eisner, F., *Ber. Dtsch. Chem. Ges.* **40**, 2926 (1907).
94. Ruff, O., Eisner, F., and Heller, W., *Z. Anorg. Allg. Chem.* **52**, 256 (1907).
95. Cady, G. H., and Hargreaves, G. B., *J. Chem. Soc.* 1568 (1961).
96. Ward, B. G., and Stafford, F. E., *Inorg. Chem.* **7**, 2569 (1968).
97. Von Schmitz-Dumont, O., Bruns, I., and Heckmann, I., *Z. Anorg. Allg. Chem.* **271**, 347 (1953).
98. Burns, R. C., O'Donnell, T. A., and Waugh, A. B., *J. Fluorine Chem.* **12**, 505 (1978).
99. Selig, H., Sunder, W. A., Schilling, F. C., and Falconer, W. E., *J. Fluorine Chem.* **11**, 629 (1978).
100. Zmbov, K. F., Uy, O. M., and Margrave, J. L., *J. Phys. Chem.* **73**, 3008 (1969).
101. Tebbe, F. N., and Muetterties, E. L., *Inorg. Chem.* **7**, 172 (1968).
102. Webb, A. D., and Young, H. A., *J. Am. Chem. Soc.* **72**, 3356 (1950).
103. Noble, A. M., and Winfield, J. M., *J. Chem. Soc. (A)* 501 (1970).
104. Alexander, L. E., Beattie, I. R., Bukovszky, A., Jones, P. J., Marsden, C. J., and van Schalkwyk, G. J., *J. Chem. Soc. Dalton Trans.* 81 (1974).
105. Edwards, A. J., Jones, G. R., and Steventon, B. R., *J. Chem. Soc., Chem. Commun.* 462 (1967).
106. Alexeichuk, I. S., Ugarov, V. V., Rambidi, N. G., Legasov, V. A., and Sokolov, V. B., *Dokl. Akad. Nauk SSSR* **257**, 625 (1981).
107. Vovna, V. I., Dudin, A. S., Kleshchevnikov, A. M., Lopatin, S. N., and Rakov, E. G., *Koord. Khim.* **7**, 575 (1981).
108. Paine, R. T., and McDowell, R. S., *Inorg. Chem.* **13**, 2366 (1974).
109. Levason, W., Narayanaswamy, R., Ogden, J. S., Rest, A. J., and Turff, J. W., *J. Chem. Soc. Dalton Trans.* 2501 (1981).
110. Edwards, E. J., and Steventon, B. R., *J. Chem. Soc. (A)* 2503 (1968).
111. Edwards, A. J., Jones, G. R., and Sills, R. J. C., *J. Chem. Soc., Chem. Commun.* 1177 (1968).
112. Beattie, I. R., Livingstone, K. M. S., Reynolds, D. J., and Ozin, G. A. S., *J. Chem. Soc. (A)* 1210 (1970).
113. Edwards, A. J., and Jones, G. R., *J. Chem. Soc. (A)* 2074 (1968).
114. Beattie, I. R., and Reynolds, D. J., *J. Chem. Soc., Chem. Commun.* 1531 (1968).
115. Asprey, L. B., Ryan, R. R., and Fukushima, E., *Inorg. Chem.* **11**, 3122 (1972).
116. Bougon, R., Bui Huy, T., and Charpin, P., *Inorg. Chem.* **14**, 1822 (1975).
117. Paine, R. T., and Quarterman, L. A., *Inorg. Nucl. Chem., H. H. Hyman Mem. Vol.* 85 (1976).
118. Meinert, H., Friedrich, L., and Kohl, W., *Z. Chem.* **15**, 492 (1975).
119. Wilson, W. M., and Christe, K. O., *Inorg. Chem.* **20**, 4139 (1981).
120. Hargreaves, G. B., and Peacock, R. D., *J. Chem. Soc.* 4390 (1958).
121. Hargreaves, G. B., and Peacock, R. D., *J. Chem. Soc.* 2170 (1958).
122. Buslaev, Y. A., Kokunov, Y. V., and Bochkareva, V. A., *Russ. J. Inorg. Chem.* **16**, 1393 (1971).
123. Buslaev, Y. A., Kokunov, Y. V., Bochkareva, V. A., and Shustorovich, E. M., *J. Struct. Chem.* **13**, 491 (1972).
124. Buslaev, Y. A., Kokunov, Y. V., and Bochkareva, V. A., *J. Struct. Chem.* **13**, 570 (1972).

125. Buslaev, Y. A., Kokunov, Y. V., and Bochkareva, V. A., *Russ. J. Inorg. Chem.* **17,** 1774 (1972).
126. Blokhina, G. E., Belyaev, I. N., Opalovskii, A. A., and Belan, L. I., *Russ. J. Inorg. Chem.* **17,** 1113 (1972).
127. Buslaev, Y. A., Kokunov, Y. V., Bochkareva, V. A., and Shustorovich, E. M., *Dokl. Akad. Nauk SSSR* **201,** 925 (1971).
128. Buslaev, Y. A., Kokunov, Y. V., Bochkareva, V. A., and Shustorovich, E. M., *J. Inorg. Nucl. Chem.* **34,** 2861 (1972).
129. Buslaev, Y. A., Kuznetsova, A. A., Bainova, S. V., and Kokunov, Y. V., *Koord. Khim.* **3,** 216 (1977).
130. Buslaev, Y. A., Kokunov, Y. V., Bochkareva, V. A., and Gustyakova, M. P., *Koord. Khim.* **2,** 921 (1976).
131. Buslaev, Y. A., and Kokunov, Y. V., *Dokl. Acad. Nauk. SSSR* **233,** 357 (1977).
132. Holloway, J. H., Schrobilgen, G. J., and Taylor, P., *J. Chem. Soc., Chem. Commun.* 40 (1975).
133. Tucker, P. A., Taylor, P. A., Holloway, J. H., and Russell, D. R., *Acta Crystallogr. Sect. B* **31,** 906 (1975).
134. Holloway, J. H., and Schrobilgen, G. J., *Inorg. Chem.* **19,** 2632 (1980).
135. Atherton, M. J., and Holloway, J. H., *J. Chem. Soc., Chem. Commun.* **6,** 254 (1978).
136. Holloway, J. H., and Schrobilgen, G. J., *Inorg. Chem.* **20,** 3363 (1981).
137. Fawcett, J., Holloway, J. H., and Russell, D. R., *J. Chem. Soc. Dalton Trans.,* 1212 (1981).
138. Nikolaev, N. S., Vlasov, S. V., Buslaev, Y. A., and Opalovskii, A. A., *Izv. Sibir. Otdel Akad. Nauk SSSR* **10,** 47 (1960).
139. Burgess, J., Peacock, R. D., and Taylor, P., unpublished observations.
140. Calves, J. Y., Kergoat, R., and J. E. Guerchais, *C.R. Acad. Sci. Paris, Ser. C* **275,** 1423 (1972).
141. Buslaev, Y. A., Petrosyants, S. P., and Chagin, V. I., *Russ. J. Inorg. Chem.* **17,** 368 (1972).
142. Buslaev, Y. A., and Petrosyants, S. P., *Russ. J. Inorg. Chem.* **16,** 702 (1971).
143. Buslaev, Y. A., Kokunov, Y. V., and Bochkareva, V. A., *Russ. J. Inorg. Chem.* **20,** 495 (1975).
144. Buslaev, Y. A., and Bochkareva, V. A., *Izv. Akad. Nauk SSSR, Neorg. Mater.* **1,** 316 (1965).
145. Nikolaev, N. S., and Opalovskii, A. A., *Izv. Akad. Nauk SSSR* **12,** 49 (1959).
146. Grandjean, D., and Weiss, R., *C.R. Acad. Sci. Paris, Ser. C* **262,** 1864 (1966).
147. Marinina, L. K., Rakov, E. G., Gromov, B. V., Minaev, V. A., and Karanov, S. A., *Tr. Mosk. Khim. Tekhnol. Inst.* 83 (1970).
148. Müller, A., Ruck, A., Dartmann, M., and Reinsch-Vogell, U., *Angew. Chem. Int. Ed. Engl.* **20,** 483 (1981).
149. Müller, A., Jostes, R., and Cotton, F. A., *Angew. Chem. Int. Ed. Engl.* **19,** 875 (1980).
150. Mattes, R., and Mennemann, K., *Z. Anorg. Allg. Chem.* **437,** 175 (1977).
151. Mennemann, K., and Mattes, R., *Angew. Chem. Int. Ed. Engl.* **19,** 72 (1980).
152. Engelbrecht, A., and Grosse, A. V., *J. Am. Chem. Soc.* **76,** 2042 (1954).
153. Aynsley, E. E., *J. Chem. Soc.* 2425 (1958).
154. Varetti, E. L., and Müller, A., *Z. Anorg. Allg. Chem.* **442,** 230 (1978).
155. Edwards, A. J., Hugill, D., and Peacock, R. D., *Nature (London)* **200,** 672 (1963).
156. Hugill, D., and Peacock, R. D., *J. Chem. Soc. (A)* 1339 (1966).

157. Edwards, A. J., and Jones, G. R., *J. Chem. Soc. (A)* 1651 (1969).
158. Edwards, A. J., Jones, G. R., and Sills, R. J. C., *J. Chem. Soc (A)* 2521 (1970).
159. Sites, J. R., Baldock, C. R., and Gilpatrick, L. O., *U.S.A.E.C. Rep. ORNL* 1327 (1952).
160. Selig, H., and Malm, J. G., *J. Inorg. Nucl. Chem.* **25**, 349 (1963).
161. Franklin, K. J., Lock, C. J. L., Sawyer, B. G., and Schrobilgen, G. J., *J. Am. Chem. Soc.* **104**, 5303 (1982).
162. Ruff, O., and Kwasnik, W., *Z. Anorg. Allg. Chem.* **219**, 65 (1934).
163. Ruff, O., and Kwasnik, W., *Angew. Chem.* **47**, 480 (1934).
164. Hargreaves, G. B., and Peacock, R. D., *J. Chem. Soc.* 1099 (1960).
165. Dudin, A. S., Vovna, V. I., Rakov, E. G., and Lopatin, S. N., *Izv. Vyssh. Uchebn. Zaved. Khim. Khim. Tekhnol.* **21**, 1564 (1978).
166. Paine, R. T., *Inorg. Chem.* **12**, 1457 (1973).
167. Yagodin, G. A., Opalovskii, A. A., Rakov, E. G., and Dudin, A. S., *Dokl. Akad. Nauk SSSR* **256**, 1400 (1980).
168. Holloway, J. H., Selig, H., and Claassen, H. H., *J. Chem. Phys.* **54**, 4305 (1971).
169. Aynsley, E. E., Peacock, R. D., and Robinson, P. L., *J. Chem. Soc.* 1622 (1950).
170. Selig, H., and El-Gad, U., *J. Inorg. Nucl. Chem.* **35**, 3517 (1973).
171. Alekseichuk, I. S., Ugarov, V. V., Sokolov, V. B., and Rambidi, N. G., *J. Struct. Chem.* **22**, 795 (1981).
172. Yagodin, G. A., Dudin, A. S., Rakov, E. G., and Opalovskii, A. A., *Zh. Neorg. Khim.* **25**, 170 (1980).
173. Selig, H., and Karpas, Z., *Isr. J. Chem.* **9**, 53 (1971).
174. Holloway, J. H., and Raynor, J. B., *J. Chem. Soc. Dalton Trans.* 737 (1975).
175. Münch, V., Selig, H., and Ebert, L. B., *J. Fluorine Chem.* **15**, 223 (1980).
176. Hagenmuller, P., Portier, J., Cadiou, J., and de Pape, R., *C.R. Acad. Sci. Paris Ser. C* **260**, 4768 (1965).
177. Holloway, J. H., and Peacock, R. D., *J. Chem. Soc.* 527 (1963).
178. Sakurai, T., and Takahashi, A., *J. Inorg. Nucl. Chem.* **39**, 427 (1977).
179. Hepworth, M. A., and Robinson, P. L., *J. Inorg. Nucl. Chem.* **4**, 24 (1957).
180. Hargreaves, G. B., and Peacock, R. D., *J. Chem. Soc.* 2618 (1960).
181. Falconer, W. E., Burbank, R. D., Jones, G. R., Sunder, W. A., and Vasile, M. J., *J. Chem. Soc., Chem. Commun.* 1080 (1972).
182. Selig, H., Sunder, W. A., Disalvo, F. A., and Falconer, W. E., *J. Fluorine Chem.* **11**, 39 (1978).
183. Bartlett, N., Jha, N. J., and Trotter, J., *Proc. Chem. Soc.* 277 (1962).
184. Falconer, W. E., Disalvo, F. J., Griffiths, J. E., Stevie, F. A., Sunder, W. A., and Vasile, M. J., *J. Fluorine Chem.* **6**, 499 (1975).
185. Burbank, R. D., *J. Appl. Crystallogr.* **7**, 41 (1974).
186. Nguyen-Nghi, and Bartlett, N., *C.R. Acad. Sci. Paris Sec. C* **269**, 756 (1969).
187. Ruff, O., and Fischer, J., *Z. Anorg. Allg. Chem.* **179**, 161 (1929).
188. Bartlett, N., and Lohmann, D. H., *Proc. Chem. Soc.* 14 (1960).
189. Bartlett, N., and Lohmann, D. H., *J. Chem. Soc.* 629 (1964).
190. Klemm, W., and Klein, H. A., *Z. Anorg. Allg. Chem.* **248**, 167 (1941).
191. Batsanova, L. R., and Kustova, G. N., *Zh. Neorg. Khim.* **9**, 330 (1964).
192. Bedford, R. G., and Catalano, E., *J. Solid State Chem.* **2**, 585 (1970).
193. De Kozak, A., Samouel, M., and Chretien, A., *Rev. Chim. Mineral.* **10**, 259 (1973).
194. Ruchkin, E. D., Travkina, L. N., and Kopaneva, L. I., *Zh. Neorg. Khim.* **19**, 1969 (1974).

195. Podberezskaya, N. V., Batsanova, L. R., and Egorova, L. S., *Zh. Strukt. Khim.* **6**, 850 (1965).
196. Weigel, F., and Scherer, V., *Radiochim. Acta* **7**, 40 (1967).
197. Popov, A. I., and Knudsen, G. E., *J. Am. Chem. Soc.* **76**, 3921 (1954).
198. Wendlandt, W. W., *Science* **129**, 842 (1959).
199. Zachariasen, W. H., *Acta Crystallogr.* **4**, 231 (1951).
200. Mergault, P., and Duffault-Devred, M., *C.R. Acad. Sci. Paris. Ser. C* **261**, 4392 (1965).
201. Bannert, M., Blumenthal, G., Sattler, H., Schoenherr, M., and Wittrich, H., *Z. Anorg. Allg. Chem.* **446**, 251 (1978).
202. Finkelnburg, W., and Stein, A., *J. Chem. Phys.* **18**, 1296 (1950).
203. Pannetier, J., and Lucas, J., *C.R. Acad. Sci. Paris Ser. C* **268**, 604 (1969).
204. Kustova, G. N., Obzherina, K. F., and Batsanova, L. R., *Zh. Prikl. Spektrosk.* **9**, 43 (1968).
205. Fried, S., Hagemann, F., and Zachariasen, W. H., *J. Am. Chem. Soc.* **72**, 771 (1950).
206. Peterson, J. R., and Burns, J. H., *J. Inorg. Nucl. Chem.* **30**, 2955 (1968).
207. Lucas, J., and Rannou, J. P., *C.R. Acad. Sci. Paris Ser. C* **266**, 1056 (1968).
208. Zachariasen, W. H., *Acta Crystallogr.* **2**, 388 (1949).
209. Zachariasen, W. H., *Natl. Nucl. Energy Ser. (Div. IV, 14B) Transuranium elements* Pt. 11, 1454 (1949).
210. D'Eye, R. W. M., Booth, G. W., and Harper, E. A., *At. Energy Res. Estab. (Gt. Brit.)* C/R 1735, 3 (1955).
211. D'Eye, R. W. M., *J. Chem. Soc.* 196 (1958).
212. Dimmick, G. L., *U.S.* 2,427,592 Sept. 16 (1947).
213. Satpathy, K. C., and Sahoo, B., *Indian J. Chem.* **5**, 278 (1967).
214. Vdovenko, V. M., Romanov, G. A., and Solntseva, L. V., *Radiokhimiya* **9**, 727 (1967).
215. Brown, D., "Halides of the Lanthanides and Actinides." Wiley, New York, 1968.
216. Rampey, G. A., *U.S. Energy Comm.* **GAT-265,** 30 (1959).
217. Brown, D., and Easy, J. F., *J. Chem. Soc. (A)* 3378 (1970).
218. Stein, L., *Inorg. Chem.* **3**, 995 (1964).
219. Kemmler-Sack, S., *Z. Anorg. Allg. Chem.* **364**, 88 (1969).
220. Bagnall, K. W., Brown, D., and Easey, J. F., *J. Chem. Soc. (A)* 2223 (1968).
221. Jouniaux, B., Legoux, Y., Merinis, J., and Bouissieres, G., *Radiochem. Radioanal. Lett.* **39**, 129 (1979).
222. Laser, M., and Merz, E., *J. Inorg. Nucl. Chem.* **31**, 349 (1969).
223. Wilson, P. W., *J. Chem. Soc., Chem. Commun.* 1241 (1972).
224. Wilson, P. W., *J. Inorg. Nucl. Chem.* **36**, 303 (1974).
225. Paine, R. T., Ryan, R. R., and Asprey, L. B., *Inorg. Chem.* **14**, 1113 (1975).
226. Wilson, P. W., *J. Inorg. Nucl. Chem.* **36**, 1783 (1974).
227. Levy, J. H., Taylor, J. C., and Wilson, P. W., *J. Inorg. Nucl. Chem.* **39**, 1989 (1977).
228. Taylor, J. C., and Wilson, P. W., *J. Chem. Soc., Chem. Commun.* 232 (1974).
229. Jacob, E., and Polligkeit, W., *Z. Naturforsch.* **B28**, 120 (1973).
230. Bagnall, K. W., du Preez, J. G. H., Gellatly, B. J., and Holloway, J. H., *J. Chem. Soc. Dalton Trans.* 1963 (1975).
231. Joubert, P., and Bougon, R., *C.R. Acad. Sci. Paris Ser. C* **280**, 193 (1975).
232. Joubert, P., Bougon, R., and Gaudreau, B., *Can. J. Chem.* **56**, 1874 (1978).
233. Joubert, P., Weulersse, J. M., Bougon, R., and Gaudreau, B., *Can. J. Chem.* **56**, 2546 (1978).

234. Bougon, R., Fawcett, J., Holloway, J. H., and Russell, D. R., *C.R. Acad. Sci. Ser. C* **287,** 423 (1978).
235. Bougon, R., Fawcett, J., Holloway, J. H., and Russell, D. R., *J. Chem. Soc. Dalton Trans.* 1881 (1979).
236. Holloway, J. H., Laycock, D., and Bougon, R., *J. Chem. Soc. Dalton Trans.* 2303 (1983).
237. O'Hare, P. A. G., and Malm, J. G., *J. Chem. Thermodyn.* **14,** 331 (1982).
238. Peacock, R. D., and Edelstein, N., *J. Inorg. Nucl. Chem.* **38,** 771 (1976).
239. Burns, R. C., and O'Donnell, T. A., *Inorg. Nucl. Chem. Lett.* **13,** 657 (1977).
240. Drobyshevskii, Y. V., Serik, V. F., Sokolov, V. B., and Tul'skii, M. N., *Radiokhimiya* **20,** 238 (1978).
241. Drobyshevskii, Y. V., Serik, V. F., and Sokolov, V. B., *Dokl. Akad. Nauk SSSR* **225,** 1079 (1975).
242. Hoekstra, H. R., *Inorg. Chem.* **2,** 492 (1963).
243. Chakravosti, M. C., and Bharadwaj, P. K., *J. Inorg. Nucl. Chem.* **40,** 1643 (1978).
244. Knacke, O., Lossmann, G., and Mueller, F., *Z. Anorg. Allg. Chem.* **371,** 32 (1969).
245. Kolditz, L., Feist, M., Wilde, W., and Proesch, U., *Z. Chem.* **17,** 194 (1977).
246. Zachariasen, W. H., *Acta Crystallogr.* **1,** 277 (1948).
247. Atoji, M., and McDermot, M. J., *Acta Crystallogr. Sect. B* **26,** 1540 (1970).
248. Ohwada, K., *J. Inorg. Nucl. Chem.* **33,** 1615 (1971).
249. Barraclough, C. G., Cockman, R. W., and O'Donnell, T. A., *Inorg. Nucl. Chem. Lett.* **17,** 83 (1981).
250. Yahata, T., and Iwasaki, M., *J. Inorg. Nucl. Chem.* **26,** 1863 (1964).
251. Brown, D., personal communication.
252. Bohinc, M., and Frlec, B., *J. Inorg. Nucl. Chem.* **34,** 2942 (1972).
253. Goekcek, C., *Ger. Offen.* 2,626,427, 22 Dec. (1977).
254. Shrewsberry, R. C., and Williamson, E. L., *J. Inorg. Nucl. Chem.* **28,** 2535 (1966).
255. Luce, M., and Hartmanshenn, O., *J. Inorg. Nucl. Chem.* **28,** 2535 (1966).
256. Ellis, J. F., and Forrest, C. W., *J. Inorg. Nucl. Chem.* **16,** 150 (1960).
257. Midgley, A., Phillips, T. R., and Hamlin, A. G., *J. Inorg. Nucl. Chem.* **28,** 3047 (1966).
258. Reynes, J., Bethuel, L., and Aubert, J. *Nucl. Sci. Abstr.* **27,** 27446 (1973).
259. Tsvetkov, A. A., Seleznev, V. P., Sudarikov, B. N., and Gromov, B. V., *Zh. Neorg. Khim.* **18,** 783 (1973).
260. Seleznev, V. P., Tsvetkov, A. A., Sudarikov, B. N., and Gromov, B. V., *Zh. Neorg. Khim.* **17,** 2587 (1972).
261. Vdovenko, V. M., Skoblo, A. I., Suglobov, D. N., Shcherbakova, L. L., and Shcherbakova, V. A., *Zh. Neorg. Khim.* **12,** 2863 (1967).
262. Vdovenko, V. H., Kovaleva, T. V., Kanevskaya, N. A., and Suglobov, D. N., *Radiokhimiya* **16,** 66 (1974).
263. Schchelokov, R. N., Tsivadze, A. Y., Orlova, I. M., and Podnebesnova, G. V., *Inorg. Nucl. Chem. Lett.* **13,** 367 (1977).
264. Baker, A. E., and Haendler, H. M., *Inorg. Chem.* **1,** 127 (1962).
265. Zaitseva, L. L., Lipis, L. V., Fomin, V. V., and Chebotarev, N. B., *Russ. J. Inorg. Chem.* **7,** 795 (1962).
266. Sokolov, I. P., and Seleznev, V. P., *Russ. J. Inorg. Chem.* **22,** 1854 (1977).
267. Nguyen Quy Dao, Sadok Chourou, and Heckly, J., *J. Inorg. Nucl. Chem.* **43,** 1835 (1981).
268. Galkin, N. P., Veryatin, U. D., and Karpov, V. I., *Russ. J. Inorg. Chem.* **7,** 1043 (1962).

269. Tsvetkov, A. A., Seleznev, V. P., Sudarikov, B. N., and Gromov, B. V., *Russ. J. Phys. Chem.* **45,** 563 (1971).
270. Fawcett, J., Holloway, J. H., Laycock, D., and Russell, D. R., *J. Chem. Soc. Dalton Trans.* 1355 (1982).
271. Holloway, J. H., and Laycock, D., *J. Chem. Soc. Dalton Trans.* 1635 (1982).
272. Mangleberg, C. J., Rae, H. K., Hurst, R., Long, G., Davies, D., and Francis, K. E., *J. Inorg. Nucl. Chem.* **2,** 358 (1956).
273. Florin, A. E., Tannenbaum, I. R., and Lemons, J. F., *J. Inorg. Nucl. Chem.* **2,** 368 (1956).
274. Alenchikova, I. F., Zaitseva, L. L., Lipis, L. V., Nikolaev, N. S., Fomin, V. V., and Chebotarev, N. T., *Zh. Neorg. Khim.* **3,** 951 (1958).
275. Otey, M. G., and Le Doux, R. A., *J. Inorg. Nucl. Chem.* **29,** 2249 (1967).
276. Buslaev, Y. A., Nikolaev, N. S., and Tananaev, I. V., *Dokl. Akad. Nauk SSSR* **148,** 832 (1963).
277. Rannou, J. P., and Lucas, J., *Mater. Res. Bull.* **4,** 443 (1969).
278. Kemmler-Sack, S., *Z. Naturforsch. B* **22,** 597 (1967).

CHEMICAL EFFECTS OF NUCLEAR TRANSFORMATIONS

G. A. BRINKMAN

Department of Chemistry, National Institute of Nuclear Physics and High-Energy Physics, Amsterdam, The Netherlands

I. Introduction 101
II. Reactions of Thermalized Recoil Atoms 102
   A. Thermal Reactions of ^{18}F Atoms 102
   B. Thermal Reactions of ^{38}Cl Atoms 108
III. Stereochemistry in Substitution Reactions 112
   A. Tritium 112
   B. Chlorine 114
   C. Fluorine, Bromine, and Iodine 117
   D. Conclusions 118
IV. Muonium Chemistry 119
   A. Gaseous Phase 119
   B. Liquid Mixtures 120
   C. Formation and Reactions of Muonic Radicals 122
   References 130
   Notes Added in Proof 133

I. Introduction

The chemical effects of nuclear transformations are mainly the chemical reactions of energetic (hot), electronically excited, and thermal radioactive recoil atoms, produced by nuclear reactions and of hot, excited, and thermal ions, produced by nuclear decay (α, β^-, β^+, IT, EC). The study of the reactions of recoil particles began in the 1930s, when Szilard and Chalmers (1) showed that after neutron irradiation of C_2H_5I, the majority of the ^{128}I activity—formed by the ^{127}I(n, γ) ^{128}I reaction—could be extracted as ^{128}I$^-$ ions. Obviously, the C—I bond was broken after the nuclear reaction. The literature on hot (or recoil) chemistry is so extensive that only some topics can be discussed in this article. The selection is such that there is little overlap with existing review articles. General reviews can be found in references (2–7). More specialized articles reviewed the reactions of radioactive T (8), F

(9–11), Cl (12), I (13), N (14), Si (9), other polyvalent atoms (15), and of muonium (16). Reviews have also been published on the reactions of recoil atoms with arenes (17), (halo)ethylenes (18), and (halo)-methanes (19). The capture of π^- in hydrogenated species is sometimes considered as a part of recoil chemistry (20), and so also are reactions of species formed after decay of multiply labeled (T, ^{14}C) molecules (21–23), for example,

$$CT_4 \xrightarrow{\beta^-} [CT_3{}^3He]^+ \xrightarrow{fast} CT_3{}^+ + {}^3He$$

II. Reactions of Thermalized Recoil Atoms

The thermalization of energetic recoil atoms in excess moderator is a useful tool to measure kinetic parameters for abstraction, substitution, and addition reactions. For thermal experiments, the bulk (>90%) of the sample must consist of a compound that is (1) inert for hot and thermal reactions with the recoil atom and (2) able to supply the radioactive atom. For example, Ne, CF_4, C_2F_6, SF_6 + Ar, CF_2Cl_2, and CF_3Cl meet these requirements for radioactive recoil F and Cl atoms.

A. Thermal Reactions of ^{18}F Atoms

The compounds SF_6, CF_4, and C_2F_6 are mainly used for the production and moderation of ^{18}F atoms. These compounds are not absolutely inert for reactions with hot ^{18}F atoms, but the product yields are low: 2% (24), 7.4%, and 14% (11), respectively.

1. Hydrogen Abstraction

The first experiments were carried out by Williams and Rowland (25), using SF_6 as the bath gas, with tracers of C_2H_2 and HI. For these mixtures, the following reactions were considered.

$$^{18}F + C_2H_2 \longrightarrow H^{18}F + C_2H \qquad (1)$$

$$^{18}F + C_2H_2 \longrightarrow CH=CH^{18}F \qquad (2)$$

$$CH=CH^{18}F + HI \longrightarrow CH_2=CH^{18}F + I \qquad (3)$$

$$^{18}F + HI \longrightarrow H^{18}F + I \qquad (4)$$

The following equation can be derived for the fractional yield Y of $CH_2=CH^{18}F$:

$$\frac{Y_{\text{total}}}{Y_{\text{CH}_2=\text{CH}^{18}\text{F}}} = \frac{k_1 + k_2}{k_2} + \frac{k_4[\text{HI}]}{k_2[\text{C}_2\text{H}_2]}$$

If a third reactant (RH) is present, an additional reaction can take place:

$$^{18}\text{F} + \text{RH} \longrightarrow \text{H}^{18}\text{F} + \text{R} \tag{5}$$

This leads to the equation:

$$\frac{Y_{\text{total}}}{Y_{\text{CH}_2=\text{CH}^{18}\text{F}}} - \frac{k_4[\text{HI}]}{k_2[\text{C}_2\text{H}_2]} = \frac{k_1 + k_2}{k_2} + \frac{k_5[\text{RH}]}{k_2[\text{C}_2\text{H}_2]}$$

Both equations predict straight lines for graphs of $Y^{-1}_{\text{CH}_2=\text{CH}^{18}\text{F}}$ versus [HI]/[C_2H_2] with slopes of k_4/k_2 and k_5/k_2, respectively.

The second group of experiments was performed by Root and coworkers (26–29), using C_2F_6 as the bath gas and C_3F_6 as the reference compound. From the reaction

$$^{18}\text{F} + \text{C}_3\text{F}_6 \longrightarrow \text{C}_3\text{F}_6{}^{18}\text{F} \tag{6}$$

and a total hot yield of 14% for pure C_2F_6, the following equation can be derived:

$$\frac{0.86}{Y_{\text{H}^{18}\text{F}}} = \frac{k_5 - k_6}{k_5} + \frac{k_6}{k_5}\left(\frac{1}{1 - [\text{C}_3\text{F}_6]}\right)$$

The formed H^{18}F is quantitatively absorbed on the sample vessel walls, from which it is removed by extraction with a K_2CO_3 solution (28, 30).

Some of the absolute rate constants for H abstraction, measured by both groups, are given in Table I. In the cases of H_2 and CH_4, the agreement with recommended literature survey values for ^{19}F atoms is very good. For C_2H_6, only one ^{19}F value has been published. The results for the deuterated compounds are compared with ^{19}F isotopic ratios.

2. Reactions with CH_3X and CF_3X

Thermal ^{18}F-for-X substitution yields have been measured in SF_6–CH_3X mixtures. Extrapolated to zero CH_3X concentration, the absolute yields are found to increase with decreasing C—X bond energies (34): 0.11 ± 0.2, 0.27 ± 0.02, and $0.45 \pm 0.15\%$ for X = Cl, Br, and I,

TABLE I

Rate Constants for Thermal H Abstraction by ^{18}F and ^{19}F Atoms at 283 K[a]

Compound	^{18}F (ref. 25)[b]	^{19}F (ref. 29)[c]	^{19}F
H_2	1.3 ± 0.2	1.29 ± 0.05	1.35[d]
D_2	0.7 ± 0.2	0.56 ± 0.02	—
H_2/D_2	1.8 ± 0.6	2.30 ± 0.12	2.00 ± 0.04[e]
CH_4	3.8 ± 0.4	4.0 ± 0.2	4.3[d]
CD_4	2.2 ± 0.4	1.94 ± 0.11	—
CH_4/CD_4	1.7 ± 0.4	2.06 ± 0.16	1.5 ± 1.0[f]
C_2H_6	12.9 ± 1.1	14.9 ± 2.1	12.9[g]
C_2D_6	—	9.3 ± 0.6[h]	—

[a] Absolute constants in 10^{10} liters mol^{-1} sec^{-1}.

[b] Relative to ^{18}F addition to C_2H_2: $k^{283} = (9.2 \pm 0.7) \times 10^{10}$ liters mol^{-1} sec^{-1} (31).

[c] Relative to ^{18}F addition to C_3F_6: $k^{283} = (6.0 \pm 0.3) \times 10^{10}$ liters mol^{-1} sec^{-1}.

[d] Calculated from recommended literature survey data of H_2 (1.5×10^{10} liters mol^{-1} sec^{-1}) and CH_4 (4.8×10^{10} liters mol^{-1} sec^{-1}) at room temperature (32).

[e] Recommended literature survey data (32).

[f] Calculated from $(1.0 \pm 0.3) \exp(0.96 \pm 0.84)/RT$. At 300 K, a ratio of 1.8 was measured (32).

[g] Relative to H abstraction from CH_4: $k^{283} = 4.3 \times 10^{10}$ liters mol^{-1} sec^{-1} (33).

[h] At 300 K.

respectively. The remaining ^{18}F activity—about 98%, when corrected for a contribution of 2% for hot reactions with SF_6—is the result of thermal H abstraction from CH_3X. Rate constants for the substitution reactions were calculated from the rate constants of the H-abstraction reactions (Table II). Since no thermal ^{18}F-for-X substitution was observed for CH_3Br and CF_3I, although their bond energies are similar to those of CH_3Br and CH_3I, respectively, the absolute substitution yields apparently do not depend only on bond energies. This effect can be understood if the substitution proceeds via an inversion mechanism: the H atom in CH_3 can relax rapidly enough to permit formation of the necessary trigonal pyrimidal intermediate, with ^{18}F and X in the apical positions (36).

TABLE II

ABSOLUTE RATE CONSTANTS[a] FOR THERMAL ^{18}F ATOMS
(31, 34, 35)

Reaction	k^{283}
$^{18}\text{F} + \text{CH}_3\text{Br} \rightarrow \text{HF} + \text{CH}_2\text{Br}$	$(3.7 \pm 0.4) \times 10^{10b}$
$^{18}\text{F} + \text{CH}_3\text{I} \rightarrow \text{HF} + \text{CH}_2\text{I}$ $^{18}\text{F} + \text{CH}_3\text{I} \rightarrow \text{IF} + \text{CH}_3$	$(10.5 \pm 0.9) \times 10^{10b}$
$^{18}\text{F} + \text{CF}_3\text{I} \rightarrow \text{IF} + \text{CF}_3$	$(9.8 \pm 1.0) \times 10^{10c}$
$^{18}\text{F} + \text{CH}_3\text{F} \rightarrow \text{CH}_3\text{F} + \text{F}$	$(0.66 \pm 0.24) \times 10^7$
$^{18}\text{F} + \text{CH}_3\text{Cl} \rightarrow \text{CH}_3\text{F} + \text{Cl}$	$(2.2 \pm 0.8) \times 10^{7d}$
$^{18}\text{F} + \text{CH}_3\text{Br} \rightarrow \text{CH}_3\text{F} + \text{Br}$	$(10.2 \pm 1.8) \times 10^7$
$^{18}\text{F} + \text{CH}_3\text{I} \rightarrow \text{CH}_3\text{F} + \text{I}$	$(48 \pm 18) \times 10^7$

[a] In liters mol^{-1} sec^{-1}.
[b] Relative to ^{18}F addition to C_2H_2 with $k^{283} = (9.2 \pm 0.7) \times 10^{10}$ liters mol^{-1} sec^{-1}.
[c] Literature values for ^{19}F are: (8 ± 3) and $(11 \pm 5) \times 10^{10}$ liters mol^{-1} sec^{-1} at 296 and 293 K, respectively (32).
[d] Relative to H abstraction from CH_3Cl by ^{19}F atoms: $k = (2.0 \pm 0.6) \times 10^{10}$ liters mol^{-1} sec^{-1} (35).

3. Addition

Addition reactions of ^{18}F atoms with alkenes and alkynes were reviewed in 1978 by Rowland et al. (37). In the case of C_2H_4 as the reactant in excess CF_4, the excited $C_2H_4{}^{18}$F radicals can either decompose (to $C_2H_3{}^{18}$F) or become collisionally stabilized and react with added HI (to $C_2H_5{}^{18}$F) (38, 39). A plot of the decomposition/stabilization ratio (D/S = $Y_{C_2H_3{}^{18}F}/Y_{C_2H_5{}^{18}F}$) versus the inverse pressure results in a straight line. The half stabilization pressure (D/S = 1) found is 19.2 ± 1.3 kPa, corresponding to a lifetime of the excited $C_2H_4{}^{18}$F radical of 1 nsec. The total yield for thermal addition of ^{18}F to C_2H_4 is about 65%, the remaining yield being due to thermal H abstraction. Rate constants measured by this method are given in Table III.

In the case of C_2H_2, addition accounts for 86% of the total activity (41). The yield of $C_2H_3{}^{18}$F in the SF_6–C_2H_2–HI system does not vary in the pressure range of 33–530 kPa, indicating that decomposition of the $(C_2H_2{}^{18}F)^*$ radical to $C_2H{}^{18}$F is negligible.

Using C_3H_6 as the bath gas, the ratio of terminal to central attack (resulting in $CH_3CH^{18}FCH_3$ and $CH_3CH_2CH_2{}^{18}$F, respectively) is 1.4, regardless of pressure (65–530 kPa) (42). The radical formed after

TABLE III

Absolute Rate Constants^a for Thermal ^{18}F Addition Reactions at 283 K

Compound	Ref. (37, 40)		Ref. (29)	
	k^{283}	Standard	k^{283}	Standard
C_2H_2	10.2	C_2H_4	9.4 ± 0.7	H_2, D_2, CH_4, CD_4, C_2H_6
C_2H_4	8.4	C_2H_2	7.7 ± 0.6	C_2H_2
CHCl=CHCl	10.2	CH_4	9.2 ± 0.7	RH, RD, C_3F_6
CFCl=CFCl	1.7	C_2H_2	1.5	C_2H_2
C_3F_6	1.2	CH_4	0.94 ± 0.05	H_2, CH_4, CHF_3
CH≡CCH$_3$	10.2	C_2H_2	—	—
CH≡CCF$_3$	3.6	C_2H_2	—	—

^{a} In liters mol^{-1} sec^{-1} × 10^{-10}.

central addition can undergo unimolecular decomposition at lower pressures:

$$(CH_3CH^{18}FCH_2)^* \longrightarrow CH_3 + CH^{18}F=CH_2 \quad (7)$$

The terminal/central attack ratios for $H_2C=C=CH_2$, $H_2C=CHC_2H_5$ (37), and HC≡CCH$_3$ (40) are 1.9, 1.4, and 3.7, respectively.

With C_2F_4 in excess SF_6, the excited $C_2F_4{}^{18}F$ radicals decomposed by C—C bond scission (43):

$$(C_2F_4{}^{18}F)^* \longrightarrow CF_2 + CF_2{}^{18}F \quad (8a)$$

$$CF_2{}^{18}F + HI \longrightarrow CHF_2{}^{18}F + I \quad (8b)$$

The half-stabilization pressure is 30 kPa, indicating a lifetime of 2 nsec. Selectivities for addition to other fluoroethylenes are given in Table IV (44). The product yields reflect the behavior of C_2H_4 and C_2F_4, i.e., (1) C—C scission as a decomposition mode of an excited $C_2H_nF_{4-n}{}^{18}F$ radical is important if a CF_2 group is present, because the C—F bond energy in CF_2 (522 kJ mol^{-1}) is high compared with the ethylenic C—F bond energy (480 kJ mol^{-1}); and (2) the next most energetic decomposition mode of these radicals is the loss of an H atom, which is also an exothermic reaction.

The CH_2 end of CH_2=CHCl is 2.5 times as reactive toward addition of thermal ^{18}F as is the CHCl end (45). The decomposition of the excited $CH_2CHCl^{18}F$ radical is extremely rapid (Cl loss), with no stabilization at 500 kPa. Excited CHCl^{18}FCHCl radicals, formed after the

TABLE IV (44)

SELECTIVITY IN THERMAL ^{18}F ADDITION TO ETHYLENES

Olefin	Relative yield per C atom			Relative total addition yield
	CH_2	CHF	CF_2	
CH_2=CH_2	1.0	—	—	1.0
CH_2=CHF	0.8	0.6	—	0.70
CH_2=CF_2	1.1	—	0.2	0.65
trans-CHF=CHF	—	0.3	—	0.30
CHF=CF_2	—	0.4	0.1	0.25
CF_2=CF_2	—	—	0.1	0.20

addition of ^{18}F to cis- and trans-CHCl=CHCl, also decompose very rapidly (98% at 270 kPa), or some 50 times as fast as CFCl^{18}FCFCl (50% at 270 kPa) (37).

4. Reactions with Organometallic Compounds

Rowland and co-workers have investigated the reactions of thermal ^{18}F atoms with organometallic compounds (Sn, Ge, Hg) (31, 46–50). In Table V, product yields and rate constants are given for organotin compounds and $(CH_3)_2$Hg. The yields of CH_3^{18}F from $(CH_3)_4$Sn and of CH_2=CH^{18}F from $(CH_2$=CH$)_4$Sn are independent of the pressure in the range between 65 and 400 kPa, indicating that the abstraction reactions take place in a time that is considerably shorter than 0.1 nsec (47). In both cases, the abstraction is a direct thermal reaction, in contrast with neopentane, from which CH_3^{18}F is only formed via a hot reaction (49). Apart from CH_2=CHCH$_2^{18}$F, produced from $(CH_2$=CH-CH$_2)_4$Sn, CH_2=CH^{18}F was also detected. The yield of this product decreased linearly as a function of $1/P$ from 0.5% to 65 kPa to zero at infinite pressure, which proves that it is formed through decomposition of excited SnC$_{12}$H$_{20}^{18}$F radicals (51). The decomposition rate of these radicals is 10^3 times as high as that calculated with the RRKM theory. This is attributed to the lack of internal energy equilibration beyond the central C—Sn—C bonding in a time range of 0.1 1 nsec. There seems to be a bottleneck of energy transfer because of that bonding. A similar non-RRKM behavior was also found for $(CH_2$=CHCH$_2)_4$Ge (49), but no CH_2=CH^{18}F was observed from $(CH_2$=CHCH$_2)_2$Si$(CH_3)_2$ (50).

TABLE V (46–50)

Yields (%) and Rate Constants[a] for Reactions of Thermal ^{18}F Atoms with Organometallic Compounds

Compound	Labeled product	Yield	k
$(CH_3)_3SnH$	CH_3F	5	—
$(CH_3)_4Sn$	CH_3F	8.5	2.3
	HF	—	24
$(CH_3)_2Hg$	CH_3F	9.6	2.8 ± 0.2
	Total[b]	—	28 ± 3
$(C_2H_5)_4Sn$	C_2H_5F	2	—
$(CH_2=CH)_4Sn$	$CH_2=CHF$	16	13 ± 2
	Total[c]	—	72 ± 12
$(n\text{-}C_3H_7)_4Sn$	C_3H_7F	0.8	—
$(CH_2=CHCH_2)_4Sn$	$CH_2=CHCH_2F$	0.5	—
$(CH_3)_3SnC_6H_5$	CH_3F	5.3	—

[a] In 10^{10} liters mol^{-1} sec^{-1}.
[b] $HF + CH_3F + CH_3HgF$.
[c] Addition + H abstraction + $CH_2=CH_2$ abstraction and substitution.

B. Thermal Reactions of ^{38}Cl Atoms

For the production and moderation of recoil ^{38}Cl atoms, CF_3Cl and CF_2Cl_2 are in use as bath gases. Hot reactions with both gases account for only 3 and 6% of the total yields, respectively (51, 52).

1. Abstraction

Stevens and Spicer (53) measured the yields for abstraction from H_2 (3.7% H^{38}Cl) and from D_2 (1.3% D^{38}Cl) in excess CF_2Cl_2. In purely thermal systems, with nonradioactive Cl atoms, an HCl–DCl isotope effect in the range 9–10 has been measured at 300 K. Based upon extrapolation of thermal data, the observed ^{38}Cl isotope effect of 2.8 indicates that the effective temperature in the recoil experiment is 800–900 K. Lee and Rowland (54) measured rate constants for thermal H abstraction from CH_4 and C_2H_6, in competition with addition to C_2H_3Br. At 243 K they found these constants to be $(1.9 \pm 0.4) \times 10^7$ and 2.7×10^{10} liters mol^{-1} sec^{-1}, respectively, in agreement with average values obtained with thermal nonradioactive Cl atoms of 2.4×10^7 and 3.6×10^{10} liters mol^{-1} sec^{-1}, respectively (55). Abstraction of CH_3 and C_2H_5 was observed from $(CH_3)_4Pb$ and $(C_2H_5)_4Pb$, respectively (56). In the former case, the yield of $CH_3^{38}Cl$ was 18% [the rate constant is $(1.8 \pm 0.3) \times 10^{10}$ liters mol^{-1} sec^{-1}], whereas H^{38}Cl accounts

TABLE VI
RELATIVE REACTION RATES FOR THERMAL ^{38}Cl ADDITION

Reactant	Relative rate	Reference
^{38}Cl + HI → H^{38}Cl	(1.0)	
^{38}Cl + H$_2$S → H^{38}Cl	0.75 ± 0.10	57
H$_2$C=CH$_2$	1.7 ± 0.1	57
H$_2$C=CHCH$_3$	1.2 ± 0.1	58
HC≡CH	1.6 ± 0.3	59
HC≡CCH$_3$	1.9 ± 0.1	60
H$_2$C=CHF	1.3	45
H$_2$C=CHBr	1.2; 1.7^a	61

^{a} The limiting total reactivity at high pressure.

for 75% of the total activity [rate constant $(7.8 \pm 1.4) \times 10^{10}$ liters mol^{-1} sec^{-1}].

2. Addition

Rates of addition to alkenes and alkynes, relative to H abstraction from HI, are given in Table VI. The reactions of thermal ^{38}Cl atoms with traces of C_2H_4 in a CF_2Cl_2 or $CFCl_3$ matrix lead to high yields (90%) of $C_2H_4{}^{38}$Cl (57, 62, 63). The excited radical can decompose by Cl loss, but that causes the ^{38}Cl atom to be available again for reaction. A minor reaction channel is the elimination of H^{38}Cl. Extrapolation of the H^{38}Cl–$C_2H_4{}^{38}$ClI (I$_2$ scavenged experiment) to $1/P = 0$ results in a ratio of 0.06, indicating that H abstraction also takes place: the addition/abstraction ratio is 15.5 ± 0.5 (62). Relative to the rate of H abstraction from C_2H_6, evaluated as $(3.4 \pm 0.4) \times 10^{10}$ liters mol^{-1} sec^{-1} at 298 K, the rate constant for removal of Cl atoms by addition to C_2H_4 is $(1.0 \pm 0.1) \times 10^{11}$ liters mol^{-1} sec^{-1} (63). Correction for back reaction of C_2H_4Cl leads to an absolute rate constant of $(1.14 \pm 0.12) \times 10^{11}$ liters mol^{-1} sec^{-1}. Thermal reactions of ^{38}Cl with C_3H_6 result in the formation of 86% $C_3H_6{}^{38}$Cl (58). The terminal/central addition ratio depends upon the HI concentration and varies between 6.5 and 12.3. This is explained by ^{38}Cl migration from $CH_3CH^{38}ClH_2$ to $CH_3CHCH_2{}^{38}Cl$, with a rate constant of 10^7 sec^{-1}. The original terminal/central ratio is 6. Allylic H abstraction accounts for no more than 14% of the total number of reactions. Thermal ^{38}Cl atoms react almost quantitatively with C_2H_2 (59). The addition takes place in not more than two to five collisions. The reaction with propyne also proceeds almost entirely by addition, with less than 5% H abstraction (60). The

TABLE VII

Relative ^{38}Cl-for-Cl Substitution Yields in 1:1:1
Mixtures of o-, m-, and p-$C_6H_4ClX^a$

	Ortho	Meta	Para
$C_6H_4Cl_2$ (69)			
Hot	35 ± 1	25 ± 1	40 ± 1
Thermal	42 ± 2	2 ± 2	56 ± 2
C_6H_4ClF (70)			
Hot	38 ± 1	26 ± 1	36 ± 2
Thermal	55 ± 4	1 ± 2	44 ± 3
$C_6H_4ClCH_3$ (67)			
Hot	46 ± 4	20 ± 4	34 ± 1
Thermal	55 ± 6	6 ± 2	39 ± 4

^{a} Dose: 7.5 kGy.

terminal/central ratio is 8 ± 1, but there is no indication of a 1,2 shift as observed with C_3H_6.

The addition of ^{38}Cl to CH_2=CHF occurs preferentially at the CH_2 end by a factor of two (45). A major proportion of both $C_2H_3F^{18}Cl$ radicals are stabilized at a pressure of 500 kPa, their lifetimes being 1 nsec.

In the presence of HI, the reactions of ^{38}Cl with CH_2=CHBr produce both CH_2=$CH^{38}Cl$ and $CH_2{}^{38}ClCH_2Br$ (no $CH_3CH^{38}ClBr$) in pressure-dependent yields, indicating a long-lived (0.1–1 nsec) excited $C_2H_3{}^{38}ClBr$ radical (54, 61). The proposed mechanism shows very little preference for addition to the CH_2 site versus CHBr. The observation that no $CH_3CH^{38}ClBr$ is found is explained by a 1,2-Br shift after the formation of $CH_2CH^{38}ClBr$, the "anti-Markovnikov" product.

cis- and trans-CHCl=CHCl were used in hot-atom chemistry experiments as scavengers for thermal ^{38}Cl atoms (52, 64, 65). The lifetime of the CHClCHCl^{38}Cl radical is 0.5–0.7 nsec. With either isomer as the reactant, the loss of a Cl atom from this radical leads to CHCl=$CH^{38}Cl$ with a trans/cis ratio of 0.50. Similar experiments with C_2Cl_4 as a scavenger resulted in yields of $C_2Cl_3{}^{38}Cl$ between 9 and 92%, owing to radiation-induced reactions (52).

3. ^{38}Cl-for-Cl *Exchange*

The hot substitution of a Cl atom in liquid chlorobenzenes by recoil ^{34m}Cl and ^{38}Cl atoms accounts for 4–6% of the total activity. Apart

TABLE VIII

Hot and Thermal ^{34m}Cl-for-Cl Substitution Yields (%) in Equimolar Mixtures of o-Dichlorobenzenes (71)

1:1 Mixtures		Hot yields		Thermal yields	
A	B	A	B	A	B
$C_6H_4Cl_2$	C_6H_4ClF	15 ± 1	6 ± 1	11 ± 3	6 ± 2
$C_6H_4Cl_2$	$C_6H_4ClCH_3$	8 ± 1	8 ± 1	0	13 ± 1
$C_6H_4Cl_2$	$C_6H_4ClCF_3$	12 ± 1	4 ± 1	22 ± 4	1 ± 1
C_6H_4ClF	$C_6H_4ClCH_3$	3 ± 1	5 ± 1	6 ± 1	13 ± 1
C_6H_4ClF	$C_6H_4ClCF_3$	9 ± 2	7 ± 2	31 ± 2	0
$C_6H_4ClCH_3$	$C_6H_4ClCF_3$	4 ± 1	11 ± 1	29 ± 1	0

from the substitution by energetic Cl atoms, there is a high yield of a thermal ^{38}Cl-for-Cl exchange reaction that can be completely suppressed by 1–2 mol % Br_2 or I_2. In C_6H_5Cl, this thermal yield is 30% and increases to 50% at high radiation doses (66). Comparable high exchange yields were observed in $C_6H_4Cl_2$ (70), $C_6H_4ClCH_3$, and C_6F_5Cl (67, 68). The thermal exchange reaction rates are influenced by the position of the second substituent. Table VII gives relative hot and thermal substitution yields for equimolar mixtures of the three isomers of some substituted chlorobenzenes: the rate constants for the thermal reactions are substantially lower for the meta isomers. Not only the position of the second substituent affects the thermal exchange rates, but also the nature of that substituent, as can be seen in Table VIII, for equimolar mixtures of two ortho-substituted chlorobenzenes. The "hot" yields are those measured in the presence of 2 mol % I_2, the "thermal" yields are the differences between the yields measured in mixtures without and with I_2. From the thermal data it can be concluded that the sequence of the rate constants for thermal Cl-for-Cl exchange is $CH_3 > Cl > F > CF_3$. Similar values were found for equimolar mixtures of the meta isomers (72). The effect of the substituent X on the thermal exchange rate in C_6H_4ClX compounds is in the same order as the σ_m^+ and σ_p^+ Hammett constants, indicating the electrophilic nature of the reaction. The thermal exchange yield in C_6F_5Cl is (15 ± 1)%, but in equimolar mixtures with C_6H_5Cl thermal exchange is only found with C_6H_5Cl, the thermal yield of $C_6F_5{}^{34m}Cl$ being zero (72). The strong electron withdrawal properties will decrease the rate of the formation of a π-complex in C_6F_5Cl.

4. Discussion–Conclusions

The accurate determination of rate constants for the reactions of ^{19}F atoms is often hampered by the presence of reactive F_2 and by the occurrence of side reactions. The measurement of the absolute concentration of F atoms is sometimes a further problem. The use of thermalized ^{18}F atoms is not subject to these handicaps, and reliable and accurate results for abstraction and addition reactions are obtained. The studies of the reactions of ^{18}F atoms with organometallic compounds are unique, inasmuch as such experiments have not been performed with ^{19}F atoms. In the case of addition reactions, the fate of the excited intermediate radical can be studied by pressure-dependent measurements. The non-RRKM behavior of tetraallyltin and -germanium compounds is very interesting inasmuch as not many other examples are known. The next phase in the ^{18}F experiment should be the determination of Arrhenius parameters for selected reactions, i.e., those occurring in the earth's atmosphere, since it is expected that the results will be more precise than those obtained with ^{19}F atoms.

As a consequence of the occurrence of cage reactions in the liquid phase, which cannot be suppressed by small amounts of scavengers, studies of the reactions of thermalized recoil atoms are, in general, not possible. The investigations of thermal exchange reactions of ^{34m}Cl and ^{38}Cl atoms with chlorobenzenes are an exception to this rule. It must also be noted that X-for-X exchange reactions can only be studied with radioactive (or enriched) isotopes. In order to gain more information regarding the role of a second substituent on the rate of exchange, further experiments with varying substitutents will have to be conducted. Similar types of experiments should also be performed with recoil Br and I atoms. The observation (73) that recoil T atoms do not react with liquid CCl_4 indicates the viability of the study of reactions of thermal T atoms with compounds present in low concentrations in CCl_4.

III. Stereochemistry in Substitution Reactions

A. TRITIUM

An important aspect of hot-atom chemistry concerns the stereochemistry of substitution reactions, particularly if the reactions proceed via retention or (Walden) inversion. Cross sections for the reactions of energetic T atoms with CH_4, calculated by trajectory studies, show that T-for-H substitution with inversion is a feasible but minor process at low energies (74). Chou et al. (75) investigated the reactions of photolytically produced T atoms (1.8–3.2 eV) with CH_3F. The threshold energy is much lower for the T-for-F substitution (1.3 eV) than for the T-for-H substitution (1.8 eV). Furthermore, the ratio of the yields of both reactions (0.6 at 2.7 eV) is far smaller than unity, as has been measured for energetic T atoms originating from a nuclear reaction (76). It was suggested that at low energies the T-for-F substitution involves the preferential loss of F along a linear T—C—F axis, leading to Walden inversion (75). Similar experiments with CHF_3 resulted in higher threshold energies (1.9 eV for both H and F substitution) and lower reactivity for the T-for-F substitution ($CTF_3/CHTF_2 \simeq 7$), indicating the absence of inversion during the F-substitution reaction (77). Such a reaction involves two heavy F substituents, which would not readily adjust to changes in configuration. Retention of configuration is expected to be even more efficient for the T-for-H substitution reaction in CHF_3.

Substitution reactions at asymmetric C atoms in optically active molecules were first studied with glucose, galactose, and alanine, and it was proved that retention was preserved during the substitution of H atoms bound at these asymmetric atoms (78–80). However, it was not possible from the results of these solid-phase experiments to establish the mechanism. More direct information was obtained with gas-phase experiments (Table IX). Only for $C_2H_5CHOHCH_3$ and $(CHFCl)_2$ was substitution at the asymmetric C atom studied; for the other compounds, the total T-for-H (all H atoms) substitution yields were measured. Even if all the T activity in meso-$(CH_3CHCl)_2$, formed from reactions with dl-$(CH_3CHCl)_2$, were in the asymmetric position, the gas-phase preference for retention of configuration is more than 93% (84). In gaseous 1,3-dimethylcyclobutane, the yield of the inverted isomer is less than 1% (85). Since only 2 out of the 12 H atoms are bonded in asymmetric positions, more than 94% of the H substitution at the asymmetric C atom occurs with retention of configuration (if it is assumed that T-for-H substitution yields are equal for all 12 H atoms). It is even probable that this small inversion yield arises from secondary isomerization of the excited product and not from the primary event itself. In the condensed phases of $(CHFCl)_2$ and $(CH_3CHCl)_2$, the inversion yields are higher than in the (scavenged) gas phases. This is thought to be due to the decomposition of excited, labeled isomers by Cl loss, followed by caged recombination, in competition with racemization of the radical (82, 83).

TABLE IX

INVERSION FOR T-FOR-H SUBSTITUTION REACTIONS[a]

Sample	Phase[b]	Inversion (%)	Sample	Phase	Inversion (%)
d-$C_2H_5CHOHCH_3$ (81)	G	10 ± 3	dl-$(CH_3CHCl)_2$ (84)	G	1.3 ± 0.1
l-$C_2H_5CHOHCH_3$	G	9 ± 4		L	3.1 ± 0.8
				$L(I_2)$	2.3 ± 0.6
dl-$(CHFCl)_2$ (82)	G	15.3 ± 1.1	$meso$-$(CH_3CHCl)_2$	$S(I_2)$[c]	4.3 ± 1.0
	$G(O_2)$	2.1 ± 1.4		$G(O_2)$	0
	$L(I_2)$	16.4 ± 0.3	cis-$(CH_3)_2$-c-C_4H_8 (85)		
$meso$-$(CHFCl)_2$	G	8.6 ± 0.5		$G(O_2)$	1.0 ± 0.1
	$G(O_2)$	0.7 ± 0.4		L	6.7
	$L(I_2)$	5.4 ± 0.2	$trans$-$(CH_3)_2$-c-C_4H_6[d]	$L(I_2)$	0.9
dl-$(CHFCl)_2$ (83)	$G(I_2)$	4		$G(O_2)$	0.8 ± 0.1
	$L(I_2)$	20		L	6.7
$meso$-$(CHFCl)_2$	$G(I_2)$	4		$L(I_2)$	0.2
	$L(I_2)$	8			

[a] Inversion + Retention = 100%.
[b] G: gas, L: liquid. (O_2), (I_2): O_2 or I_2 scavenger present.
[c] S: solid.
[d] $(CH_3)_2$-c-C_4H_6: 1,3-dimethylcyclobutane.

B. Chlorine

1. Gaseous Phase

The ^{38}Cl and ^{39}Cl-for-Cl substitution in gaseous *meso*- and *dl*-$(CH_3CHCl)_2$ (scavenged with butadiene) proceeds with almost complete (≥93%) retention of configuration (86). In the absence of butadiene, (26 ± 2)% inversion was measured for dl-$(CH_3CHCl)_2$ and (15 ± 2)% for the meso compound. Decomposition of labeled products will occur if they possess over 6 eV of excitation energy. Radical reactions appear then in the formation of both diastereomers, with some preference for the more stable meso compound. Virtually complete retention of configuration was also found in gaseous dl-$(CHFCl)_2$ (91%) and $meso$-$(CHFCl)_2$ (92%) when scavenged with I_2 (91).

In contrast with these experiments, a high degree of inversion (81%) was observed for the substitution by recoil ^{38}Cl and ^{39}Cl atoms of the Cl atom bound at the asymmetric C atom in gaseous d- and l-$CH_3CHClCOCl$ (88). In the gas phase, the gauche prime configuration is present in high concentrations. This conformation provides a relatively unhindered approach to an attack of the asymmetric C atom from the rear (with respect to the 2-Cl atom), resulting in Walden inversion upon substitution. This theory is upheld by the finding that in $(CH_3)_2CHCHClCOCl$, where the approach is sterically hindered, the amount of inversion is only $(41 \pm 1)\%$. In d- and l-$CH_3CHClCH_2OH$, the inversion decreases from 80% at a pressure of 38 kPa to 42% at 100 kPa (89). Infrared spectroscopy shows that an increase in pressure results in the appearance of a 3470 cm^{-1} absorption band (O—H stretching in a dimer) and in a decrease of the 3600 cm^{-1} band (O—H stretching in the monomer). As the extent of aggregation by hydrogen bonding increases, the chances for unhindered rear attack, leading to inversion, decrease (see Notes Added in Proof, p. 133).

2. Condensed Phases

The ^{38}Cl-for-Cl substitution yields in dl- and $meso$-$(CH_3CHCl)_2$ are 10 times as high in the condensed phases as in the gas phase (86). The inversion in the liquid phase for both isomers is 28% at 298 K and 30% at 217 K, independent of the presence of scavengers, whereas it is 50 and 38% in the solid phase for the dl and meso isomers, respectively (87). These findings confirm the recombination of a ^{38}Cl atom with a $CH_3CHClCHCH_3$ radical in a solvent cage. Such a process will not be affected by normal concentrations of scavengers. The increase of inversion at lower temperatures is consistent with a slightly higher activation energy for the combination process than for the racemization of the radicals. Caged radical recombination reactions could also explain the observations that in d- and l-$CH_3CHClCOCl$ the total substitution yield increases (from 1.2% to 4.3%) and the inversion decreases (from 80% to 50%) when going from the gaseous to the liquid phase (88). In dl- and $meso$-$(CHFCl_2)$, the ^{38}Cl-for-Cl substitution yields increased by a factor of about three, and the inversion increases from 9 to 30%, when going from the gaseous to the liquid phase (87). These effects were not explained by caged recombination between a ^{38}Cl atom and a CHFClCHF radical but by the formation of a caged complex: "i.e., an electronically unstable intermediate, which is held together by the surrounding solvent molecules for a time sufficient for configurational changes to occur" (87).

3. Solutions

An interesting behavior of the retention/inversion ratio in dl- and $meso$-$(CH_3CHCl)_2$ was observed upon dilution with several compounds (Fig. 1) (91). This effect was assigned to the relative concentrations of

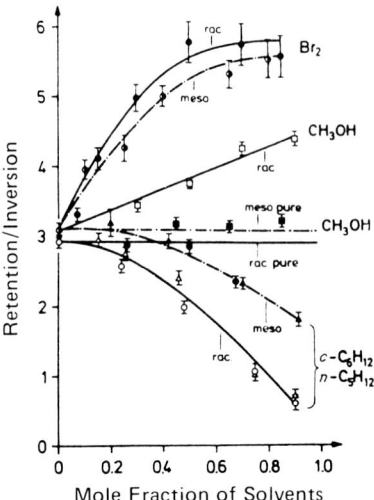

FIG. 1. Solvent effect on the stereochemical course (ratio retention/inversion) of ^{38}Cl-for-Cl substitution in rac- and meso-2,3-DCB. Key: ⊗: rac-DCB-Br$_2$; □: rac-DCB-CH$_3$OH; △: rac-DCB-n-C$_5$H$_{12}$; ○: rac-DCB-o-C$_5$H$_{12}$; ◐: meso-DCB-Br$_2$; ■: meso-DCB-CH$_3$OH; ▲: meso-DCB-n-C$_5$H$_{12}$; ●: meso-DCB-o-C$_5$H$_{12}$. [Reprinted with permission from ref. (91). Copyright 1972 American Chemical Society.]

the three possible conformers (RT, RG, and RG′), these concentrations depending on the type and amount of the additive. A similar effect on the retention/inversion ratio was found for (CH$_3$CHCl)$_2$CH$_2$ (92), but in this case each of the dl and meso compounds has only one preferred conformation, with minor amounts of the others, whereas the conformation population is barely affected by the nature of the solvent. Furthermore, only small changes in the retention/inversion ratio were observed in the case of (CH$_3$CHCl)$_2$ when going from the liquid to the solid phase (dl from 2.78 to 2.57, meso from 2.45 to 2.21), whereas distinct changes in the relative conformer concentrations were to be expected. Alterations in the retention/inversion ratios were similarly observed in dl- and meso-(CHFCl)$_2$ and in d- and l-CH$_3$CHClCH$_2$OH (93, 94). It was proposed that the dielectric properties of the solvent, or more precisely the quantity $(\varepsilon - 1)/(2\varepsilon + 1)$, causing differences in the solute–solvent interactions, control the substitution mechanism to a large degree. A strong interaction prevents the intermediate radical from obtaining planarity, maintaining the configuration that is obtained in the primary substitution step (94). For liquid cis- and trans-1,2-dichlorohexafluorocyclobutane, the retention is 76%

(95) and the addition of 80 mol % n-C_7H_{16}, c-C_6H_{12}, or n-$C_5H_{11}OH$ increases this value to 80%. In contrast with the other experiments, no dependence was found on the dielectric constants of the various hydrogen-containing solvents. This was attributed to a much higher activation energy being required for achieving planarity in the case of the c-C_4F_6Cl radical than for the other radicals. The addition of 80 mol % n-C_7F_{16} decreases the retention from 76 to 55 and 72% for the cis and trans isomers, respectively. The different behavior of C_6F_{16}, which has the same dielectric constant as C_7H_{16} and c-C_6H_{12}, has been attributed to the self-scavenging of ^{38}Cl by H abstraction from the hydrogen-containing solvents.

C. FLUORINE, BROMINE, AND IODINE

No inversion of configuration was found in the case of ^{18}F-for-F substitution in gaseous dl- and $meso$-$(CHFCl)_2$, in accord with the results obtained with the same compounds for recoil T and ^{38}Cl atoms (96).

The isomeric transitions $^{80m}Br \xrightarrow{IT} {}^{80}Br$ and $^{125}Xe \xrightarrow{IT} {}^{125}I$ result in highly positively charged ^{80}Br and ^{125}I ions. In a study with gaseous $(CH_3CHCl)_2$, it was safely assumed that both species react as singly charged Br^+ and I^+ ions (97). Electrophilic ^{80}Br and ^{125}I-for-Cl substitution leads to $erythro$- and $threo$-2-bromo(iodo)-3-chlorobutanes. In the pure systems, the retention/inversion ratios were 2.5 for ^{80}Br (for both the dl and meso isomers) and 1.9 for ^{125}I (for the dl isomer). Extrapolation of these ratios to 100 mol % moderator (Ar for ^{80}Br and Xe for ^{125}I) result in the following ratios: 3.3 (^{80}Br/meso), 0.3 (^{80}Br/dl), and 0.5 (^{125}I/dl); the attack of the thermal ions proceeds preferentially from the front and results in the formation of a halocarbocation with a three-centered bond structure and which retains the original configuration:

$$CH_3-\underset{\underset{H}{|}}{\overset{\overset{Cl}{|}}{C}}-\underset{\underset{H}{|}}{\overset{+}{C}}-CH_3 \quad \text{with} \quad Cl\cdots Br \text{ bridge}$$

After the formation of the carbocation, two competing processes occur in the moderated systems: racemization and Cl^+ transfer. In the case of the meso compound, front attack leads to the thermodynamically stable erythro form, whereas in the dl system, the less stable threo diastereomer is formed, which readily leads to racemization.

Dilution of liquid $(CH_3CHCl)_2$ with several additives results, in the

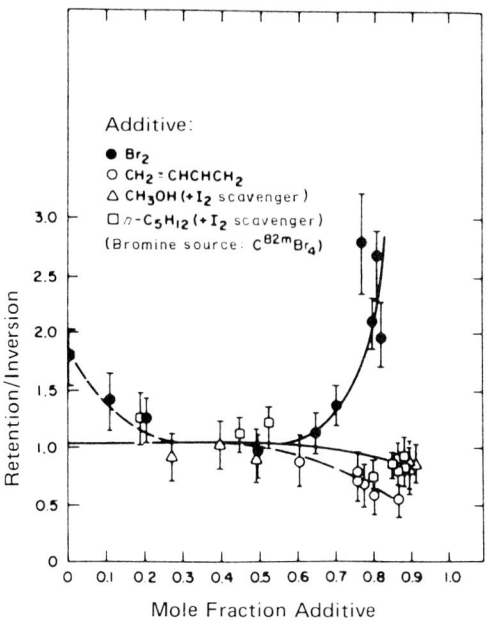

FIG. 2. ^{82}Br-for-Cl exchange in *dl*-2,3-dichlorobutane solutions following ^{82m}Br(T) ^{82}Br. [Reprinted with permission from ref. (*98*). Copyright 1976 American Chemical Society.]

case of Cl substitution by neutral ^{80m}Br atoms [produced by the ^{79}Br(n,γ)^{80m}Br reaction], in curves similar to those plotted for ^{38}Cl (Fig. 1), indicating that direct hot reactions are involved (*98*). In Fig. 2, results are shown for ^{82}Br, produced via the ^{82m}Br $\xrightarrow{IT}$ ^{82}Br decay. Auger radiolysis leads to the formation of CH$_3$CHClCHCH$_3$ radicals, which can react with a neutral ^{82}BR atom. The stereochemistry depends on the time needed to obtain a planar configuration of the radical, this time interval depending on the neutralization time of the ^{82}Br$^+$ ion and the density of the solvent.

D. CONCLUSIONS

Hot T-for-H substitution occurs mainly with retention of configuration. Inversion seems theoretically possible only when very light substituents are bound to that C atom at which the substitution takes place. Unfortunately, it appears to be impossible to prove inversion experimentally. During Cl substitution in gaseous compounds by recoil Cl atoms, the configuration is mainly retained in the case of simple molecules, but for larger molecules the situation becomes more complex. If unhindered approach can take place from the rear of the atom that is to be substituted, inversion is the main reaction channel. Such a situation can occur if one of the possible conformations is significantly more abundant than the others. However, the degree of inversion can be affected by steric hindrance. Many more experiments are needed in order to gain further insight into the significance of the parameters predicting the substitution process, such as conformational effects and steric hindrance. In the condensed phase, more experiments are needed to obtain information regarding the importance of cage reactions. In particular, experiments in liquid mixtures can provide more information about the time scale of the reactions and the interactions of radicals with the surrounding molecules. As noted in the previous section on thermal Cl-for-Cl substitution, hot X-for-X substitution reactions as discussed in this article can be studied only with radioactive atoms. Although not further discussed, the electrophilic substitution reactions by thermal $^{80}Br^+$ and $^{125}I^+$ ions are of special importance in these cases where other methods, such as ion cyclotron resonance and high-pressure mass spectroscopy, cannot provide the necessary information on the stereochemical course of the reaction.

IV. Muonium Chemistry

Muonium (Mu) is the lightest hydrogen-like atom ($m_{Mu} = 0.11\ m_H$) available for chemical research; it has a positive muon (μ^+, $\tau = 2.2$ μsec) as the nucleus. The muon spin resonance (μSR) technique is described in several review articles (*16, 99–102*). Most of the research is performed in the condensed phases, but because of the development of the "surface muon beams" (*103, 104*), experiments in the gaseous phase have received more attention. At present three muonic fractions can be detected: (1) f_{Mu}, free muonium; (2) f_D, free μ^+, or Mu bound in a diamagnetic compound; and (3) f_R, Mu bound in a paramagnetic compound. In liquid phases, there is quite often a missing fraction, $f_L = 1 - f_{Mu} - f_D - f_R$.

A. Gaseous Phase

During the deceleration process in matter, a μ^-, formed through the decay of a π^-, generally captures an atomic electron, resulting in Mu formation. In compounds where the ionization potentials are higher than that of Mu (13.6 eV), no (or only partial) neutralization takes place: $f_{Mu} = 0$ for He, 0.07 for Ne, 0.74 for Ar, and 1 for Kr and Xe (*105*).

TABLE X

REACTION RATES AND ISOTOPIC RATIOS FOR GAS-PHASE REACTIONS OF Mu (111)

Reactant	Reaction type	k_{Mu} (300 K) ($\times 10^{10}$ liters mol^{-1} sec^{-1})	k_{Mu}/k_H
F_2	Abstraction	1.46 ± 0.11	9.2 ± 3.1
Cl_2	Abstraction	5.30 ± 0.15	3.5 ± 0.8
Br_2	Abstraction	2.4 ± 3	5.3 ± 1.5
HBr	H abstraction / H exchange	0.91 ± 0.15	3.0 ± 1.0
C_2H_4	Addition	0.4 ± 0.05	5.8 ± 0.8
O_2	Spin exchange	15.8 ± 2.4	2.5 ± 0.4
NO	Spin exchange	18.3 ± 2.0	2.7 ± 0.3

The addition of 0.09 mol % Xe to He increases f_{Mu} to 0.75 (106). Similar effects were found on the addition of small amounts of Xe, CH_4, and NH_3 to Ne (105). This demonstrates the importance of the neutralization process right down to thermal energies. Neutralization may also proceed through a reaction with $Ne\mu^+$:

$$Xe + Ne\mu^+ \rightarrow Xe^+ + Ne + Mu \quad (107)$$

In gaseous N_2, H_2O, NH_3, n-C_6H_{14}, c-C_6H_{12}, $(CH_3)_4Si$, CH_2Cl_2, and $CHCl_3$, $f_D = 0.1$–0.25 and $f_{Mu} = 0.9$–0.75 (an exception in CCl_4: $f_D = f_{Mu} = 0.5$). The f_D fraction is supposed to be formed by hot reactions of μ^+ or of Mu (108). In collisions with paramagnetic molecules, fast spin exchange can take place. Cross sections at room temperature are reported for O_2: (5.9 ± 0.6) (109) and (7.8 ± 0.4) (110) × 10^{-16} cm^2; while for NO: (7.1 ± 1.0) (109) and (10.3 ± 0.4) (110) × 10^{-16} cm^2. In Table X, the rate constants are listed for the reactions of Mu with several gaseous compounds and also the isotopic k_{Mu}/k_H ratios. If the reactions are diffusion controlled, an isotope effect of $k_{Mu}/k_H = (m_H/M_\mu)^{1/2} = 3$ is to be expected. Ratios higher than 3, as for F_2 and C_2H_4, point to tunneling effects (111), as has been corroborated (for F_2) by theoretical calculations (112).

B. LIQUID MIXTURES

In order to gain more information about (1) relative reaction rates of Mu, (2) occurrence of hot Mu reactions, and (3) the high diamagnetic yield in CCl_4 ($f_D = 1$), several experiments have been performed in liquid mixtures.

In several mixtures, no preferential interaction with one of the two compounds was observed. The linear increase of f_D as a function of additive concentration between 0 and 100 mol % [from 0.56 to 0.85 in CH_3OH–$CHCl_3$, from 0.16 to 0.56 in C_6H_6–CH_3OH (*100*), and from 0.16 to 0.61 in C_6H_6–c-C_6H_{12} (*113*)] was taken as evidence for hot Mu reactions. In binary mixtures of C_6H_6 with C_6H_5Br, $C_6H_5NH_2$, and p-$C_6H_4F_2$, the values of f_D and of f_R (C_6H_6Mu and the isomeric C_6H_5XMu) were measured as a function of the relative concentration (*114*). The relative reaction rates of Mu do not differ to a large extent from those measured for thermal H atoms (*115*); these Mu results do not contribute much to the discussion on hot/thermal reactions. The partial rate factors, relative to C_6H_6, differ for $C_6H_5NH_2$ (ortho 1.9, meta 1.2, and para 1.7) from those measured with thermal T atoms [4.7, 1.36, and 2.0, respectively (*116*)].

In mixtures of C_6H_6 and CH_3I, the values of f_D and f_R deviate significantly from the proposed linearity for hot reactions (*117*). The results indicate that both compounds compete in reactions with thermal Mu, CH_3I being the more efficient. More information was obtained by investigations of Roduner (*118*) on binary mixtures of C_6H_6 with c-C_6H_{12}, DMBD (2,3-dimethyl-1,3-butadiene), and CCl_4.

1. From experiments with C_6H_6–c-C_6H_{12}, the rate constant for addition of Mu to C_6H_6 was found to be $(8.9 \pm 0.6) \times 10^9 \, M^{-1} \, \mathrm{sec}^{-1}$, which is considerably below the diffusion-controlled limit, proving that Mu is not hot when it adds.

2. From experiments with C_6H_6–DMBD, the rate constant for addition of Mu to DMBD was deduced as $4 \times 10^{10} \, M^{-1} \, \mathrm{sec}^{-1}$, which is close to the diffusion-controlled limit. The selectivity for addition to DMBD over that to C_6H_6 (by a factor of 4.5) is much lower than for thermal H atoms. This effect was attributed to tunneling rather than to reactions of hot Mu.

3. In former experiments with C_6H_6–CCl_4 mixtures, only f_D values were measured (*113*). Roduner (*118*) has also measured f_R values, in particular at low CCl_4 concentrations (Fig. 3). Since it was proved that Mu atoms are the direct radical precursors for addition to C_6H_6, it was concluded that CCl_4, an excellent electron scavenger, inhibits Mu formation by scavenging spur electrons before their combination with μ^+. This means that thermal Mu is formed in an end-of-track process: $\mu^+ + e^- \rightarrow$ Mu. The rate constant of $2.7 \times 10^{12} \, M^{-1} \, \mathrm{sec}^{-1}$ for the reaction of CCl_4 with electrons shows that Mu is formed within a picosecond after the creation of the last spur.

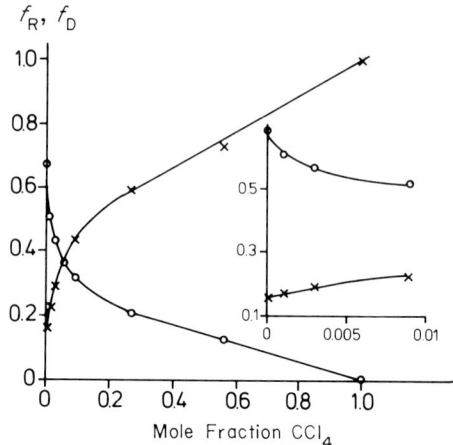

FIG. 3. f_R(O) and f_D(X) in C_6H_6–CCl_4 mixtures. [By permission of E. Roduner, ref. (118).]

For most of the hydrocarbons it has been found that f_D = 0.6–0.7, f_{Mu} = 0.1–0.2, and f_L = 0.1–0.2 (100, 119). These data were taken as evidence that hot abstraction (MuH) is of comparable efficiency for all of these compounds and that only 10–20% of the hot Mu atoms become thermalized and contribute to f_M. The data for f_D are not much different from those obtained for the reactions of recoil T atoms with C_4H_{10}: HT (40%), C_4H_9T (20%) (120). In n-C_6F_{14}, similar yields were found: f_D = 0.64 and f_{Mu} = 0.20, whereas F abstraction is a less probable process (121).

C. Formation and Reactions of Muonic Radicals

Before the first muonic radicals were observed in 1978 by Roduner et al. (122), rate constants had been measured for the addition of Mu to unsaturated compounds in aqueous solution (Table XI). If k_{Mu}/k_H is larger than 3, tunneling may be important, since otherwise differences in the vibrational zero-point energy in the transition state can decrease this ratio (128). In the case of maleic acid, the Arrhenius parameters were determined as $A = (2.3 \pm 0.2) \times 10^{13}\ M^{-1}\ \text{sec}^{-1}$ and $E = 18.8 \pm 1.7$ kJ mol^{-1}. Rate constants were also measured for addition to the CN triple bond (127).

TABLE XI

REACTION CONSTANTS[a] FOR ADDITION OF Mu IN AQUEOUS SOLUTIONS AT 298 K

Compound (123, 124, 125)	k_{Mu}	k_{Mu}/k_H
Maleic acid	1.1×10^{10}	18
Fumaric acid	1.4×10^{10}	16
Ascorbic acid	1.8×10^{9}	16
Dihydroxyfumaric acid	4.5×10^{7}	0.5
Acrylamide	1.9×10^{10}	1.1
Acrylic acid	1.6×10^{10}	
Acrylonitrile	1.1×10^{10}	2.8
Methyl metacrylate	1.0×10^{10}	

[a] In mol^{-1} sec^{-1}.

1. Alkenes

Roduner et al. (128) studied the formation of muonic radicals in 24 monoolefins and in 9 dienes. Hyperfine coupling constants (A_μ) were measured for 44 radicals (11 compounds gave two radicals each). For comparison with the corresponding protonic constants A_p, the values of A_μ must be multiplied by the ratio of the magnetic moments: $A'_\mu = (\mu_p/\mu_\mu)A_\mu = 0.3141 A_\mu$. The isotopic effect A'_μ/A_p is found to be of the order of 1.4 (see also comments below in Section C,2). The terminal and nonterminal olefins yield primary, secondary, and tertiary alkyl radicals with A_μ = 330, 300, and 270 MHz, respectively. Allylic radicals were usually formed from dienes with A_μ = 160–190 MHz. The assignment of the radicals is based on more extensive investigations of ethene, propene, 2-methylpropene (128), and tetramethylethylene (129). The regioselectivity of Mu addition is similar to that of H atoms, in that it occurs (1) at the unsubstituted C atom for terminal olefins, with the exception of allyl ethers, (2) preferentially at the less substituted C atom for nonterminal olefins, and (3) at the end C atoms for dienes to yield the thermodynamically more stable allyl type radicals.

In methyl-substituted dienes, the following yields were found: f_D = 0.2–0.3, f_R = 0.45–0.35, and f_L = 0.35 (130). The missing fraction of muon polarization f_L is thought to be due to muonic radicals having lost spin polarization during encounters with other paramagnetic species in the spur of the muon track. If Mu is the precursor for the formation of radicals, then its lifetime must be less than 20 psec. Comparing the selectivities, the activation energies, and the rate constants for addition between Mu and H leads to the conclusion that Mu is close to thermal energy at the moment of its addition. However, μ^+ cannot be excluded as the precursor (131), in which case its lifetime is less than 2 nsec, because the precession frequency is 100 times lower than that of Mu. The hyperfine coupling constants A_μ decrease with increasing temperature for CH_2MuCH_2, $CH_2MuCHCH_3$, and $CH_2MuC(CH_3)_2$, which is evidence of a higher barrier against rotation about the C—C bond than that for H and D atoms. This is a consequence of higher zero-point vibrational amplitudes, which result in a more effective van der Waals radius of Mu (128) (see Notes Added in Proof, p. 133).

In Table XII, A_μ values are given along the f_R, f_D and relative rates λ for a selection of chloroolefins (C_4H_9Cl). The tentative assignments of the radicals are based on the data of other olefins, e.g., the A_μ values for both radicals from 1-chloro-3-butene being similar for terminal and nonterminal Mu addition to allyl propyl and diallyl ethers (128). The relaxation rates are somewhat high, indicating slow addition rates. In the case of 2-chloro-2-butene, no radicals were observed, but this may be due to even higher relaxation rates (similarly no radicals were observed for chloroethylenes). Ring closure and ring fission were observed for muonic radicals (132). The Arrhenius parameters are in good agreement with literature values for the corresponding protonic radicals, supporting the view that the substitution of H by Mu in a CH_3 group neighboring the reactive center changes hardly the rate constants.

TABLE XII

Radicals from Chloroolefins (C_4H_7Cl) at Room Temperature (121)

Olefin	Radical	A_μ (MHz)	λ (μsec)$^{-1}$	f_R	f_D
CH_2=$CHCH_2CH_2Cl$	$CH_2Mu\dot{C}HCH_2CH_2Cl$	318.4	0.75 ± 0.01	0.13 ± 0.02 ⎫	
	$\dot{C}H_2CHMuCH_2CH_2Cl$	332.1	3.1 ± 0.6	0.23 ± 0.02 ⎭	0.52 ± 0.01
CH_2=$CHCHClCH_3$	No radical	—	—	—	0.60 ± 0.01
CH_3CH=$CHCH_2Cl$	$CH_3CHMu\dot{C}HCH_2Cl$	302.5	2.3 ± 0.1	0.056 ± 0.006 ⎫	
	$CH_3\dot{C}HCHMuCH_2Cl$	276.7	1.2 ± 0.3	0.046 ± 0.005 ⎭	0.57 ± 0.01
CH_3CH=$CClCH_3$	$CH_3CHMu\dot{C}ClCH_3$	235.0	4.4 ± 0.2	0.31 ± 0.02	0.47 ± 0.01
CH_2=$C(CH_3)CH_2Cl$	$CHMu\dot{C}(CH_3)CH_2Cl$	265.5	—	0.20 ± 0.02	0.60 ± 0.02
$CHCl$=$C(CH_3)_2$	$CClMu\dot{C}(CH_3)_2$	79.0	—	0.16 ± 0.02	0.52 ± 0.02

TABLE XIII (133, 134)

ADDITION OF Mu TO MONOSUBSTITUTED ARENES

	A_μ (MHz) and relative ortho/meta/para ratios (%)			Σf_R	f_D	f_L
$C_6H_5CH_3$	498.6	509.3	496.4			
	48	35	17	0.50	0.25	0.25
C_6H_5F	485.7	511.8	511.8			
	42	44	14	0.43	0.19	0.38
$C_6H_5CF_3$	500.3	510.5	508.7			
	37	44	19	0.42	0.24	0.34
C_6H_5Cl	487.3	509.4	485.3			
	39	44	17	0.28	0.33	0.39
$C_6H_5CCl_3$	477.5[a]	155.7	477.5[a]			
	a	36	a	0.16	0.67	0.17

[a] Ortho and para isomers possibly degenerate.

2. Arenes

The first muonic cyclohexadienyl radical detected was C_6H_6 Mu (122). Later, many others were found, e.g., those originating from all the liquid CH_3- and F-substituted arenes (133). Furthermore, 24 monosubstituted arenes were studied, and the relative yields of Mu addition at the ortho, meta, and para positions were determined (134) (Table XIII). The assignment of these three isomeric radicals is determined via reactions with the p-deutero-substituted analogs (137, 138), where (1) at a magnetic field of 0.3 T there is a slight shift in the A_μ values in the sequence para > meta > ortho (Fig. 4), and (2) at a magnetic field of 0.1 T the signals from addition at C(H) atoms are split, but not from those pertaining to C(D) atoms (Fig. 4). In most cases, ortho addition occurs at a rate somewhat higher than the statistical probability, even in the presence of a bulky substituent. It is difficult to correlate the deviations from statistics with any substituent properties. In several cases, ipso substitution was observed (Fig. 4 and reference 133). Muonic radicals were also detected in solid benzene (135) and durene (136).

For some of these compounds, the protonic hyperfine coupling constants are known. The isotopic A'_μ/A_p ratio is 1.21 for radicals produced by addition to C_6H_6 and to the ortho position of $C_6H_5CH_3$ and 1.15–1.18 for several fluorobenzenes (133). Quantum chemical calculations that include averaging over 33 vibrational modes in C_6H_7–C_6H_6Mu have shown that the dynamics account quantitatively ($A'_\mu/A_p = 1.16$) for the

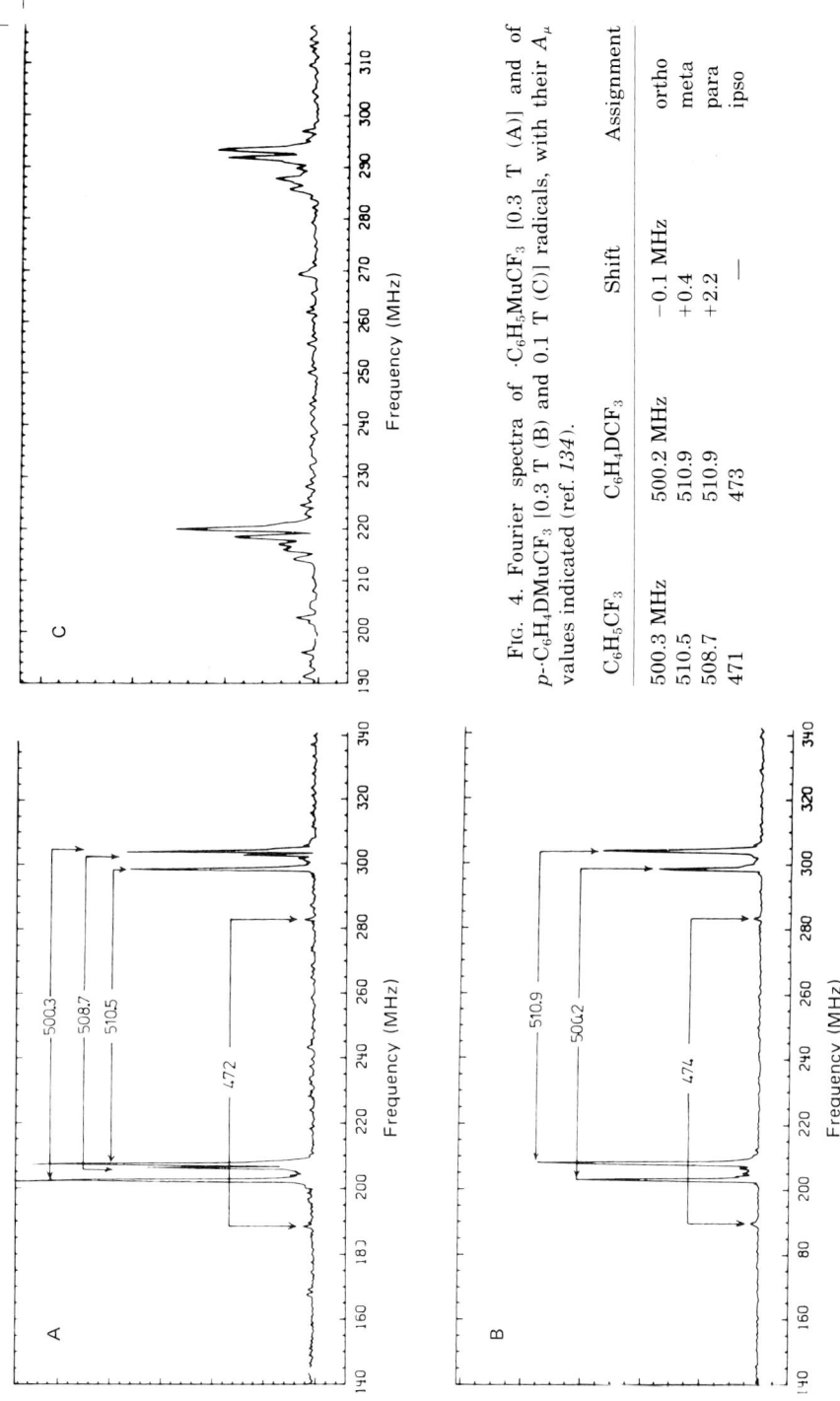

FIG. 4. Fourier spectra of ·C$_6$H$_5$MuCF$_3$ [0.3 T (A)] and of p-·C$_6$H$_4$DMuCF$_3$ [0.3 T (B) and 0.1 T (C)] radicals, with their A_μ values indicated (ref. 134).

C$_6$H$_5$CF$_3$	C$_6$H$_4$DCF$_3$	Shift	Assignment
500.3 MHz	500.2 MHz	−0.1 MHz	ortho
510.5	510.9	+0.4	meta
508.7	510.9	+2.2	para
471	473	—	ipso

isotope effect (137). The normal C—Mu stretching frequency and the two out of plane vibration frequencies are, in particular, responsible for the isotopic effect.

The value of f_D for monosubstituted arenes range between 0.15 (C_6H_6) and 0.76 (C_6H_5SH). In the series of halobenzenes, it increases from 0.19 (F) to 0.52 (I) (134). As discussed in Section IV,B, radical formation in mixtures of C_6H_6 involves the thermalization of μ^+ and subsequent combination with end-of-track electrons ($\leq$ 1 psec), followed by addition (10 psec) (118). The compounds with high f_D values undergo efficient dissociative electron capture, increasing from C_6H_5F to C_6H_5I, inhibiting the formation of Mu and of muonic radicals.

3. Reactions of Muonic Radicals

On addition of small amounts of benzoquinone to benzene, the relaxation rates increase linearly with concentration, in accord with a pseudo-first-order reaction between C_6H_6Mu and BQ (136). In the case of $C_6H_5CH_3$, the rate constants for the reactions of the o- and m-$C_6H_5MuCH_3$ radicals with BQ are comparable, but they differ considerably for the three isomers formed from $C_6H_5OCH_3$ (138) (Table XIV). The rate constant for the reaction of C_6H_6Mu with 2,3-dimethyl-1,3-butadiene is 7×10^5 mol^{-1} sec^{-1} at room temperature (139), whereas for C_6H_7 it is less than 12 mol^{-1} sec^{-1}. The large isotope effect is due to direct Mu transfer from C_6H_6 to C_6H_{10}, a reaction that is considerably faster than H transfer from C_6H_7, owing to the much higher zero-point vibrational energies in the muonic radical. For the dimerization reaction of the $C_6H_5CHCH_2Mu$ radical with styrene, a rate constant of 1.3×10^5 mol^{-1} sec^{-1} at room temperature has been reported (131). (Mu adds at the C=C bond in styrene, and not at the ring.)

4. Miscellaneous

Apart from addition to alkenes and arenes, other addition reactions have also been observed:

Styrene	$C_6H_5\dot{C}HCH_2Mu$	A_μ = 213.3 (134)
Phenylacetylene	$C_6H_5\dot{C}$=CHMu	421.2 (134)
Furane	c-$\dot{C}_4H_8OMu$	379.2 (140)
Acetone	$(CH_3)_2\dot{C}OMu$	26.0 (122)
Methyl methacrylate	$CH_2Mu\dot{C}(CH_3)COOCH_3$	270 (140)
Cyclohexanone	c-$\dot{C}_6H_{10}OMu$	21.4 (140)
Nitrobenzene	$C_6H_5\dot{N}OOMu$	38.7 (134)
Azobenzene	$C_6H_5NMu\dot{N}C_6H_5$	21.2 (121)

TABLE XIV (138)

RATE CONSTANTS FOR THE REACTION OF MUONIC CYCLOHEXADIENYL-TYPE RADICALS WITH QUINONES AT 293 K

Compound	Substitution of the radical	k_{BQ} ($10^8\ M^{-1}\ sec^{-1}$)	k_{DQ} ($10^8\ M^{-1}\ sec^{-1}$)	k_{BQ}/k_{DQ}
Benzene	—	2.6 ± 0.4^a	0.62 ± 0.09	4.2 ± 0.9
Benzene-d_6	—	—	0.56 ± 0.08	—
Toluene	ortho	3.4 ± 0.9	0.52 ± 0.05	6.5 ± 1.8
	meta	3.5 ± 1.5	0.79 ± 0.13	4.4 ± 2.2
Anisole	ortho	10.5 ± 1.3	1.2 ± 0.3	8.8 ± 2.5
	meta	4.2 ± 0.5	0.66 ± 0.08	6.4 ± 1.1
	para	20	—	—

^{a} See ref. 117.

5. Conclusions

Despite its short mean life of 2.2 μsec, many chemical reactions of the hydrogen-like Mu atom can be studied. From investigations with recoil T atoms, it was suggested that Mu atoms could also react while possessing an excess of kinetic energy (hot reactions). Later investigations in C_6H_6–c-C_6H_{12} mixtures showed that addition to C_6H_6 is a thermal reaction. The relative addition rates to the ortho, meta, and para positions in monosubstituted benzenes, although not as conclusive, also point in the same direction. As free μ^+, MuCl, and CMuCl$_3$ cannot be distinguished in liquid CCl$_4$, it could not be deduced whether the value of $f_D = 1$ was due to hot abstraction (MuCl), hot substitution (CMuCl$_3$), or to nonneutralization of the μ^+. From experiments with CCl$_4$–C_6H_6 mixtures, it is now believed that CCl$_4$ scavenges spur electrons, preventing the formation of Mu.

The discovery of muonic radicals in 1978 by Roduner opened up a broad field of interesting experiments. Aromatic substitution is normally studied by the measurement of the yields of stable products. This includes not only the site of addition of the reactant, but also the splitting off of the atom that is substituted. With the MuSR technique only the first step is studied exclusively. Another development concerns the measurements of the reactions of muonic radicals: dimerization, ring opening, and ring closure. The observations of different reaction rates of the three isomeric muonic cyclohexadienyl radicals from anisole with quinones also opens a new field of investigation.

Absolute and relative reaction rates of Mu were measured for several compounds. Some of the results can be explained only by the acceptance of tunneling reactions.

Obviously the field of MuSR reaction studies is still in its infancy. Many more experiments must be performed in order to reach firm conclusions.

ACKNOWLEDGMENTS

I am very grateful to Professor Dr. F. S. Rowland (Irvine) and Dr. E. Roduner for their permission to use some experimental data prior to publication.

This work is part of the research program of the National Institute for Nuclear Physics and High-Energy Physics (NIKHEF, Section K), made possible by financial support from the Foundation for Fundamental Research on Matter (FOM) and the Netherlands Organization for the Advancement of Pure Research (ZWO).

REFERENCES

1. Szilard, L., and Chalmers, T. A., *Nature (London)* **134,** 462 (1934).
2. Wolfgang, R., *Prog. React. Kinet.* **3,** 97 (1965).
3. Stöcklin, G., "Chemie heisser Atome." Verlag Chemie, Weinheim, 1969.
4. Urch, D. S., *Int. Rev. Sci. Radiochem. Inorg. Chem. Ser. Two* **8,** 49 (1975).
5. "Hot Atom Chemistry Status Report." IAEA, Vienna (1975).
6. "Chemical Effects of Nuclear Transformations in Inorganic Systems" (G. Harbottle and A. G. Maddock, eds.). North-Holland Publ., Amsterdam, 1979.
7. Tominaga, T., and Tachikawa, E., "Modern Hot-Atom Chemistry and Its Applications." Springer-Verlag, Berlin and New York, 1981.
8. Tang, Y. N., *in* "Isotopes in Organic Chemistry" (E. Buncel and C. C. Lee, eds.), Vol. 4, p. 85. Elsevier, Amsterdam, 1978.
9. Gaspar, P. P., and Root, J. W., *Radiochim. Acta* **28,** 191 (1981).
10. *ACS Symp. Ser.* **66** (1978).
11. Root, J. W., and Manning, R. E., *Adv. Chem. Ser.* **197,** 79 (1981).
12. Brinkman, G. A., *Int. J. Appl. Radiat. Isot.* **34,** 985 (1983).
13. Rack, E. P., *Radiochim. Acta* **28,** 221 (1981).
14. Tilbury, R. S., *Adv. Chem. Ser.* **197,** 261 (1981).
15. Gaspar, P. P., *Adv. Chem. Ser.* **197,** 3 (1981).
16. Walker, D. C., "Muon and Muonium Chemistry." Cambridge Univ. Press, London and New York, 1983.
17. Brinkman, G. A., *Chem. Rev.* **81,** 267 (1981).
18. Brinkman, G. A., *Chem. Rev.* **82,** 245 (1982).
19. Brinkman, G. A., *Chem. Rev.* (in press).
20. Horváth, D., *Radiochim. Acta* **28,** 241 (1981).
21. Cacace, F., *Adv. Chem. Ser.* **197,** 143 (1981).
22. Raadschelders-Buyze, C., Ph.D. Thesis, Free University of Amsterdam, 1974.
23. Raadschelders-Buyze, C., and van Zanten, B., *Radiochim. Acta* **22,** 71 (1975).
24. Williams, R. L., and Rowland, F. S., *J. Phys. Chem.* **76,** 3509 (1972).
25. Williams, R. L., and Rowland, F. S., *J. Phys. Chem.* **75,** 2709 (1971); Williams, R. L., and Rowland, F. S., *J. Phys. Chem.* **77,** 301 (1973).
26. Grant, E. R., and Root, J. W., *Chem. Phys. Lett.* **27,** 484 (1974).

27. Manning, R. G., Grant, E. R., et al., Int. J. Chem. Kin. **7,** 39 (1975).
28. Mo, S. H., Grant, E. R., et al., ACS Symp. Ser. **66,** 59 (1978).
29. Root, J. W., Mathis, C. A., et al., Adv. Chem. Ser. **197,** 207 (1981).
30. Parks, N. J., Krohn, K. A., et al., J. Chem. Phys. **55,** 2690 (1971).
31. McKeown, F. P., Iyer, R. S., et al., to be published (1984).
32. Foon, R., and Kaufman, M., Prog. React. Kin. **8,** 81 (1975).
33. Smith, D. J., Setser, D. W., et al., J. Phys. Chem. **81,** 898 (1977).
34. Iyer, R. S., and Rowland, F. S., J. Phys. Chem. **85,** 2488 (1981).
35. Iyer, R. S., and Rowland, F. S., J. Phys. Chem. **85,** 2493 (1981).
36. Iyer, R. S., and Rowland, F. S., Chem. Phys. Lett. **21,** 346 (1973).
37. Rowland, F. S., Rust, F., et al., ACS Symp. Ser. **66,** 26 (1978).
38. Williams, R. L., and Rowland, F. S., J. Phys. Chem. **76,** 3509 (1972).
39. Frank, J. P., and Rowland, F. S., J. Phys. Chem. **78,** 850 (1974).
40. Concannon, C., and Rowland, F. S., J. Phys. Chem. **85,** 89 (1981).
41. Williams, R. L., and Rowland, F. S., J. Am. Chem. Soc. **94,** 1047 (1972).
42. Williams, R. L., Iyer, R. S., et al., J. Am. Chem. Soc. **94,** 7192 (1972).
43. Smail, T., Miller, G. E., et al., J. Phys. Chem. **74,** 3464 (1970).
44. Smail, T., Iyer, R. S., et al., J. Am. Chem. Soc. **94,** 1041 (1972).
45. Rowland, F. S., Lawrence, B., et al., Hot At. Chem. Symp. 11th, Davis X.7 (1982).
46. Cramer, J. A., Iyer, R. S., et al., J. Am. Chem. Soc. **95,** 643 (1973).
47. Kikuchi, M., Cramer, J. A., et al., J. Phys. Chem. **86,** 2677 (1982).
48. Rogers, P., Montague, D. C., et al., Chem. Phys. Lett. **89,** 9 (1982).
49. Rogers, P. J., Selco, J. I., et al., Chem. Phys. Lett. **97,** 313 (1983).
50. Rowland, F. S., personal communication.
51. Lee, F. S. C., and Rowland, F. S., J. Phys. Chem. **75,** 2685 (1971).
52. Lee, F. S. C., and Rowland, F. S., J. Phys. Chem. **81,** 1229 (1977).
53. Stevens, D. J., and Spicer, L. D., J. Chem. Phys. **64,** 4798 (1976).
54. Lee, F. S. C., and Rowland, F. S., J. Phys. Chem. **81,** 86 (1977).
55. Watson, R. T., J. Phys. Chem. Ref. Data **6,** 871 (1977).
56. Kikuchi, M., Lee, F. S. C., et al., J. Phys. Chem. **84,** 84 (1981).
57. Lee, F. S. C., and Rowland, F. S., J. Phys. Chem. **81,** 1235 (1977).
58. Lee, F. S. C., and Rowland, F. S., J. Phys. Chem. **81,** 1222 (1977).
59. Lee, F. S. C., and Rowland, F. S., J. Phys. Chem. **81,** 684 (1977).
60. Lee, F. S. C., and Rowland, F. S., J. Phys. Chem. **84,** 1876 (1980).
61. Iyer, R. S., and Rowland, F. S., submitted (1984).
62. Stevens, D. J., and Spicer, L. D., J. Phys. Chem. **81,** 1217 (1977).
63. Iyer, R. S., Rogers, P. J., et al., J. Phys. Chem. **87,** 3799 (1983).
64. Wai, C. M., and Rowland, F. S., J. Am. Chem. Soc. **90,** 3638 (1968).
65. Wai, C. M., and Rowland, F. S., J. Am. Chem. Soc. **91,** 1053 (1969).
66. Brinkman, G. A., Kaspersen, F. M., et al., Radiochim. Acta **28,** 61 (1981).
67. Bhave, R. N., Brinkman, G. A., et al., Radiochim. Acta **31,** 185 (1982).
68. Veenboer, J. Th., and Brinkman, G. A., Radiochim. Acta **33,** 7 (1983).
69. van Halteren, B. W., Veenboer, J. Th., et al., Radiochem. Radioanal. Lett. **51,** 373 (1982).
70. van Halteren, B. W., and Brinkman, G. A., unpublished results.
71. Veenboer, J. Th., and Brinkman, G. A., Proc. Int. Congr. Radiat. Res., 7th Amsterdam A6-07 (1983).
72. Veenboer, J. Th., and Brinkman, G. A., unpublished results.
73. Veenboer, J. Th., and Brinkman, G. A., Radiochim. Acta **31,** 7 (1982).
74. Valencich, T., and Bunker, D. L., J. Chem. Phys. **79,** 671 (1983).

75. Chou, C. C., Wilkey, D. D., et al., Chem. Phys. Lett. **20**, 53 (1973).
76. Lee, E. K. C., Miller, G., et al., J. Am. Chem. Soc. **87**, 190 (1965).
77. Min, B. K., Yeh, C. T., et al., J. Phys. Chem. **82**, 971 (1978).
78. Rowland, F. S., Turton, C. N., et al., J. Am. Chem. Soc. **78**, 2354 (1956).
79. Keller, H., and Rowland, F. S., J. Phys. Chem. **62**, 1373 (1958).
80. Kay, J. G., Malsan, R. P., et al., J. Am. Chem. Soc. **81**, 505 (1959).
81. Henchman, M., and Wolfgang, R., J. Am. Chem. Soc. **83**, 2991 (1961).
82. Palino, G. F., and Rowland, F. S., J. Phys. Chem. **78**, 1299 (1971).
83. Machulla, H. J., and Stöcklin, G., J. Phys. Chem. **78**, 658 (1974).
84. Tang, Y. N., Ting, C. T., et al., J. Phys. Chem. **74**, 675 (1970).
85. Ting, C. T., and Rowland, F. S., J. Phys. Chem. **74**, 445 (1970).
86. Wai, C. M., and Rowland, F. S., J. Phys. Chem. **74**, 434 (1970).
87. Machulla, H. J., and Stöcklin, G., J. Phys. Chem. **78**, 658 (1974).
88. Wolf, A. P., Schueler, P., et al., J. Phys. Chem. **83**, 1237 (1979).
89. To, K. C., Rack, E. P., et al., J. Phys. Chem. **74**, 1499 (1981).
90. Wai, C. M., Ting, C. T., et al., J. Am. Chem. Soc. **86**, 2525 (1964).
91. Vasáros, L., Machulla, H. J., et al., J. Phys. Chem. **76**, 501 (1972).
92. Wu, J., and Ache, H. J., J. Am. Chem. Soc. **99**, 6021 (1977).
93. Acciani, T. R., Su, Y., et al., J. Phys. Chem. **82**, 975 (1978).
94. Wu, J., Booth, T. E., et al., J. Phys. Chem. **68**, 5285 (1978).
95. Acciani, T. R., and Ache, H. J., J. Phys. Chem. **82**, 1465 (1978).
96. Palino, G. F., and Rowland, F. S., Radiochim. Acta **15**, 57 (1971).
97. Daniel, S. H., Ache, H. J., et al., J. Phys. Chem. **78**, 1043 (1974).
98. Su, Y., and Ache, H. J., J. Phys. Chem. **80**, 659 (1976).
99. Brewer, J. H., and Crowe, K. M., Annu. Rev. Nucl. Part. Sci. **28**, 239 (1978).
100. Fleming, D. G., Garner, D. M., et al., Adv. Chem. Ser. **175**, 279 (1979).
101. Percival, P. W., Radiochim. Acta **26**, 1 (1979).
102. Walker, D. C., J. Phys. Chem. **85**, 3960 (1981).
103. Pifer, A. E., Bowen, T., et al., Nucl. Instrum. Methods **135**, 39 (1976).
104. Oram, C. J., Warren, J. B., et al., Nucl. Instrum. Methods **179**, 95 (1981).
105. Fleming, D. G., Mikula, R. J., et al., Phys. Rev. **A26**, 2527 (1982).
106. Stambaugh, R. D., Casperson, D. E., et al., Phys. Rev. Lett. **83**, 568 (1974).
107. Garner, D. M., personal communication.
108. Fleming, D. G., and Arseneau, D. J., Hyperfine Interact. **17(19)**, 655 (1984).
109. Mobley, R. M., Amato, J. J., et al., J. Chem. Phys. **47**, 3074 (1974).
110. Senba, M., Garner, D. M., et al., Hyperfine Interact. **17(19)**, 703 (1984).
111. Fleming, D. G., Garner, D. M., et al., Hyperfine Interact. **8**, 337 (1981).
112. Connor, J. N. L., Hyperfine Interact. **8**, 423 (1981).
113. Jean, Y. C., Ng, B. W., et al., J. Phys. Chem. **85**, 451 (1981).
114. Roduner, E., Brinkman, G. A., et al., Hyperfine Interact. **17(19)**, 803 (1984).
115. Henderson, R. W., and Pryor, W. A., J. Am. Chem. Soc. **97**, 7437 (1975).
116. Pryor, W. A., Lin, T. H., et al., J. Am. Chem. Soc. **95**, 6993 (1973).
117. Roduner, E., Hyperfine Interact. **8**, 561 (1981).
118. Roduner, E., Hyperfine Interact. **17(19)**, 785 (1984).
119. Walker, D. C., Hyperfine Interact. **8**, 329 (1981).
120. Wolfgang, R., Prog. React. Kin. **3**, 1 (1965).
121. Roduner, E., Brinkman, G. A., et al., unpublished results.
122. Roduner, E., Percival, P. W., et al., Chem. Phys. Lett. **57**, 37 (1978).
123. Percival, P. W., Roduner, E., et al., Chem. Phys. Lett. **47**, 11 (1977).
124. Percival, P. W., Roduner, E., et al., Adv. Chem. Ser. **175**, 333 (1979).

125. Stadlbauer, J. M., Ng, B. W., et al., Can J. Chem. **59**, 3261 (1981).
126. Ng, B. W., Jean, Y. C., et al., J. Phys. Chem. **85**, 454 (1981).
127. Stadlbauer, J. M., Ng, B. W., et al., Hyperfine Interact. **17(19)**, 715 (1984).
128. Roduner, E., Strub, W., et al., Chem. Phys. **67**, 275 (1982).
129. Roduner, E., and Fischer, H., Chem. Phys. **54**, 261 (1981).
130. Roduner, E., and Webster, B. C., J. Chem. Soc. Faraday Trans. I **79**, 1939 (1983).
131. Cox, S. F. J., Hill, A., et al., J. Chem. Soc. Faraday Trans. I **78**, 2975 (1982).
132. Burkhard, P., Roduner, E., et al., J. Phys. Chem. **88**, 773 (1984).
133. Roduner, E., Brinkman, G. A., et al., Chem. Phys. **73**, 117 (1982).
134. Roduner, E., Brinkman, G. A., et al., Chem. Phys., to be published (1984).
135. Roduner, E., Chem. Phys. Lett. **81**, 191 (1981).
136. Roduner, E., Hyperfine Interact. **8**, 561 (1981).
137. Münger, K., Diplome thesis, Univ. of Zürich (1980).
138. Roduner, E., Brinkman, G. A., et al., Hyperfine Interact. **17(19)**, 797 (1984).
139. Roduner, E., and Münger, K., Hyperfine Interact. **17(19)**, 793 (1984).
140. Hill, A., Allen, G., et al., J. Chem. Soc. Faraday Trans. I **78**, 2959 (1982).

NOTES ADDED IN PROOF

(Page 115): Not only does steric hindrance predict the retention/inversion ratio, but also the nature of the leaving group. In halopropionyl halides the degree of inversion for ^{34m}Cl-for-X substitution is

72% (X = F), 79% (X = Cl), and 78% (X = Br)

Furthermore, the mass of the incoming atom is also important, as can be seen by the extent of inversion for ^{18}F-for-X substitution for the propionyl halides:

30% (X = F), 35% (X = Cl), 39% (X = Br)

[Reference: To, K. C., Wolf, A. P., et al., J. Phys. Chem. **87**, 4929 (1984).]

(Page 124): The height of the barriers for internal rotation has been determined as 2.71 kJ mol^{-1} for both the CMuH$_2$CH$_2$ and CMuD$_2$CD$_2$ radicals, but is only 0.35 and 0.38 kJ mol^{-1} for the CDH$_2$CH$_2$ and CHD$_2$CD$_2$ radicals, respectively.

[Reference: Ramos, M. J., McKenna, D., et al., J. Chem. Soc. Faraday Trans. I **80**, 255, 267 (1984).]

HOMOCYCLIC SELENIUM MOLECULES AND RELATED CATIONS

RALF STEUDEL and EVA-MARIA STRAUSS

Institut für Anorganische und Analytische Chemie, Technische Universität Berlin, Berlin, Federal Republic of Germany

I.	General	135
II.	Neutral Selenium Ring Molecules	136
	A. Cyclopentaselenium, Se_5	136
	B. Cyclohexaselenium, Se_6	137
	C. Cycloheptaselenium, Se_7	142
	D. Cyclooctaselenium, Se_8	144
	E. Other Ring Molecules	150
	F. Thermodynamic Properties of Selenium Allotropes	151
III.	Homocyclic Selenium Cations	152
	A. Mass Spectra of Selenium Vapors	154
	B. Dications Se_n^{2+} in Solids and Solutions	155
	C. Other Dications	160
IV.	Selenium Iodide Cations, Se_nI^+	161
V.	Conclusions and Outlook	162
	References	163

I. General

The heavier chalkogens are known for their tendency toward homonuclear catenation, which rivals that of phosphorus and silicon and is exceeded only by that of carbon. Especially sulfur forms a large number of rings S_n (n = 6, 7, ...) (1) and of chainlike compounds of type R—S_n—R (R = inorganic or organic group) (2), which have been extensively studied. In the case of selenium the number of known cyclic compounds with more than one Se—Se bond, that is, compounds with cumulated Se—Se bonds, was formerly quite small, and for a long time the Se_8 molecule was the only example. However, developments have shown that selenium is able to form rings Se_n of various sizes as well as cyclic, bicyclic, and cagelike dications Se_n^{2+}, and cyclic derivatives like

TABLE I

PRESENTLY KNOWN HOMOCYCLIC SELENIUM COMPOUNDS

	Neutral rings	Monocations	Dications	Ring derivatives
As pure materials:	Se_6, Se_8	Se_n^+ ($n = 5 ... 10$)	Se_n^{2+}	$[Se_6I][AsF_6]$
In solution:	Se_n ($n = 6-8$)	(in the vapor	($n = 4, 8,$ or 10)	
In the gas phase:	Se_n ($n = 5-10$)	phase only)		

Se_6I^+, as does sulfur. In addition, a number of presumably cyclic monocations Se_n^+ have been studied in the vapor phase.

The presently known homocyclic selenium compounds are summarized in Table I. All these compounds are interesting owing to their chemical simplicity—the allotropes of an element and its binary derivatives should interest every chemist—and their bond properties. Furthermore, homocyclic selenium species are related to the much studied and still poorly understood semiconducting amorphous selenium, which may be a mixture of chain- and ringlike molecules similar to liquid or amorphous (polymeric) sulfur. In this connection, the small rings Se_6, Se_7, and Se_8 can serve as model species to study the chemical transformations of amorphous selenium on heating or on irradiation, as well as the changes in its chemical and physical properties on doping or because of the presence of impurities. Selenium–selenium bonds are of lower bond energy and consequently of higher reactivity than comparable sulfur–sulfur bonds. The preparation and characterization of compounds containing cumulated Se–Se bonds therefore require more sophisticated preparative and analytical methods. The later results in this field are reviewed in this article; for earlier reviews, see refs. (3–7).

II. Neutral Selenium Ring Molecules

Neutral selenium rings Se_n have been studied in the vapor phase by mass spectroscopy, in solution by high-pressure liquid chromatography, and in the solid state by Raman spectroscopy and X-ray crystallography.

A. CYCLOPENTASELENIUM, Se_5

The five-membered ring Se_5 is one of the major components of selenium vapor. However, despite several extensive investigations, there is still no agreement between different authors regarding the quantitative molecular composition of the saturated vapor above metallic ("gray") or liquid selenium (mp 221°C, 494 K). The Se_5 concentration in the vapor has been derived from vapor-pressure measurements as well as from EI mass-spectrometric studies, using 10-eV electrons whose energy is sufficient for ionization but not for fragmentation plus ionization of molecules larger than Se_5. Near 470 K, Se_5 concentrations of between 15 and 50% and at 1000 K between 20 and 55% have been found (8–11) (see also Fig. 1). For the vapor leaving the surface of metallic selenium ("freely subliming selenium") a concentration of ≤29% Se_5 has been determined at 448 K, using 40-eV electrons (12). The thermodynamic data for Se_5 (see Section II,F) indicate that the molecule is cyclic rather than a chain, but the accurate structure is unknown since pure Se_5, like pure S_5, has never been prepared. There is also no evidence for the presence of Se_5 in solutions.

B. CYCLOHEXASELENIUM, Se_6

Cyclohexaselenium is the main component of the vapor above freely subliming metallic selenium, accounting for as much as 58% of the vapor at 448 K (12, 13). In the saturated vapor above metallic or liquid selenium, Se_6 is also one of the main species at temperatures up to about 800–850 K, but there is no close agreement between the quantitative data published by different authors (8–11) (see also Fig. 1). By angular distribution mass spectrometry, a concentration of 40% Se_6 has been determined at 463 K (11).

By high-pressure liquid chromatography, using a UV absorbance detector, cyclohexaselenium has been detected in various selenium solutions in equilibrium with Se_7 and Se_8 (see also Section II,D) (14). Cyclohexaselenium is also formed by thermal decomposition of Se_7 in inert organic solvents like CS_2 according to the equation

$$2Se_7 \longrightarrow Se_6 + Se_8$$

(see Section II,C) (15).

1. *Preparation*

Crystalline Se_6 has been prepared. The synthesis consists of some kind of recrystallization of red amorphous selenium, prepared from highly purified selenium, which is extracted with carbon disulfide. By a special crystallization method, two types of single crystals have been

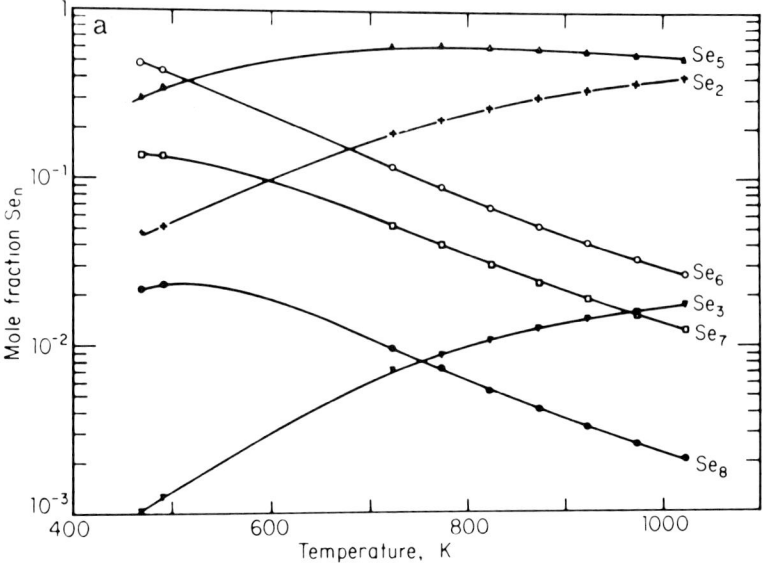
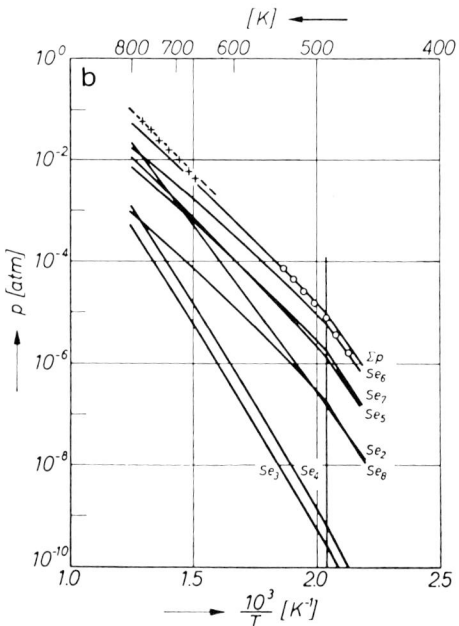

FIG. 1. Molecular composition of saturated selenium vapor: (a) according to Berkowitz et al. (8) [cited in (4a)]; (b) according to Keller et al. (9) (1 atm = 1.013 bar).

obtained from the solution and the needlelike cyclohexaselenium was separated under a microscope from the well-known α-monoclinic Se$_8$ (*16*), and characterized by X-ray crystallography, vibrational spectroscopy, and differential scanning calorimetry (DSC) (*16–19*).

The only known *chemical* synthesis of Se$_6$ is based on the reactions

$$Se_2Cl_2 + 2KI \xrightarrow[20°C]{CS_2} Se_2I_2 + 2KCl$$

$$nSe_2I_2 \xrightarrow{20°C} Se_{2n} + nI_2 \ (n = 6 \text{ or } 8)$$

which are analogous to the synthesis of S$_6$ from S$_2$Cl$_2$ (*20*). A crystalline mixture of Se$_6$ and Se$_8$ is obtained in which the two components can be detected by low-temperature Raman spectroscopy (*15*).

2. Molecular Structure and Bonding

Cyclohexaselenium crystallizes in the rhombohedral space group $R\bar{3}$ and therefore represents a second form of "trigonal selenium" besides the well-known gray or "metallic" or "hexagonal" or "trigonal" selenium (space group $P3_121$), which is polymeric and represents the thermodynamically stable allotrope at 25°C/1.013 bar. In order to avoid misunderstandings, in this review, cyclohexaselenium will simply be termed Se$_6$ or rhombohedral selenium, and gray selenium (Se$_\infty$) will be called trigonal or metallic selenium. Other presumably polymeric forms of selenium are the "glassy" or "vitreous" selenium, obtained by quenching of liquid selenium, and the two forms of "red amorphous" selenium (*5*), obtained either by quenching of selenium vapor in liquid nitrogen (*21*) or by chemical reduction of water-soluble selenium compounds (*22*) (e.g., SeO$_2$) with SO$_2$ or hydrazine. In this article, the latter two forms will be termed as "red amorphous Se(vap)" and "red amorphous Se(redn)." Since red amorphous Se(redn) turns irreversibly black at 30°C (*19*) the "black amorphous Se" obtained in this way is sometimes regarded as another form of elemental selenium.

According to a complete X-ray diffraction analysis, Se$_6$ consists of ring molecules with the molecular symmetry of D_{3d}; the crystal and molecular parameters are listed in Table II (*17*) and the crystal structure is shown in Fig. 2. Refinement by the least squares method resulted in the following atomic parameters of the single atom in the asymmetric unit: $x = 0.1602 \pm 0.00048$, $y = 0.20227 + 0.00047$, $z = 0.12045 \pm 0.00120$; calculated density, 4.71 g/cm^3. An earlier investigation of selenium vapor by electron diffraction led to an internuclear distance of 234 ± 1 pm and an average bond angle of 102 ± 0.5° for the chairlike cyclic Se$_6$ molecule (*23*).

TABLE II

Crystal and Molecular Parameters of Se_6, α-, β-, and γ-Se_8, and Metallic Selenium (Se_∞)

	Se_6 [17][a]	α-Se_8 [31][a]	β-Se_8 [28][a]	γ-Se_8 [29][a]	Se_∞ [32, 33][a]
Space group	$R\bar{3}$	$P2_1/n$	$P2_1/a$	$P2_1/c$	$P3_121$
Crystal data (20°C) (a,b,c in pm)	$a = 1136.2(1)$ $c = 442.9(8)$	$a = 905.4 \pm 0.3$ $b = 908.3 \pm 0.5$ $c = 1160.1 \pm 0.6$ $\beta = 90.81 \pm 0.05°$	$a = 1285$ $b = 807$ $c = 931$ $\beta = 93.75$	$a = 1501.8(1)$ $b = 1471.3(1)$ $c = 878.9(1)$ $\beta = 93.61(1)$	$a = 436.6$ $c = 495.8$
Z (number of molecules in unit cell)	3	4	4	8	3 (atoms)
Calculated density (g/cm³)	4.71	4.400	4.352	4.33	4.807
Site symmetry of the molecules in the crystal structure	D_{3d}	C_1	C_1	C_1	—
Average Se—Se bond length (pm)	235.6 ± 0.9	$233.6(6)$	234 ± 1.4	233.4 ± 0.5	237.4 ± 0.5
Average Se—Se—Se bond angle	$101.1 \pm 0.3°$	$105.7(1.6)°$	105.7 ± 0.8	105.8 ± 1.4	103.1 ± 0.2
Average torsional angle	$76.2 \pm 0.4°$	$101.3°$	$101.4°$	$101 \pm 2°$	$100.7 \pm 0.1°$
Shortest intermolecular distance (pm)	341.4	347.6	340	334.6	343.6

[a] Numbers in brackets indicate standard deviations.

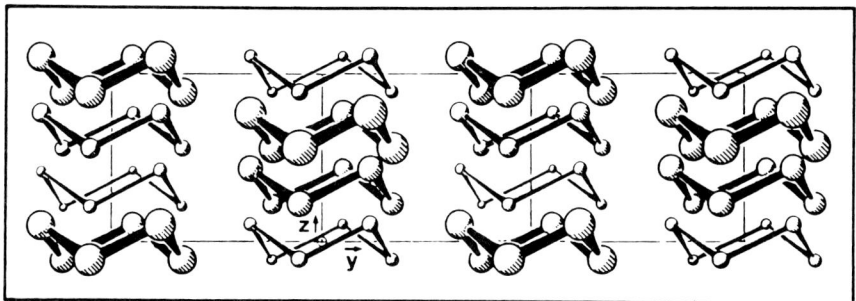

FIG. 2. Molecular and crystal structure of cyclohexaselenium, Se_6.

3. Molecular Spectra

Infrared and Raman spectra of Se_6 have been measured, and the observed wave numbers have been assigned to the eight fundamental vibrations by comparison with the wave numbers calculated from force constants taken from Se_8 and using a modified Urey–Bradley force field. The spectra of S_6 and Se_6 are completely analogous (see Table III). The four g vibrations are Raman active, the three a_{2u} and e_u vibrations are active in the infrared while the a_{1u} mode cannot be observed directly.

By adjusting the calculated to the observed wave numbers, the following five Urey–Bradley force constants have been obtained for Se_6 (in N/cm): $K = 1.188$ (bond stretching), $H = 0.102$ (bond-angle bending), $F = 0.082$ (next nearest atom repulsion), $Y = 0.242$ (torsion), and $P = 0.207$ (bond–bond interaction) (18, 19).

4. Chemical Properties and Reactions

Since pure Se_6 has so far been prepared only in minute amounts, its chemical properties have hardly been explored. Differential scanning

TABLE III

Observed and Calculated Wave Numbers[a] of the Eight Fundamental Modes of Cyclohexaselenium, Se_6 (19)

	ν_1	ν_2	ν_3	ν_4	ν_5	ν_6	ν_7	ν_8
Symmetry	a_{1g}	a_{1g}	a_{1u}	a_{2u}	e_g	e_g	e_u	e_u
Observed	247	129	—	151	221	102	253	103
Calculated	264	126	216	154	236	102	257	83

[a] In cm^{-1}.

TABLE IV
APPEARANCE POTENTIALS OF SELENIUM MOLECULAR IONS[a]

	Reference (8)	Reference (12)
Se_5^+	8.6 ± 0.2	9.2 ± 0.2
Se_6^+	8.9 ± 0.2	9.08 ± 0.05
Se_7^+	8.4 ± 0.2	8.87 ± 0.05
Se_8^+	8.6 ± 0.2	8.97 ± 0.05

[a] In eV.

calorimetry measurements on Se_6 crystals from 373 K up to the melting point of metallic selenium (494 K) have been carried out. They show a small endothermic peak near 393 K and an exothermic peak at 408 K, which have been attributed in terms of melting and recrystallization, respectively, to the polymeric metallic phase (17).

C. CYCLOHEPTASELENIUM, Se_7

The seven-membered ring Se_7 has been established as part of the vapor above freely subliming selenium in an amount of 11.4% at 448 K (12, 13). The equilibrium concentration in the saturated vapor above metallic or liquid selenium is of the same order of magnitude but decreases rapidly with temperature (see Fig. 1) (8–11).

Since the ionization potential of Se_7 is lower than that of both Se_6 and Se_8 (see Table IV), as has also been observed in the case of the corresponding sulfur rings, the structure of Se_7 should be analogous to that of S_7, which forms a chairlike molecule of C_s symmetry:

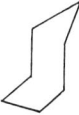

The torsional angle of 0° at the unique bond results in a strong repulsion of lone electron pairs, leading to a relatively low ionization energy (24).

Cycloheptaselenium been detected in various solutions by HPLC, using a column with octadecylsilane as a stationary phase in connection with a UV absorbance detector. The retention time of the Se_7 molecule is between those of Se_6 and Se_8, in complete analogy to the

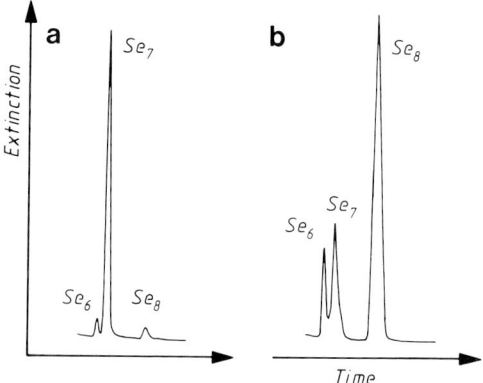

FIG. 3. Chromatogram (HPLC) of a solution of Se_7 in CS_2 at 25°C (15); (a) freshly prepared; (b) after 4 hr at 25°C in the dark.

behavior of the corresponding sulfur rings (1, 14). Cycloheptaselenium-containing solutions are obtained by dissolution of Se_8 or of red amorphous Se(redn) since the following equilibrium is established within several minutes at 25°C (14):

$$Se_8 \rightleftharpoons 8/7\ Se_7 \rightleftharpoons 4/3\ Se_6 \tag{1}$$

However, only small relative concentrations of Se_7 (with regard to Se_8) are obtained in this way. Almost pure Se_7 solutions can be prepared from titanocenepentaselenide and dichlorodiselane (Se_2Cl_2) according to the following equation (15):

$$(C_5H_5)_2TiSe_5 + Se_2Cl_2 \longrightarrow Se_7 + (C_5H_5)_2TiCl_2 \tag{2}$$

The chromatogram of this solution is shown in Fig. 3a; after 4 hr at 25°C, equilibrium [Eq. (1)] is established, and the chromatogram shown in Fig. 3b is obtained (15). The interconversion of Se_7 to Se_8 obviously begins with the reaction

$$2Se_7 \longrightarrow Se_6 + Se_8 \tag{3}$$

which is followed by the decomposition of the Se_6 by a still unknown reaction path (presumably via intermediate formation of larger rings). The mechanism of the reaction shown in Eq. (3) very likely involves the formation of a transient intermediate with a four-coordinated selenium atom, as has also been proposed in the case of sulfur rings (1):

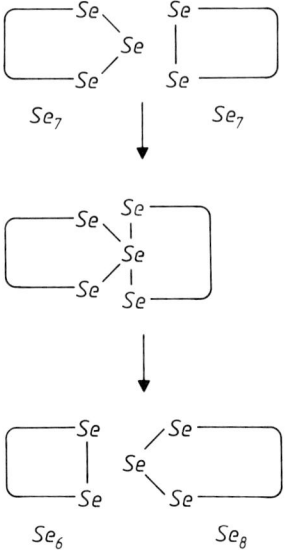

Dark-red crystals of another form of selenium have been grown from a CS_2 solution. X-ray rotating crystal and Weisenberg methods showed the material to be orthorhombic (or nearly so) with the lattice constants $a = 2632$, $b = 688$, and $c = 434$ pm, resulting in a unit-cell volume of 7.86×10^{-22} cm^3, which leads to a cell content of 28 atoms (observed density, 4.6 g/cm^3) (25). Thus the material may consist of Se_7 ring molecules.

D. CYCLOOCTASELENIUM, Se_8

1. *Preparation*

The crown-shaped molecule Se_8 has long been known to crystallize in two monoclinic lattices termed α- and β-Se_8, respectively (see Table II). Crystals of various sizes have been obtained from CS_2 solutions prepared by dissolution of either red amorphous Se (22) or of vitreous selenium in CS_2 and subsequent cooling or evaporation of the solvent (3a,b, 4a, 5). Red amorphous Se is readily soluble in CS_2, but since both vitreous and red amorphous Se do not contain more than small amounts of Se_8, the dissolution must be accompanied by a chemical interconversion. It has, however, been reported (26) that the dissolution of red amorphous Se in CS_2 requires illumination with photons of energies in excess of 2.3 eV, ambient room light levels being sufficient.

In the dark, the surface attack of CS_2 is said to leave a rubbery insoluble fraction similar to polymeric sulfur.

α- and β-Cyclooctaselenium form crystals of different shapes and with different interfacial angles and consequently can be separated under a microscope (*3b*). Slow evaporation of a saturated solution in CS_2 yields α-monoclinic Se_8, while on rapid evaporation α- and β-monoclinic Se_8 are formed (*27, 28*). Since the β form dissolves in a carbon disulfide solution saturated with the α form, the latter seems to be the more stable of the two (*29*). This observation is in agreement with the densities, which decrease in the order $\alpha\text{-}Se_8 > \beta\text{-}Se_8 > \gamma\text{-}Se_8$ (see Table II). Large single crystals of α-monoclinic Se_8 have been grown from a solution in methylene iodide (*30*).

A third monoclinic allotrope, γ-Se_8, has been prepared from a solution of dipiperidinotetraselane in carbon disulfide (*29*):

$$Se_4(NC_5H_{10})_2 + 2CS_2 \longrightarrow 3/8\ Se_8 + Se(S_2CNC_5H_{10})_2$$

γ-Cyclooctaselenium forms long prisms with eight molecules in the unit cell (see Table II) and is quite stable at 25°C (*29*).

An investigation has shown that dissolution of crystalline Se_8 or extraction of glassy or red amorphous selenium (the latter prepared from aqueous SeO_2 by reduction) by different organic solvents yields solutions containing Se_6, Se_7, and Se_8 in equilibrium. Through high-pressure liquid chromatography on reversed bonded phases, the three rings Se_n have been separated and detected by their UV absorption at 254 nm. In contrast to selenium vapor, Se_8 is the main component (*14*). The total solubility of Se in CS_2 at 25°C amounts to 0.05% by weight or 0.008 mol (Se) per liter (*34, 35*). These results have led to the conclusion that Se_6, Se_7, and Se_8 undergo a rapid interconversion in solution at 25°C, leading to an equilibrium (*14*). The equilibrium vapor above metallic or liquid selenium contains only 1–2% Se_8 at temperatures of between 440 and 720 K (*8–12*).

The spontaneous but slow crystallization of monoclonic selenium (Se_8) from red amorphous Se, prepared by quenching of selenium vapor (1000°C) in liquid nitrogen, depends on the storage temperature of the samples. Below 303 K, red amorphous Se(vap) is completely transformed into the monoclinic phase, whereas above 303 K transformation into the metallic phase takes place. (The relative Se_8 content of the material can be determined by DSC.) In contrast, chemically prepared red amorphous Se(redn) is transformed only into the metallic phase, even below 303 K (*21*).

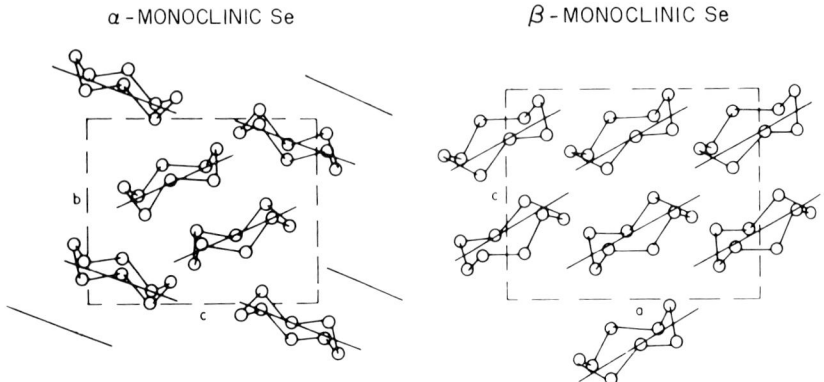

FIG. 4. Molecular and crystal structures of monoclinic α-Se$_8$ and β-Se$_8$.

2. Molecular Structure and Bonding

The crystal and molecular structure data of the three Se$_8$ forms listed in Table II have been determined by X-ray diffraction (27–29, 31). α-, β-, and γ-cyclooctaselenium crystallize in the same space group but differ in the packing of the molecules (see Fig. 4). The average bond distances, bond angles, and torsional angles of the Se$_8$ molecules are identical within the limits of the standard deviation. The torsional angle of 101° is close to the value of 99° observed in the case of S$_8$ (36) and obviously corresponds to the minimum of the torsional potential energy function. The shortest intermolecular distance has been observed in the case of γ-Se$_8$: the value of 334.6 pm is even smaller than the shortest intermolecular contact in orthorhombic cyclooctasulfur, S$_8$ [337 pm (33)].

3. Molecular Spectra

Raman and infrared spectra of α-monoclinic selenium have been reported (3c,d, 4b, 14, 37); the low-temperature Raman spectrum is shown in Fig. 5. It differs sufficiently from that of Se$_6$ for both compounds to be detected in this way in mixtures with each other (15). The observed wave numbers have been assigned to the 11 fundamental modes of the Se$_8$ molecule, assuming D_{4d} symmetry (see Table V). The b_2 and e_1 modes are infrared active, and the a_1, e_2, and e_3 vibrations are Raman active. The wave number of the inactive b_1 fundamental has been estimated from the second-order Raman spectrum.

Using an extended Urey–Bradley force field with six force constants,

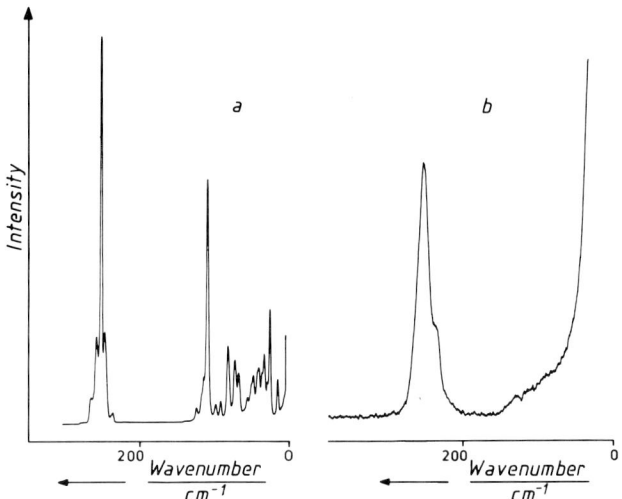

FIG. 5. Low-temperature Raman spectra ($-105 \pm 5°C$) of monoclinic α-Se$_8$ (a) and of red amorphous selenium prepared by reduction of SeO$_2$ (b) (14).

a very good agreement between the observed and calculated wave numbers has been achieved. These force constants have the following values (in N/cm) (38): K (bond stretching) = 1.341, H (bond angle bending) = 0.021, F (next nearest atom repulsion) = 0.214, Y (torsion) = 0.015, P (bond–bond interaction) = 0.175, and C (long range repulsion) = 0.070. For other values, see ref. (19); for valence force constants, see ref. (38). The bond–bond interaction constant P is unusually large, which has been rationalized by electron rearrangement during asymmetric stretching vibrations (38), as has also been found for S–S bonds (39, 40).

TABLE V

OBSERVED AND CALCULATED WAVE NUMBERS[a] OF THE 11 FUNDAMENTAL MODES OF CYCLOOCTASELENIUM, Se$_8$

	ν_1	ν_2	ν_3	ν_4	ν_5	ν_6	ν_7	ν_8	ν_9	ν_{10}	ν_{11}	Reference
Symmetry	a_1	a_1	b_1	b_2	e_1	e_1	e_2	e_2	e_2	e_3	e_3	
Observed	256	113	—	120	253	95	256	86	47	240	128	3c,d,4b,37
Calculated	255	111	231	120	255	97	255	88	46	240	127	38
Calculated	257	115	224	130	253	111	252	78	44	235	115	19

[a] In cm^{-1}.

The highly characteristic Se_8 Raman line at 113 cm^{-1} (totally symmetrical ring bending mode a_1) has occasionally been observed in the Raman spectrum of red amorphous Se(redn) (3d, 37, 41); this has been interpreted on the basis of either a certain Se_8 content of this material (37, 41) or of Se_8-like "molecular fragments" in a polymeric chain structure (42). However, the low-temperature Raman spectrum of red amorphous Se, recorded with the red line of a krypton laser (to avoid photodecomposition), does not exhibit this line (14). Instead, an almost continuous Raman scattering is observed for the region of the Se—Se—Se bending vibrations, indicating the presence of a large number of molecular species (e.g., rings) or of many different conformations of a long chain. The same holds for the Raman spectrum of freshly prepared red amorphous Se(vap), recorded at room temperature (42). The Se_8 signals in the earlier spectra are explained as photodecompositon of amorphous Se with formation of Se_8 in the laser beam (14).

The resonance Raman spectrum of a thin film of selenium exhibits 10 signals in the region 115–1400 cm^{-1}, which have been assigned to the fundamentals $\nu_2(a_1)$ and $\nu_{10}(e_3)$ of Se_8 and to their overtones and combination vibrations (44).

4. Chemical Properties and Reactions

Any external force, such as temperature or pressure, initiates the conversion of monoclinic selenium (Se_8) to the thermodynamically stable metallic modification. Both α-Se_8 and β-Se_8 are converted directly to the trigonal phase in the temperature range from 394 to 430 K, without being transformed into one another (28, 45, 46). Investigation of α-monoclinic Se_8 by DSC from 373 K up to the melting point of metallic selenium (494 K) has revealed an endothermic peak at 418 K and an exothermic peak at 430 K (heating rate 10 K/min), which have been interpreted by postulating melting of Se_8 followed by recrystallization to the metallic phase, respectively (17). Monoclinic selenium is thermodynamically unstable relative to metallic Se in the whole temperature range from 0 to 420 K (47). The conversion of Se_8 to Se_∞ depends on the quality of the crystals (21, 48) and gets complicated by premature melting of Se_8 if high heating rates are used (21, 49, 50). With single crystals of good quality investigated at 320 K/min, only one broad endothermic peak for the melting of monoclinic Se is observed [melting temperature, 429 K (49)], while at lower heating rates additional peaks for the exothermic conversion to solid metallic Se, followed by the endothermic melting of the latter, are found. The separation of the two melting peaks decreases with increasing heating rate

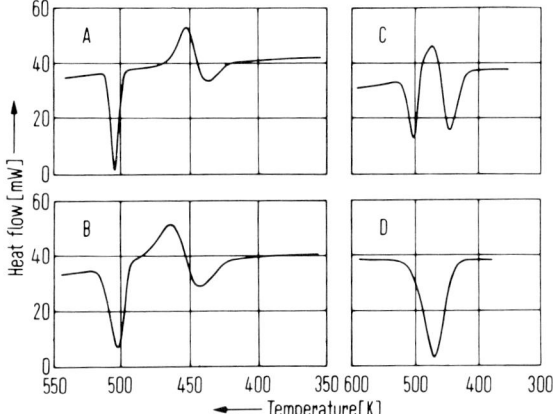

FIG. 6. DSC diagrams of single crystals of monoclinic α-Se$_8$ at different heating rates (sample mass, 4–5 mg) (49). Heating rates: A, 40; B, 80; C, 160; and D, 320 K/min.

and disappears at 320 K/min (49) (see Fig. 6). With a polycrystalline sample and a heating rate of 410 K/min, the melting point of monoclinic Se has been found to be 413 K (50), but extrapolation to zero heating rate indicates 428 K [Fig. 10 in ref. (50)].

Bombardment of monoclinic Se with fast electrons (51) or α particles (52) also results in conversion to the metallic phase. The mechanism of the conversion has been studied by X-ray diffraction, optical microscopy, and thermal analysis (28, 46, 50) and has been found to be topochemical. The formation of the crystal nuclei of the metallic phase is often related to larger lattice defects and occurs primarily at or near the crystal surface (50) and with an Arrhenius activation energy of 100 kJ/mol (50). The greater density of the metallic phase (see Table II) causes tensions that are responsible for the weakening of further Se—Se bonds. Ring opening of an Se$_8$ molecule is believed to be the rate-determining step of the metallic crystal growth though the mechanism discussed in Section III,B,3 may also be quite effective in transforming small rings into very large ones and vice versa:

$$2Se_8 \rightleftharpoons Se_{16} \xrightleftharpoons{+Se_8} Se_{24} \rightleftharpoons \xrightleftharpoons{+nSe_8} \rightleftharpoons Se_\infty$$
$$Se_6 + Se_{10} \quad Se_7 + Se_9$$

In the temperature range from 343 to 487 K, the metallic crystal growth has an activation energy of 113 kJ/mol (50) and 116.39 kJ/mol

(53), respectively. Comparison of the results of different authors makes it obvious that the process of conversion varies with temperature. To explain the effect of halogens, alkali metals, and amines, the following reactions have been proposed (54):

$$I_2 + (Se)_{ring} \longrightarrow ISe\text{—} ... \text{—}SeI + x(Se)_{ring} \longrightarrow I_2 + (Se)_{trig}$$

$$2Na + (Se)_{ring} \longrightarrow 2Na^+ + {}^-Se\text{—} ... \text{—}Se^- + x(Se)_{ring} \longrightarrow Na_2Se_y + (Se)_{trig}$$

$$R_3N + (Se)_{ring} \longrightarrow R_3N^+Se\text{—} ... \text{—}Se^- + x(Se)_{ring} \longrightarrow R_3N + (Se)_{trig}$$

Elemental selenium and elemental sulfur (any allotrope of either one) react on heating (melting) to give a large number of cyclic molecules Se_nS_m. While the eight-membered rings are much preferred, HPLC analysis has shown that both six- and seven-membered species are also formed. For a review of these cyclic selenium sufides, see ref. (55).

E. OTHER RING MOLECULES

Investigations by mass spectrometry (see also Section III,C) have shown that molecules Se_n ($n = 2$–10) are the constituents of saturated selenium vapor (Knudsen cell) as well as of the vapor above freely subliming selenium (Langmuir cell), though the latter method yields a greater amount of the smaller molecules (56). Whereas Se_4, Se_8, Se_9, and Se_{10} are of minor significance, Se_5, Se_6, and Se_7 are the main components of selenium vapor in the temperature range from 473 to 650 K. At higher temperatures (up to 1400 K), the amount of the smaller noncyclic species, especially Se_2 and Se_3, increases rapidly (9, 10, 57–60).

Despite a large number of investigations, no agreement exists concerning the molecular composition of molten, quenched (glassy or vitreous), and red amorphous selenium. The usual, simplifying model of a dynamic equilibrium (61) of Se_8 rings and Se_n chains seems to explain the behavior of molten selenium qualitatively, but there are no reasons to exclude other small and medium sized rings, as have been detected in liquid sulfur (1). Several authors have pointed out that long chains as well as small and larger ring molecules may exist in glassy and in red amorphous selenium (14, 17, 46), the population of each component depending on the preparation procedure and the storage time and temperature.

F. THERMODYNAMIC PROPERTIES OF SELENIUM ALLOTROPES

In connection with metallic selenium being a model compound for a homoatomic macromolecule, the heat capacity of selenium has been extensively studied. In a 1981 review (63) 23 investigations on the heat capacities of metallic, molten, amorphous, and monoclinic selenium have been critically evaluated, and a set of recommended data has been derived for the temperature range from 0 to 1000 K. The thermodynamic functions enthalpy, entropy, and Gibbs energy have been calculated. The heat capacity of metallic selenium is a smooth function of temperature from 0 K up to the melting point (494 K). The heat capacity of amorphous Se is also a smooth function of temperature from 0 to 270 K. An upward slope above 270 K can be associated with the glass transition. The heat capacity of monoclinic selenium below 100 K can be represented as an average of the heat capacities of metallic and amorphous Se. Above 100 K the heat capacities of monoclinic and metallic selenium are identical. Below 100 K the order of heat capacities is: amorphous > monoclinic > metallic. From 100 to 200 K the differences are minimal. Above 200 K the heat capacity of amorphous Se is higher than that of crystalline selenium. Setting the metallic selenium entropy at 0 K equal to zero, monoclinic selenium has, at 0 K, a residual entropy of about 1.7 J mol^{-1} K^{-1}, and the free enthalpy of the transition of monoclinic to metallic Se is negative; $\Delta G = -2.1$ kJ/mol(Se). At 0 K the residual entropy of amorphous (glassy) selenium is 3.63 J mol^{-1} K^{-1} and the free enthalpy of the transition of amorphous (glassy) to metallic selenium is equal to -3.8 kJ/mol. At 494.2 ± 0.1 K metallic selenium melts to an equilibrium melt with a heat of fusion of 6.20 kJ/mol and an entropy of fusion of 12.55 J mol^{-1} K^{-1}. Assuming that the equilibrium melt consists of 64% Se$_8$ rings and 36% long chains, the thermodynamic quantities for the monoclinic-to-melt transitions have been estimated as follows (50):

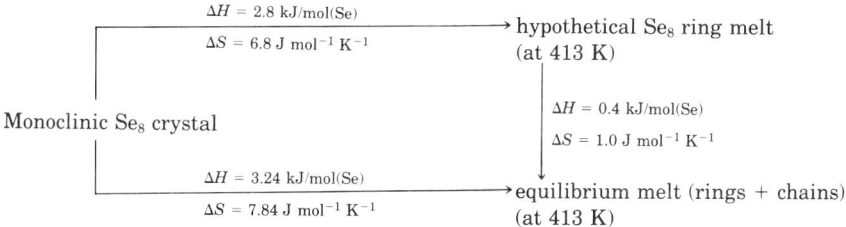

From these data it was concluded that, owing to the considerably higher enthalpy of fusion of metallic selenium, the slightly positive enthalpy for the reaction rings → chains in the melt is changed to a negative enthalpy of reaction for the ring-to-chain reaction in the crystalline state and makes the Se_8 rings metastable at all temperatures. The main reason for the higher stability of metallic selenium is its much higher packing density (see Table II) as a result of the fairly strong interchain attraction forces. According to Franz et al. (49) the enthalpy of fusion of α-monoclinic Se_8 amounts to 3.79 kJ/mol at 429 K.

The enthalpies of formation and atomization, the heat capacities, and the entropies of the selenium rings Se_n ($n = 5$–12), as derived from mass spectrometric measurements and statistical thermodynamics (8–10, 12, 13, 57), are given in Table VI.

The entropies, heat capacities, and thermodynamic functions of gaseous cyclooctaselenium have been calculated from spectroscopic and structural data for temperatures of up to 3000 K (64). Both the heat capacities and entropies of sulfur rings S_n ($n = 6, 7, 8, 12$) at a given temperature depend linearly on the ring size n (65). Therefore, it has been assumed that analogous relationships exist for the cyclic Se_n molecules, and the following equations have been derived from the data of Se_2 and Se_8 at 298 K (64):

$$C_p^\circ \text{ (J mol}^{-1}\text{ K}^{-1}) = 22.17n - 3.33$$

$$S_{298}^\circ \text{ (J mol}^{-1}\text{ K}^{-1}) = 49.6n + 135.4$$

The particular values are also listed in Table VI. It has further been found that at given temperatures $T_1 > T_2$ the function $H(T_1) - H(T_2)$ of S_n molecules depends on the number of atoms in the molecule. Provided an analogous relationship exists for Se_n molecules, corresponding equations can be derived from the data of Se_2 (66) and Se_8 (64), e.g.:

$$Se_n: H_{400} - H_{298} = 2.278n - 0.317 \text{ (kJ/mol)}$$

Equations of this type allow the functions $H(T_1) - H(T_2)$ to be calculated also for other Se_n ($n = 5$–12) molecules (64).

III. Homocyclic Selenium Cations

Cyclic selenium cations are known as monocations Se_n^+ observed in selenium vapor by mass spectrometry, and as dications Se_n^{2+} of which

TABLE VI

Formation Enthalpies $\Delta H_{\text{f}}^{\circ\,a}$, Atomization Enthalpies $\Delta H_{\text{at}}^{\circ\,a}$, Atomization Enthalpies Referred to the Number of Atoms $\Delta H_{\text{at}}^{\circ}/n^{a}$, Heat Capacities $C_{\text{p}}^{\circ\,b}$, and Entropies $S^{\circ\,b}$ of Gaseous Cyclic Selenium Molecules Se_n ($n = 5\text{–}12$)

	$\Delta H_{\text{f},298}^{\circ}$		$\Delta H_{\text{at},0}^{\circ}$	$\Delta H_{\text{at},550}^{\circ}$	$\Delta H_{\text{at}}^{\circ}/n$			$C_{\text{p},298}^{\circ}$	$\bar{C}_{\text{p}}^{\,c}$	S_{298}°		S_{600}°
	10^{d}	8	9	57	9	57		64	12,13	64	10	9
Se_5	148.7	170 ± 6	1042	946 ± 21	208	189 ± 4		107.5	100	383.4	405.9	458.5
Se_6	138.4	157 ± 6	1283	1183 ± 25	214	197 ± 4		129.7	124	433.0	444.7	523.9
Se_7	153.5	174 ± 6	1509	1390 ± 33	216	199 ± 5		151.9	147	482.6	507.9	592.9
Se_8	169.4	186 ± 6	1737	1624 ± 38	217	203 ± 5		174.0	170	532.2	565.6	655.6
Se_9	—	—	—	1796 ± 42	—	200 ± 5		196.2	—	581.8	—	—
Se_{10}	—	—	—	2010 ± 50	—	201 ± 5		218.4	—	631.4	—	—
Se_{11}	—	—	—	—	—	—		240.5	—	681.0	—	—
Se_{12}	—	—	—	—	—	—		262.7	—	730.6	—	—

[a] In kJ/mol.
[b] In J mol^{-1} K^{-1}.
[c] Average heat capacity in the temperature region 298–418 K.
[d] Reference.

those with $n = 4$, 8, or 10 have been isolated as salts with various anions. In general, the removal of electrons from neutral selenium molecules Se_n increases the total bond energy since the highest occupied molecular orbitals are always antibonding π orbitals. The ions formed therefore contain additional σ or π bonds as can be seen from the structures of Se_4^{2+}, Se_8^{2+}, and Se_{10}^{2+}. The structures of the monocations Se_n^+ are unknown.

A. Mass Spectra of Selenium Vapors

The mass spectra of evaporated elemental gray metallic selenium (Se_∞) have been investigated, using different techniques. Conventional heating to 102–187°C combined with ionization by electron impact (40 eV) leads to cations Se_n^+ with $n = 1$–10, Se_6^+ being the most abundant ion. Se_5^+, Se_6^+, Se_7^+, and Se_8^+ arise mainly from the corresponding cyclic molecules and not from fragmentation of larger molecular ions (12). For appearance potentials, see Table IV.

Heating of HgSe or of metallic selenium in a Knudsen cell (544–566 K) or surface evaporation of metallic or α-monoclinic Se, followed by electron impact ionization (75 eV), yields Se_n^+ ions with $n = 1$–8, Se_2^+ being the most abundant species (8). An investigation of the angular distribution of the ions Se_n^+ ($n = 1$–8) at different ionizing energies (8.5–75 eV) and 210°C (Knudsen cell temperature) provided special information concerning the fragmentation process within the ion source (11).

Evaporation of amorphous, metallic, or monoclinic Se by heating with a pulsed laser (10^7 W/cm^2, 800 μsec) also produces ions of type Se_n^+ ($n = 1$–9), Se_5^+ being the most abundant one. The analysis was carried out, using a time-of-flight mass spectrometer (58, 67).

Rather different spectra are obtained when the equilibrium vapor above metallic selenium (160°C) is ionized by a field ion source (2.5–8.0 kV), resulting in the ions Se_n^+ ($n = 2$ and 5–8); neither Se^+ nor Se_3^+ and Se_4^+ have been observed (these species usually result from fragmentation processes), but occasionally traces of Se_9^+ have been found. Using a Knudsen cell, Se_6^+ is the most abundant species, followed by Se_5^+ and Se_7^+, which all originate from the corresponding neutral, cyclic molecules (56). Field evaporation (20–100°C) of whiskers of selenium (prepared by field condensation of Se vapor) produces mainly Se_5^+, but small amounts of molecular ions with one, two, and four positive charges up to Se_{33} have been identified. However, these species may have chainlike structures (56).

TABLE VII

Electronic Absorption and Diffuse-Reflectance Spectra of the Se_4^{2+} Ion in Different Environments (70)

Sample	Color	$\pi \to \pi^*$ (nm)	$\pi \to \pi^*$ or $\pi \to n^*$ (nm)
Se_4^{2+} in solution	Yellow	410	320
$Se_4(Sb_2F_4)(Sb_2F_5)(SbF_6)_5$	Golden yellow	413	382 (shoulder)
$Se_4(AsF_6)_2$	Yellow	440	385
$Se_4(AlCl_4)_2$	Yellow orange	488	430
$Se_4(HS_2O_7)_2$	Orange	524	428

B. Dications Se_n^{2+} in Solids and Solutions

1. The Se_4^{2+} Ion

The yellow, diamagnetic cation Se_4^{2+} was first detected by spectrophotometric and conductometric measurements on solutions of selenium in highly acidic and oxidizing solutions like $H_2S_2O_7$ (20°C), H_2SO_4 (95%, with SeO_2 or $K_2S_2O_8$ as oxidants), and HSO_3F (with $S_2O_6F_2$ as an oxidant) (68). Se_4^{2+} salts are prepared by reactions like:

$$4Se + S_2O_6F_2 \longrightarrow Se_4^{2+} + 2SO_3F^-$$

$$7Se + SeO_2 + 6H_2SO_4 \longrightarrow 2Se_4^{2+} + 6HSO_4^- + 2H_3O^+$$

$$4Se + 6H_2S_2O_7 \longrightarrow Se_4^{2+} + SO_2 + 2HS_3O_{10}^- + 5H_2SO_4$$

The formation of Se_4^{2+} proceeds via the green Se_8^{2+} as an intermediate (see below).

In solution, Se_4^{2+} exhibits an intense absorption at 410 nm and a very weak one at 330 nm (68). By dissolution of selenium in 65% oleum, in a mixture of HSO_3F with $S_2O_6F_2$, in liquid SO_3, or in excess liquid SbF_5, respectively, yellow to orange solids of compositions $Se_4(HS_2O_7)_2$, $Se_4(SO_3F)_2$, $Se_4(S_4O_{13})_2$, and $Se_4(Sb_2F_{11})_2$ can be prepared (68). AsF_5 reacts with Se in liquid SO_2 to give $Se_4(AsF_6)_2$ (69). The electronic transitions of these salts are shown in Table VII (70).

The shifts of the dipole-allowed $\pi \to \pi^*$ transition and of the weak band at lower wavelengths are caused by the varying polarizability of the anions and the fairly strong cation–anion interactions (71).

Mixtures of $SeCl_4$ and $AlCl_3$ show a melting point maximum of 203 ± 2°C at a mixing ratio of 2:1; the orange material is believed to consist of $Se_4(AlCl_4)_2$ on grounds of its IR spectrum (72).

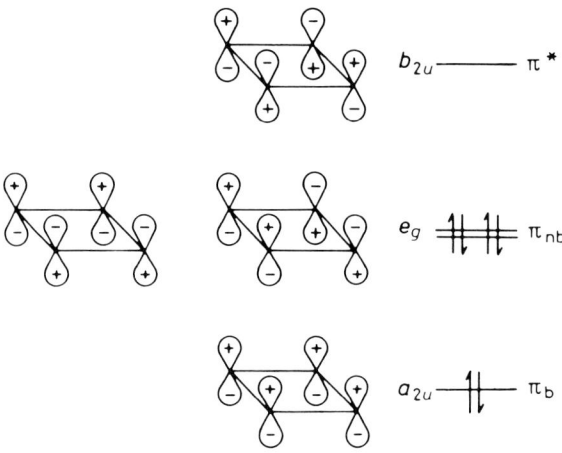

FIG. 7. Molecular orbitals of the π bond in Se_4^{2+} (π_b bonding, π_{nb} nonbonding, π^* antibonding).

Reaction of liquid SO_3 with metallic Se at ambient temperatures produces yellow solids indicated by analysis to be $Se_4S_nO_{3n+1}$ ($n = 2, 3, 4$) whose composition depends on the reaction time. UV spectra of solutions of these compounds in either H_2SO_4 or HSO_3F show the presence of Se_4^{2+} (73).

Crystals of $Se_4(HS_2O_7)_2$ have been examined by X-ray diffraction and were found to be monoclinic (space group $P2_1/c$) containing strictly square–planar Se_4^{2+} cations with an Se—Se internuclear distance of 228.3 pm (site symmetry C_i), significantly less than that of 234 pm found in the Se_8 molecule (see Table II). The shorter bond length has been explained by a delocalized π bond (74, 75). Each Se atom possesses six localized electron pairs. In addition, the four 3p orbitals perpendicular to the molecular plane form one bonding, two nonbonding, and one antibonding molecular orbitals, occupied by six electrons as shown in Fig. 7 (76). For a detailed MO theoretical study of Se_4^{2+}, see ref. (77).

X-ray structural studies were also performed on single crystals of $Se_4(Sb_2F_4)(Sb_2F_5)(SbF_6)_5$, prepared from Se and SbF_5 in liquid SO_2, and of $Se_4(AlCl_4)_2$ (71, 78). In both cases planar, centrosymmetrical cations with average bond distances of 226.0 and 228.6 pm, respectively, were found (valence angles, 90.0 ± 0.1°).

Extensive infrared, Raman, and resonance Raman (rR) spectroscopic investigations of solid and dissolved Se_4^{2+} salts led to the following wave numbers of the four fundamental vibrations of the square–planar cation (70, 76, 79):

$$a_{1g} = 323, \ b_{1g} = 184, \ b_{2g} = 324\text{–}327, \ e_u = 302 \text{ cm}^{-1}$$

Eight combination vibrations and overtones, respectively, have been observed in the rR spectrum of Se_4^{2+} in 25% oleum, and these have been used to calculate the harmonic wave number $\omega_1 = 321.8 \pm 0.5$ cm^{-1} and the anharmonicity constants $x_{11} = -0.55$ and $x_{12} = -1.3$ cm^{-1} (76). Because of the high molecular symmetry, the spectroscopic data are insufficient to calculate accurate force constants, but when most of the interaction constants are neglected (i.e., set equal to zero) the following force constants for Se_4^{2+} can be deduced (in N/cm) (70, 76):

f_r (bond stretching) = 2.09

f_α (angle bending) = 0.20

f_{rr} (bond–bond interaction) = -0.02

f'_{rr} (nonneighboring bond interaction) = 0.36

Reactions. Salts containing the Se_4^{2+} ion are extremely hygroscopic, selenium being precipitated with excess water (7):

$$2Se_4^{2+} + 6H_2O \longrightarrow 7Se + SeO_2 + 4H_3O^+$$

Oxidation of Se_4^{2+} by peroxodisulfate, $S_2O_8^{2-}$, or peroxodisulfurylfluoride, $S_2O_6F_2$, leads to SeO_2, while reduction by either elemental selenium or hydrazine, N_2H_4, results in formation of the green Se_8^{2+} cation (68). Reaction of either $Se_4(AsF_6)_2$ or $Se_4(Sb_2F_{11})_2$ with S_4N_4 in liquid SO_2 produces blue-green crystals of $(Se_4S_2N_4)(AsF_6)_2$ and $(Se_4S_2N_4)(SbF_6)_2$, respectively, which contain the dimeric cation $Se_2SN_2^+$ shown in Fig. 8 (69).

2. The Se_8^{2+} Ion

The diamagnetic Se_8^{2+} ion was discovered by Gillespie et al. (68) by spectrophotometric and conductometric measurements on solutions of selenium in highly acidic and oxidizing solutions like H_2SO_4 [100% at 50–60°C, or 95% with $K_2S_2O_8$ or $Ce(SO_4)_2$ as an oxidant], $H_2S_2O_7$ (20°C), or HSO_3F (with SO_3 as an oxidant). In solution, Se_8^{2+} is dark green and is formed in reactions like the following:

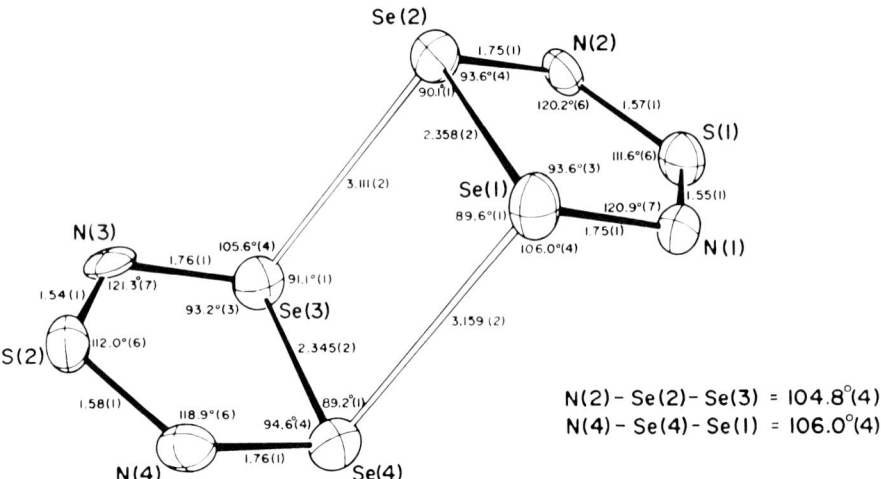

FIG. 8. Molecular structure of the cation in $(Se_4S_2N_4)$ $(AsF_6)_2$ prepared from S_4N_4 and either Se_4^{2+} or Se_8^{2+} (69). Bond distances in angstroms. Numbers in parentheses are standard deviations.

$$8Se + 2H_2S_2O_7 \longrightarrow Se_8^{2+} + 2HSO_4^- + SO_2 + H_2SO_4$$

$$8Se + S_2O_8^{2-} \longrightarrow Se_8^{2+} + 2SO_4^{2-}$$

The green species has an intense absorption at 295 nm and weak ones at 470 and 685 nm (68).

Black prismatic crystals of $Se_8(AlCl_4)_2$ have been obtained from a stoichiometric mixture of $SeCl_4$, Se, and $AlCl_3$ after fusion at 250°C for 3 hr (80).

$$15Se + SeCl_4 + 4AlCl_3 \longrightarrow 2Se_8(AlCl_4)_2$$

The melting point of this compound was determined from the phase diagram of the $Se-SeCl_4-AlCl_3$ system to be 192°C (72).

An X-ray structural analysis of $Se_8(AlCl_4)_2$ revealed a bicyclic cation structure whose characteristic feature is an endo-exo conformation of the eight-membered ring. The molecular symmetry is approximately C_s (see Fig. 9).

The most striking detail of this structure is the transannular bond Se-3—Se-7 whose length of 284 pm is significantly less than the van der Waals distance (400 pm) and even less than the nonbonding contacts Se-2–Se-8 of 336 pm and Se-4–Se-6 of 330 pm. The other bond lengths vary between 229 and 236 pm and average to 231.8 pm (80).

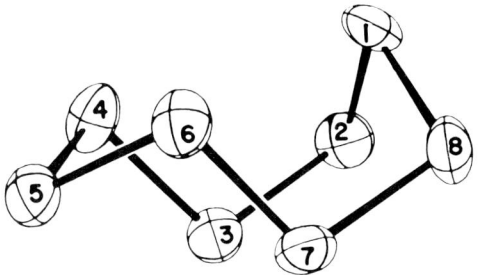

FIG. 9. Molecular conformation of the Se_8^{2+} ion in Se_8 $(AlCl_4)_2$ (80). Bond distances (pm): 1–2, 229; 2–3, 233; 3–4, 230; 4–5, 233; 5–6, 231; 6–7, 231; 7–8, 236; 8–1, 231. Distance of Se-3 from Se-7 is 284 pm.

$Se_8(AlCl_4)_2$ can also be prepared according to the equation (81):

$$5Se_2Cl_2 + 4AlCl_3 \longrightarrow Se_8(AlCl_4)_2 + 2SeCl_3(AlCl_4)$$

The compounds $Se_8(AsF_6)_2$ and $Se_8(Sb_2F_{11})_2$ have been obtained as follows:

$$8Se + 3AsF_5 \xrightarrow[-23°C]{SO_2(l)} Se_8(AsF_6)_2 + AsF_3$$

$$8Se + 5SbF_5 \xrightarrow[0°C]{HF} Se_8(Sb_2F_{11})_2 + SbF_3$$

These green materials melt near 180°C and darken rapidly on exposure to moist air. When added to water they instantly decompose to red selenium and presumably SeO_2. Dissolution in H_2SO_4 (100%), oleum ($H_2SO_4 + SO_3$), or HSO_3F produces green solutions with the characteristic absorption maxima of Se_8^{2+} (see above). The formulas given above are based on analytical data and IR as well as NMR spectra (82). For a molecular orbital study of Se_8^{2+}, see ref. (77).

Reactions. $Se_8(AsF_6)_2$ reacts with S_4N_4 in liquid SO_2 at 20°C to give blue-green crystals of $(Se_4S_2N_4)(AsF_6)_2$, which contain the dimeric cyclic $Se_2SN_2^+$ cation shown in Fig. 8 (69). Oxidation of Se_8^{2+} in solution results in formation of the yellow Se_4^{2+} (see Section III,B,1), while reduction with elemental selenium produces the Se_{10}^{2+} cation (see Section III,B,3).

Solid $Se_8(AsF_6)_2$ reacts with C_2F_4 at room temperature to give AsF_3 and $(C_2F_5)_2Se_n$ with $n = 2$ or 3. $Se_8(Sb_2F_{11})_2$ yields similar products on reaction with C_2F_4 at 100°C. In SO_2 solution, $Se_8(AsF_6)_2$ and C_2F_4 yield mainly $(C_2F_5)_2Se_2$ and $F_5C_2SeSeCF_2COF$ (83).

3. The Se_{10}^{2+} Ion

Excess metallic selenium reacts with either AsF_5 or SbF_5 in SO_2 at 20–50°C to give the saltlike compounds $Se_{10}(AsF_6)_2$ and $Se_{10}(SbF_6)_2$, respectively, which have been obtained as deep-red crystals (brown when finely powdered) from the brownish-green solution. The reaction proceeds via Se_8^{2+}. An X-ray structural analysis of $Se_{10}(SbF_6)_2$ (space group $P2_1/n$, $Z = 8$) revealed a cation structure of the bicyclo[4.2.2.]decane type (81). The asymmetric unit contains two structurally similar molecules, one of which is shown in Fig. 10.

The Se—Se bond lengths alternate on moving away from the three-coordinated atoms: the six bonds adjacent to these three-coordinated atoms are the longest, having lengths of 241–246 pm; the bonds adjacent to these long bonds are shorter with lengths of 223–227 pm, and the unique central bond Se-13—Se-14 has an intermediate length of 236 pm (81). There are also some unusually short intramolecular Se ⋯ Se contacts between atoms that are not directly bonded, e.g., Se-18–Se-110 = 330 pm, Se-13–Se-19 = 339 pm, and Se-14–Se-17 = 349 pm indicating that Se_{10}^{2+} may be regarded as some kind of cluster molecule with partly delocalized valence electrons (81).

Considerable differences between the solution and solid-state colors and electronic absorption spectra have been observed that indicate that the stucture of the Se_{10}^{2+} cation is not retained in solution. $Se_{10}(MF_6)_2$ can be dissolved without decomposition in H_2SO_4 (95.5%), while in H_2SO_4 (100%) fairly rapid oxidation to Se_8^{2+} takes place (M = As or Sb) (81). $Se_{10}(AlCl_4)_2$ is obtained as a brown material by disproportionation of $Se_8(AlCl_4)_2$ on washing with large quantities of liquid SO_2 (81).

C. OTHER DICATIONS

Spectrophotometric and potentiometric studies of the reduction of $SeCl_4$ by elemental selenium in eutectic $NaCl$–$AlCl_3$ melts at 150°C indicate that Se can exist in up to five different oxidation states, which have been interpreted in terms of the dications Se_n^{2+}, with $n = 2, 4, 8, 12$, or 16, but the exact nature and the structures of these species are unknown, and no pure compounds have been isolated from these mixtures (84, 85). Oxidation of $(C_2F_5)_2Se_2$ by either SbF_5 or $O_2(Sb_2F_{11})$ results in a wine-red cation of composition $(SeC_2F_5)_{2n}^{n+}$, where n is probably equal to 2. A homocyclic structure has been discussed for this species (86).

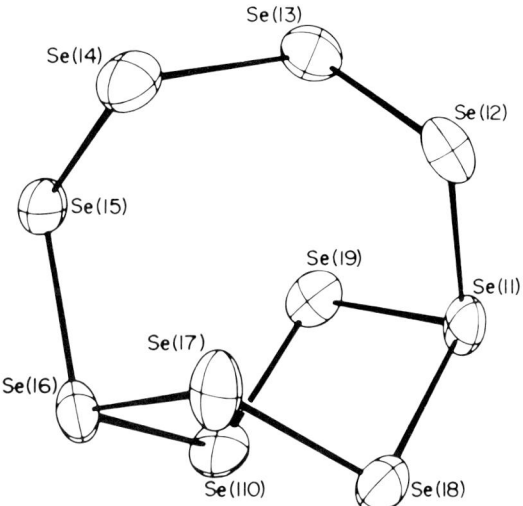

FIG. 10. Molecular conformation of the Se_{10}^{2+} ion in $Se_{10}(SbF_6)_2$ (*81*). Bond distances (pm): 11–12, 244; 12–13, 225; 13–14, 236; 14–15, 226; 15–16, 245; 16–17, 239; 17–18, 226; 18–11, 243; 11–19, 240; 19–110, 223; 110–16, 243.

IV. Selenium Iodide Cations, Se_nI^+

The first derivative of a selenium ring was prepared by Passmore *et al.* (*87*) according to the following equations:

$$32Se + 2I_2 + 9AsF_5 \longrightarrow 4(Se_6I)(AsF_6)_2 + Se_8(AsF_6)_2 + 3AsF_3$$

$$12Se + I_2 + 3AsF_3 \longrightarrow 2(Se_6I)(AsF_6) + AsF_3$$

By X-ray crystallography it was shown that the monoclinic red-black crystals of Se_6IAsF_6 contain the homocyclic Se_6I^+ cation (symmetry C_s) (*87*). The exocyclic iodine atom bridges neighboring Se_6 rings, the two SeI distances (273.6 pm) being equivalent and of bond order 0.5 (see Fig. 11).

The six-membered ring has a chairlike conformation and the Se—Se bond lengths vary between 229.2 and 237.0 pm. The longest bonds are those originating from the three-coordinated atoms, as has been observed for a number of related sulfur and selenium compounds, e.g., Se_{10}^{2+}, S_7I^+, or S_8O.

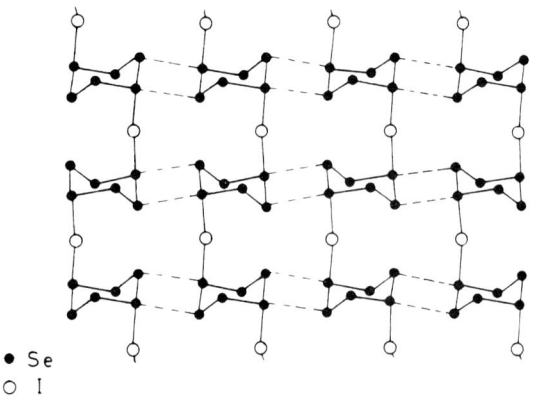

• Se
○ I

FIG. 11. Crystal structure of $(Se_6I)(AsF_6)$ (87).

Considerable intermolecular Se ⋯ Se interactions are a typical feature of the crystal structures of selenium rings, and this holds also for Se_6IAsF_6. Some of the contacts between neighboring rings are smaller than the sum of the van der Waals radii (400 pm), leading to a structure as shown in Fig. 11, in which polymeric strands of $[Se_6I^+]_n$ can be recognized. The dotted lines indicate stronger than van der Waals interactions (359.1 pm) (87).

No reactions of Se_6IAsF_6 are known.

V. Conclusions and Outlook

The structures and bond properties of the homocyclic species discussed in this article show that the selenium–selenium bond exhibits basically the same features as the well-investigated analogous sulfur–sulfur bond, which is known for its ability to adjust to a wide variety of bonding situations by large variations of the bond distances, bond angles, and dihedral angles (36). In the case of the Se—Se bonds, the fewer number of examples does not yet allow a systematic analysis, but contrary to what has been assumed in the past (when the importance of partial as well as double bonds between second- and third-row elements has been underestimated or even denied), the presently known examples show that the Se—Se bond order in homocyclic molecules can vary over a wide range, as reflected by variation of the bond distances between the 284 pm in Se_8^{2+} and 223 pm in Se_{10}^{2+}. These values should be compared with the bond length in the strain-free ring of Se_8 (234 pm) as a measure of the single-bond length and with the van der

Waals distance of 400 pm as an upper limit. A stable homocyclic compound with a selenium–selenium double bond is not known.

It can be expected that the relationships between (1) bond length and dihedral angle, (2) bond length and stretching force constant, and (3) bond length and the length of the neighboring bonds, which have been established empirically for homocyclic sulfur molecules (1, 36), will also be found in the case of selenium rings as soon as sufficiently different structures are known.

One of the major differences between two-coordinated sulfur and selenium is the tendency of the latter to increase its coordination number, as can be seen from the fairly strong intermolecular interactions between selenium rings in the various crystal structures. The shortest intermolecular distance observed (335 pm in γ-Se_8) is 65 pm shorter than the van der Waals distance (400 pm) but still 101 pm longer than the single bond distance defined above. In the case of sulfur rings the shortest observed intermolecular contact (323 pm in S_{10} (88)) is 37 pm shorter than the van der Waals distance (360 pm) but 101 pm longer than the single bond in S_8.

Many of the above-mentioned compounds have been first prepared or properly characterized after 1968. It therefore can be expected that new species of similar types will be synthesized in the near future. Their investigation by modern analytical techniques will finally help to elucidate the molecular structure of more complex systems such as amorphous, glassy, and liquid elemental selenium. In this context the already better understood homocyclic sulfur compounds and, in particular, the mixed species, the heterocyclic selenium sulfides (55), can serve as models to study the structural and chemical behavior of the selenium–selenium bond.

REFERENCES

1. Steudel, R., *Top. Curr. Chem.* **101,** 149 (1982).
2. Laur, P. H., *in* "Sulfur in Organic and Inorganic Chemistry" (A. Senning, ed.), Vol. 3, p. 91. Dekker, New York, 1972.
3. Cooper, W. Ch., ed., "The Physics of Selenium and Tellurium." Pergamon, London, 1969. (*a*) Abdullayev, G. B., Asadov, Y. G., and Mamedov, K. P., p. 179. (*b*) Izima, S., Taynai, J., and Nicolet, M. A., p. 199. (*c*) Lucovsky, G., p. 255. (*d*) Mooradian, A., and Wright, G. B., p. 269.
4. Zingaro, R. A., and Cooper, W. Ch., eds., "Selenium." Van Nostrand-Reinhold, Princeton, New Jersey, 1974. (*a*) Cooper, W. Ch., and Westbury, R. A., p. 87. (*b*) Zallen, R., and Lucovsky, G., p. 148.
5. Gmelin Handbuch der anorganischen Chemie, "Selenium," Vol. A2, p. 179. Springer-Verlag, Berlin and New York, 1980.

6. Gillespie, R. J., and Passmore, J., *Adv. Inorg. Chem. Radiochem.* **17**, 51 (1975).
7. Gillespie, R. J., *Chem. Soc. Rev.* **8**, 315 (1979).
8. Berkowitz, J., and Chupka, W. A., *J. Chem. Phys.* **45**, 4289 (1966); *J. Chem. Phys.* **48**, 5743 (1968).
9. Keller, H., Rickert, H., Detry, D., Drowart, J., and Goldfinger, P., *Z. Phys. Chem. (Frankfurt)* **75**, 273 (1971).
10. Rau, H., *J. Chem. Thermodyn.* **6**, 525 (1974).
11. Grimley, R. T., Grindstaff, Q. G., DeMercurio, T. A., and Forsman, J. A., *J. Phys. Chem.* **86**, 976 (1982).
12. Fujisaki, H., Westmore, J. B., and Tickner, A. W., *Can. J. Chem.* **44**, 3063 (1966).
13. Westmore, J. B., Fujisaki, H., and Tickner, A. W., *Adv. Chem. Ser.* **72**, 231 (1968).
14. Steudel, R., and Strauss, E.-M., *Z. Naturforsch.* **36b**, 1085 (1981).
15. Steudel, R., and Strauss, E.-M., unpublished results.
16. Miyamoto, Y., *Jpn. J. Appl. Phys.* **16**, 2257 (1977).
17. Miyamoto, Y., *Jpn. J. Appl. Phys.* **19**, 1813 (1980).
18. Nagata, K., Ishibashi, K., and Miyamoto, Y., *Jpn. J. Appl. Phys.* **19**, 1569 (1980).
19. Nagata, K., Ishibashi, K., and Miyamoto, Y., *Jpn. J. Appl. Phys.* **20**, 463 (1981).
20. Mäusle, H.-J., and Steudel, R., *Z. Anorg. Allg. Chem.* **463**, 27 (1980).
21. Gobrecht, H., Willers, G., and Wobig, D., *Z. Phys. Chem. (Frankfurt)* **77**, 197 (1972); *J. Phys. Chem. Solids* **31**, 2145 (1970).
22. Gattow, G., and Heinrich, G., *Z. Anorg. Allg. Chem.* **331**, 275 (1964).
23. Barzdain, P. P., and Alekseev, N. V., *Zh. Struct. Khim.* **9**, 520 (1968). *J. Struct. Chem.* USSR **9**, 442 (1968).
24. Steudel, R., and Schuster, F., *J. Mol. Struct.* **44**, 143 (1978).
25. Nagata, K., Tashiro, H., and Miyamoto, Y., *Jpn. J. Appl. Phys.* **20**, 2265 (1981).
26. Keezer, R. C., and Geils, R. H., unpublished results cited in ref. *43*.
27. Burbank, R. D., *Acta Crystallogr.* **4**, 140 (1951).
28. Marsh, R. E., Pauling, L., and McCullough, J. D., *Acta Crystallogr.* **6**, 71 (1953).
29. Foss, O., and Janickis, V., *J. Chem. Soc. Chem. Commun.* 834 (1977); *J. Chem. Soc. Dalton Trans.*, 624 (1980).
30. Grunwald, H. P., *Mater. Res. Bull.* **7**, 1093 (1972).
31. Cherin, P., and Unger, P., *Acta Crystallogr.* **B28**, 313 (1972).
32. Cherin, P., and Unger, P., *Inorg. Chem.* **8**, 1589 (1967).
33. Donohue, J., "The Structures of the Elements." Wiley, New York, 1974.
34. Briegleb, G., *Z. Phys. Chem. A* **144**, 321, 340 (1929).
35. Moody, J. W., and Himes, R. C., *Mater. Res. Bull.* **2**, 523 (1967).
36. Steudel, R., *Angew. Chem.* **87**, 683 (1975); *Angew. Chem. Int. Ed.* **14**, 655 (1975).
37. Lucovsky, G., Mooradian, A., Taylor, W., Wright, G. B., and Keezer, R. C., *Solid State Commun.* **5**, 113 (1967).
38. Steudel, R., *Z. Naturforsch.* **30a**, 1481 (1975).
39. Steudel, R., *Z. Naturforsch.* **30b**, 281 (1975).
40. Steudel, R., *Spectrochim. Acta* **31A**, 1065 (1975).
41. Gorman, M., and Solin, S. A., *Solid State Commun.* **18**, 1401 (1976).
42. Lucovsky, G., *in* "The Physics of Selenium and Tellurium," (E. Gerlach and P. Grosse, (eds.), p. 178. Springer-Verlag, Berlin and New York, 1979.
43. Zirke, J., Dissertation, Techn. Univ. Berlin, 1976.
44. Ohta, N., Scheuerman, W., and Nakamoto, K., *Solid State Commun.* **27**, 1325 (1978).
45. Laitinen, R., and Niinistö, L., *J. Therm. Anal.* **13**, 99 (1978).
46. Asadov, Y. G., *Kristallografiya* **14**, 356 (1969); *Sov. Phys. Crystallogr.* **14**, 292 (1969); *Dokl. Akad. Nauk SSSR* **173**, 570 (1967); *Sov. Phys. Dokl.* **12**, 199 (1967).
47. Shu, H.-C., Gaur, U., and Wunderlich, B., *Thermochim. Acta* **34**, 63 (1979).

48. Meissner, M., Tausend, A., and Wobig, D., *Phys. Status Solidi* **A49**, 59 (1978).
49. Franz, P., Gobrecht, H., Scheiba, M., Wobig, D., and Kunze, W., *Z. Phys. Chem. (Frankfurt)* **111**, 163 (1978).
50. Murphy, K. E., Altmann, M. B., and Wunderlich, B., *J. Appl. Phys.* **48**, 4122 (1977).
51. Belin, E., Bonnelle, C., and Delafosse, D., *J. Appl. Crystallogr.* **4**, 383 (1971).
52. Stech, B., *Z. Naturforsch.* **7a**, 175 (1952).
53. Asadov, Y. G., *Fiz. Svoistva Selena Selenovykh Prib.* **21**, after *Chem. Abstr.* **82** (1974); **78**, 844 (1975).
54. Krebs, H., *Z. Anorg. Allg. Chem.* **265**, 156 (1951); *Angew. Chem.* **65**, 293 (1953).
55. Steudel, R., and Laitinen, R., *Top. Curr. Chem.* **103**, 177 (1982).
56. Saure, H., and Block, J., *Int. J. Mass Spectrom. Ion Phys.* **7**, 145 and 175 (1971).
57. Hoareau, A., Reymond, J.-M., Cabaud, B., and Uzan, R., *J. Phys. (Paris)* **36**, 737 (1975).
58. Knox, B. E., *Mater. Res. Bull.* **3**, 329 (1968); *Adv. Mass. Spectrom.* **4**, 491 (1968).
59. Streets, D. G., and Berkowitz, J., *J. Electron. Spectrosc. Relat. Phenom.* **9**, 268 (1976).
60. Rau, H., *Ber. Bunsenges. Phys. Chem.* **71**, 711 (1967).
61. Eisenberg, A., and Tobolsky, A. V., *J. Polymer Sci.* **46**, 19 (1960).
62. Gobrecht, H., Willers, G., Wobig, D., and Zirke, J., *Z. Naturforsch.* **27a**, 1246 (1972).
63. Gaur, U., Shu, H.-C., Mehta, A., and Wunderlich, B., *J. Phys. Chem. Ref. Data* **10**, 89 (1981).
64. Steudel, R., *Z. Naturforsch.* **36a**, 408 (1981).
65. Steudel, R., and Mäusle, H.-J., *Z. Naturforsch.* **33a**, 951 (1978).
66. Mills, K. C., "Thermodynamic Data for Inorganic Sulfides, Selenides and Tellurides," p. 86. Butterworths, London, 1974.
67. Ban, V. S., and Knox, B. E., *Int. J. Mass Spectrom. Ion Phys.* **3**, 131 (1969).
68. Barr, J., Gillespie, R. J., Kapoor, R., and Malhotra, K. C., *Can. J. Chem.* **46**, 149 and 3607 (1968).
69. Gillespie, R. J., Kent, J. P., and Sawyer, J. F., *Inorg. Chem.* **20**, 4053 (1981).
70. Burns, R. C., and Gillespie, R. J., *Inorg. Chem.* **21**, 3877 (1982).
71. Cardinal, G., Gillespie, R. J., Sawyer, J. F., and Vekris, J. E., *J. Chem. Soc. Dalton Trans.* 765 (1982).
72. Prince, D. J., Corbett, J. D., and Garbisch, B., *Inorg. Chem.* **9**, 2731 (1970).
73. Paul, R. C., Arora, C. L., Virmani, R. N., and Malhotra, K. C., *Indian J. Chem.* **9**, 368 (1971).
74. Brown, I. D., Crump, D. B., Gillespie, R. J., and Santry, D. P., *J. Chem. Soc. Chem. Commun.* 583 (1968).
75. Brown, I. D., Crump, D. B., and Gillespie, R. J., *Inorg. Chem.* **10**, 2319 (1971).
76. Clark, R. J. H., Dines, T. J., and Ferris, L. T. H., *J. Chem. Soc. Dalton Trans.* 2237 (1982).
77. Tanaka, K., Yamabe, T., Terama-e, H., and Fukui, K., *Inorg. Chem.* **18**, 3591 (1979); *Nouv. J. Chim.* **3**, 379 (1979).
78. Brown, I. D., Crump, D. B., and Gillespie, R. J., *Inorg. Chem.* **10**, 2319 (1971).
79. Gillespie, R. J., and Pez, G. P., *Inorg. Chem.* **8**, 1229 (1969).
80. McMullan, R. K., Prince, D. J., and Corbett, J. D., *Inorg Chem.* **8**, 1749 (1971); *J. Chem. Soc. Chem. Commun.* 1438 (1969).
81. Burns, R. C., Chan, W.-L., Gillespie, R. J., Luk, W.-C., Sawyer, J. F., and Slim, D. R., *Inorg. Chem.* **19**, 1432 (1980).
82. Gillespie, R. J., and Ummat, P. K., *Can. J. Chem.* **48**, 1239 (1970).
83. Desjardins, C. D., and Passmore, J., *J. Chem. Soc. Dalton Trans.* 2314 (1973).

84. Fehrman, R., Bjerrum, N. J., and Andreasen, H. A., *Inorg. Chem.* **14,** 2259 (1975).
85. Fehrmann, R., and Bjerrum, N. J., *Inorg. Chem.* **16,** 2089 (1977).
86. Passmore, J., Richardson, E. K., and Taylor, P., *J. Chem. Soc. Dalton Trans.* 1006 (1976).
87. Shanta Nandana, W. A., Passmore, J., and White, P. S., *J. Chem. Soc. Chem. Commun.* 526 (1983).
88. Reinhardt, R., Steudel, R., and Schuster, F., *Angew. Chem.* **90,** 55 (1978); *Angew. Chem. Int. Ed.* **17,** 57 (1978); Steudel, R., Steidel, J., and Reinhardt, R., *Z. Naturforsch.* **38b,** 1548 (1983).

THE ELEMENT DISPLACEMENT PRINCIPLE: A NEW GUIDE IN p-BLOCK ELEMENT CHEMISTRY[1]

A. HAAS

Lehrstuhl für Anorganische Chemie II, Ruhr-Universität Bochum, Bochum, Federal Republic of Germany

I. Introduction	168
II. The Element Displacement Principle	169
III. Application of the Element Displacement Principle to Recent Chemical Problems	172
Equivalence between Carbon and Sulfur(IV)	172
IV. Halogen-Like Properties of Parahalogens of Different Derivation	176
A. Equivalence between CF_3 and Fluorine	176
B. Equivalence between $(CF_3S)_2N-$ and Fluorine	179
C. Equivalence between CF_3S and Chlorine	180
D. Equivalence between Tetrakis(trifluoromethylthio)pyrrolyl— and Chlorine	187
V. Interchangeability of Elements and Paraelements in Known Structures	194
A. Transition from Hexazabenzene $(-N=N-)_3$ to Borazole $(-\overset{\delta+}{N}H-\overset{\delta-}{B}H-)_3$	194
B. Exchangeability within Adamantane Structures	195
C. Interchangeability within the P_4S_3 Structure	196
VI. Electronegativities	197
References	199

[1] This review is based on an article published in 1982 entitled "A New Classification Principle: The Periodic System of Functional Groups" [Haas, A., *Chemiker Zeitung* **106**, 239 (1982)] and contains new results supporting this principle.

I. Introduction

The concept that will be discussed in this article has its origin in the hydride displacement principle published by Grimm (1) in 1925. He showed that elements situated up to four places before a noble gas change their properties by taking up 1, 2, 3, or 4 (x) hydrogen atoms so that the resulting radicals behave as pseudoatoms, which resemble the reference elements situated in the periodic system x groups to the right of them, e.g., $CH \triangleq N$, $CH_2 \triangleq O$, $CH_3 \triangleq F$, and so on. The radicals obtained by this procedure are isoelectronic and isoprotonic with the reference elements; they are termed "pseudoelements." So far this principle has been used only on elements of the second period, but it also applies without limitations to all other Group IV–VII elements. An impressive example is found in the following series of oxyacids: H_9InO_6 (as $M_3^I[In(OH)_6] \cdot 2H_2O$; $M = Na^+$, K^+), H_8SnO_6, H_7SbO_6, H_6TeO_6, H_5IO_6, and H_4XeO_6. They can be written as $H_4(InH_5)O_6$, $H_4(SnH_4)O_6$, $H_4(SbH_3)O_6$, $H_4(TeH_2)O_6$, $H_4(JH)O_6$, and $H_4(Xe)O_6$. Their similar chemical properties are due to the pseudoxenon behavior of (IH), (TeH_2), (SbH_3), (SnH_4), and (InH_5).

Birkenbach and Kellerman (2) reported also in 1925 on monovalent inorganic groups consisting of two or more electronegative atoms with chemical and physical properties partly similar to those of halogens, e.g., CN, SCN, OCN, N_3, NCO, NCS, etc. They were also termed "pseudohalogens," although they are neither isoelectronic nor isoprotonic with the reference halogens. At the beginning of the perfluorohaloorgano chemistry of the elements, Lagowski (3) drew attention in 1959 to the halogen-like behavior of the CF_3 and CF_3S radicals for, among other reasons, the fact that their electronegativities were comparable with those of the halogens [CF_3 (3.3) and CF_3S (2.7)]. The latter radicals again are neither isoprotonic nor isoelectronic with the halogens.

Incorporating these pseudoelements into the periodic system, the large differences between the chemical and physical properties of elements of different periods are appreciably alleviated. If, e.g., the electrode potentials (in volts) of pseudohalogens are arranged with those of halogens the following progression is obtained (4): F^- (2.3), $(NC)_2N^-$ (2.3), O_2NO^- (2.3), $(NC)_3C^-$ (2.1), Cl^- (2.0), NCO^- (2.0), $(O_2N)_3C^-$ (2.0), Br^- (1.7), NCS^- (1.5), I^- (1.2), $NCSe^-$ (1.1). It should be possible to eliminate sudden changes by incorporating further composite element-like groups into the periodic system, provided that a concept can be found which is able to establish a correlation between "pseudoelements," "pseudohalogens," and CF_3 or CF_3S, and which also allows the existence of new element-like groups to be deduced.

This requirement is met by the concept of *element displacement*.

II. The Element Displacement Principle

Coordination of the elements of Groups IV–VII—subsequently termed base elements—with elements or element groups—termed ligands—forming one, two, three, or four covalent bonds results in a shift of one, two, three, or four places to the right (higher atomic number) within a period. This process is termed the *element displacement principle*. In the following, fluorine is used as a representative for all other monovalent ligands. With the fluorine atom, which is isogeometric with the hydrogen atom, as a ligand the following perfluorinated groups are obtained.

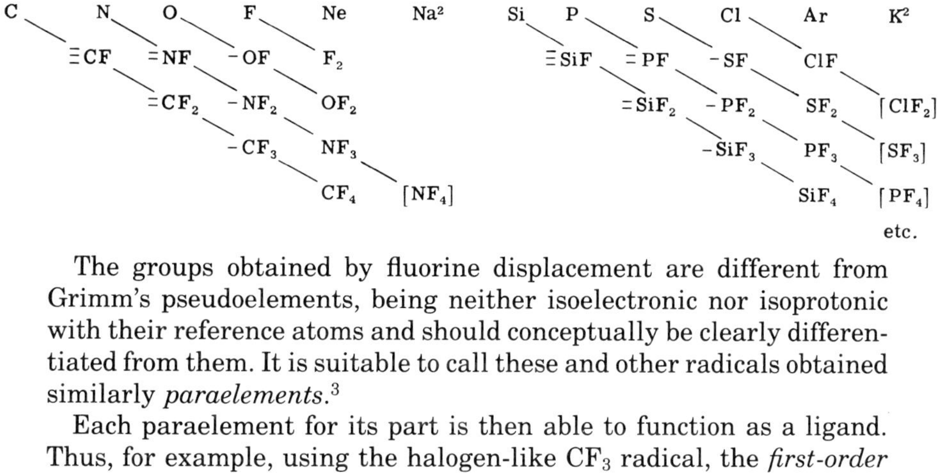

C	N	O	F	Ne		Na[2]	Si	P	S	Cl	Ar	K[2]
$\equiv CF$	$=NF$	$-OF$	F_2				$\equiv SiF$	$=PF$	$-SF$	ClF		
	$=CF_2$	$-NF_2$	OF_2					$=SiF_2$	$-PF_2$	SF_2	$[ClF_2]$	
		$-CF_3$	NF_3						$-SiF_3$	PF_3	$[SF_3]$	
			CF_4	$[NF_4]$						SiF_4	$[PF_4]$	

etc.

The groups obtained by fluorine displacement are different from Grimm's pseudoelements, being neither isoelectronic nor isoprotonic with their reference atoms and should conceptually be clearly differentiated from them. It is suitable to call these and other radicals obtained similarly *paraelements*.[3]

Each paraelement for its part is then able to function as a ligand. Thus, for example, using the halogen-like CF_3 radical, the *first-order derivative paraelements* shown below are obtained.

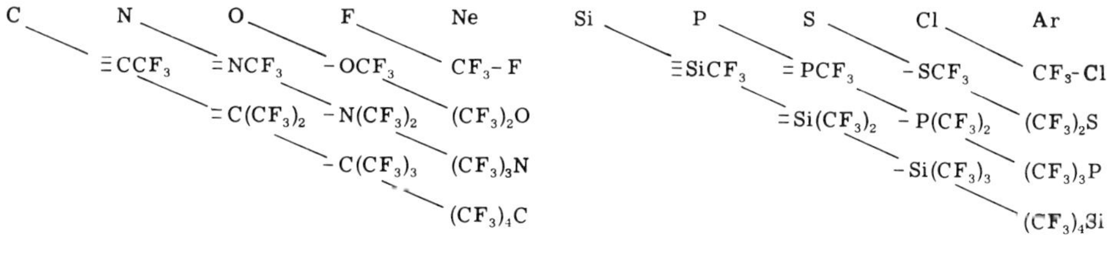

C	N	O	F	Ne	Si	P	S	Cl	Ar
$\equiv CCF_3$	$\equiv NCF_3$	$-OCF_3$	CF_3-F		$\equiv SiCF_3$	$\equiv PCF_3$	$-SCF_3$	CF_3-Cl	
	$=C(CF_3)_2$	$-N(CF_3)_2$	$(CF_3)_2O$			$=Si(CF_3)_2$	$-P(CF_3)_2$	$(CF_3)_2S$	
		$-C(CF_3)_3$	$(CF_3)_3N$				$-Si(CF_3)_3$	$(CF_3)_3P$	
			$(CF_3)_4C$					$(CF_3)_4Si$	

---

[2] Paraelements of this group are only stable as cations.

[3] παρα (para): near, secondary.

If first-order derivative paraelements are themselves used as ligands this system of nomenclature leads to second-order derivative paraelements, e.g., using CF$_3$S. This procedure can be continued in many cases with meaningful results.

```
C         N            O          F          Ne       Si           P            S           Cl         Ar
 ≡CSCF₃   ≡NSCF₃    -OSCF₃     CF₃SF              ≡SiSCF₃     ≡PSCF₃      =SSCF₃      CF₃SCl
          ≡C(SCF₃)₂  =N(SCF₃)₂  (CF₃S)₂O                    =Si(SCF₃)₂  -P(SCF₃)₂   (CF₃S)₂S
                    -C(SCF₃)₃   (CF₃S)₃N                                -Si(SCF₃)₃   (CF₃S)₃P
                                (CF₃S)₄C                                             (CF₃S)₄Si
```

Which sort of paraelements do we obtain with ligands like oxygen, sulfur, or nitrogen? When any one of these is used as the ligand a displacement of two or three groups must of course result, as follows:

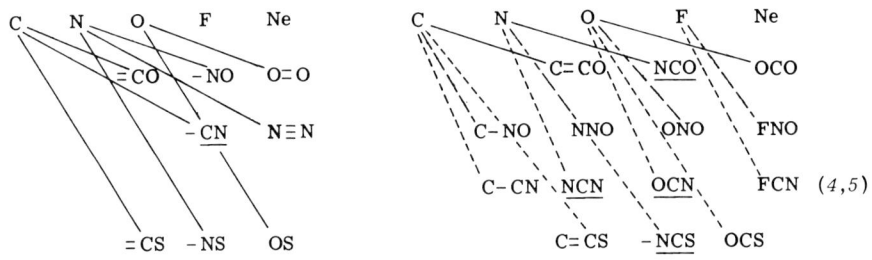

```
C     N     O     F     Ne           C     N      O     F     Ne
            ≡CO   -NO   O=O                C=CO   NCO   OCO
                  -CN   N≡N                C-NO   NNO   ONO   FNO
                                           C-CN   NCN   OCN   FCN   (4,5)
            =CS   -NS   OS                 C=CS   -NCS  OCS
```

These resulting paraelements may also appear as ligands. If, for example, the CO, NO, CN, or CS groups are coordinated above, first-order derivative paraelements are obtained.

All classical pseudohalogens can be deduced in this way. A base element and paraelement have the same number of valence electrons. This means that covalent electron pairs held in common by base atom and ligand are assigned to the base atom, which becomes electronically isovalent with the reference element.

The elements boron, aluminum, gallium, indium, and thallium similarly add ligands forming pseudo- or paraelements which are shifted to the right. Analogously, addition of an electron to an element (equivalent to a covalent bond) forms an ion which is shifted in the same direction, e.g., Tl$^-$ ≙ C (6); C$^-$ ≙ N; Si$^-$, Ge$^-$, Pb$^-$ ≙ P; Si^{2-}, As$^-$ ≙ Se; O$^-$, S$^-$ ≙ Cl (7). A shift to the left side of the periodic system is achieved by loss of electrons, e.g., N$^+$ ≙ C, C$^+$ ≙ B, O$^+$ ≙ N, Cl$^+$ ≙ S.

The elements of Groups I–III have, with the exception of boron, low ionization potentials and therefore are shifted mainly to the left when forming covalent bonds with a ligand. Covalent electron pairs between base atom and ligand are transferred in the process of heterolytic fission to the ligand, leaving a formal positive charge on the base atom, which is displaced to a reference element of lower atomic number. The elements of Group III therefore can be shifted to the right by acquiring or to the left by losing electrons. Therewith the possibility of paraelement formation is not exhausted, as elements of Groups IV–VII have at their disposal free electron pairs and consequently are able to form coordinate bonds:

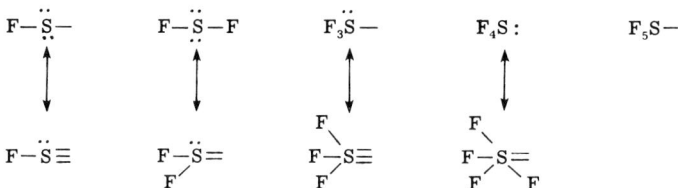

The number of unpaired valence electrons on the central atom being unchanged, base and reference element are identical.

Elements of the third to fifth periods are also able to expand their octet shells. New paraelements result in this way which have not so far been mentioned, e.g.,

$$
\begin{array}{ccccc}
F-\ddot{\underset{..}{S}}- & F-\ddot{S}-F & F_3\ddot{S}- & F_4S: & F_5S- \\
\updownarrow & \updownarrow & \updownarrow & \updownarrow & \\
F-\ddot{S}\equiv & F-\ddot{S}= & F-\underset{F}{\overset{F}{\diagup}}S\equiv & F-\underset{F}{\overset{F}{\diagup}}S\diagdown F &
\end{array}
$$

Thus when fluorine is a ligand sulfur can form the following paraelements:

$$F—\ddot{S}\equiv \text{ and } (F—)_3S\equiv \longrightarrow \text{paranitrogen}$$

$$(F—)_2\ddot{S}= \text{ and } (F—)_4S= \longrightarrow \text{paraoxygen}$$

$$F—\ddot{\underset{..}{S}}-, (F—)_3\ddot{S}-, \text{ and } (F—)_5S- \longrightarrow \text{parahalogen}$$

An analogous behavior is found for the other elements, e.g., phosphorus, arsenic, selenium, etc. Paraelements of this sort are no longer electronically isovalent with their reference elements; they do, however, remain isovalent with the reference element. Thus the criterion of isovalence is common to all paraelements.

III. Application of the Element Displacement Principle to Recent Chemical Problems

EQUIVALENCE BETWEEN CARBON AND SULFUR(IV)

An interesting relationship between carbon and s^2p^3d hybridized sulfur can be observed in a number of unsaturated functional groups containing carbon and sulfur. In these paraelements carbon and sulfur can be exchanged without altering their chemical properties significantly. The main difference between them is caused by the extra electron pair at the sulfur atom which gives rise to bent or nonplanar structures. While the following carbon-containing paraelements are linear or planar, the corresponding sulfur analogs are bent or nonplanar.

$\cdot\ddot{N}=C: \longleftrightarrow \cdot C\equiv N: \qquad \cdot\ddot{N}=C=\ddot{O} \longleftrightarrow \cdot\ddot{O}-C\equiv N:$

$\cdot\ddot{N}=\ddot{S}:^4 \longleftrightarrow \cdot\ddot{S}\equiv N: \qquad \cdot\ddot{N}=\ddot{S}=\ddot{O} \longleftrightarrow \cdot\ddot{O}-\ddot{S}\equiv N:^4$

$\cdot\ddot{N}=C=\ddot{S} \longleftrightarrow \cdot\ddot{S}-C\equiv N: \qquad \cdot\ddot{N}=C=\ddot{N}\cdot \longleftrightarrow \cdot\ddot{N}-C\equiv N:$

$\cdot\ddot{N}=\ddot{S}=\ddot{S} \longleftrightarrow \cdot\ddot{S}-\ddot{S}\equiv N:^4 \qquad \cdot\ddot{N}=\ddot{S}=\ddot{N}\cdot \longleftrightarrow \cdot\ddot{N}-\ddot{S}\equiv N:$

$X-\ddot{N}=\ddot{C}-X \longleftrightarrow \cdot\ddot{N}=CX_2$

$X-\ddot{N}=\ddot{S}-X^4 \longleftrightarrow \cdot\ddot{N}=SX_2$

This carbon–sulfur comparability extends only to sp and sp^2 hybridized carbon, meaning only to unsaturated systems. The following series of examples illustrates this.

The imines $(CF_3)_2C=NH$ and $(CF_3)_2S=NH$ are convincing examples for this carbon–sulfur exchangeability. Reactions of the two imines with RLi, RCl, ClF, as well as photolysis, hydrolysis, and their preparation are similar and take place according to Table I.

Besides these imines there are other compounds contributing evidence for the similarity between carbon and sulfur(IV). The following reactions and the compounds obtained are other conclusive arguments.

$$CF_3N=SCl_2 + Ag_2O \longrightarrow CF_3NSO \qquad (13)$$

$$CF_3N=CCl_2 + H_2O \longrightarrow CF_3NCO \qquad (14)$$

---

[4] Derivatives of these paraelements have not yet been synthesized.

TABLE I

Reaction	$(CF_3)_2C=NH$	$(CF_3)_2S=NH$
RLi	$(CF_3)_2C=NLi$	$(CF_3)_2S=NLi$ (8, 9)
R'Cl (R' = Me$_3$Si, CF$_3$S—, CF$_3$CO—)	$(CF_3)_2C=NR$ (23)	$(CF_3)_2S=NR$ (8, 9)
ClF	$(CF_3)_2C=NCl$ (11)	$(CF_3)_2S=NCl$ (10)
$h\nu$	$(CF_3)_2C=N-N=C(CF_3)_2$ (11)	$(CF_3)_2S=N-N=S(CF_3)_2$ (10)
H$_2$O	—	$(CF_3)_2S=NCl \xrightarrow{H_2O} (CF_3)_2S=NH$ (10)
H$_2$O	$(CF_3)_2C=NF + KJ + H^+$ $\longrightarrow (CF_3)_2C=NH$	—
Preparation:	$(CF_3)_2C=O + NH_3$ (21)	$(CF_3)_2SF_2 + NH_3$ (8, 12)

$$CF_3SN=SF_2 + CF_3SNH_2 \longrightarrow CF_3SN=S=NSCF_3 \quad (15)$$

$$RN=CCl_2 + RNH_2 \longrightarrow RN=C=NR \quad (R = CH_3, C_2H_5) \quad (16)$$

The reactions of isocyanidedichlorides (—N=CCl$_2$) and iminosulfurdihalides (—N=SX$_2$) are similar and the products formed show the relationship mentioned before. In recent years it was proved that in some reactions S$_2$Cl$_2$, in keeping with its isomeric structure S=SCl$_2$, forms with amines RN=S=S compounds. In its branched form S$_2$Cl$_2$ resembles, as expected, S=CCl$_2$. Both react with primary amines in a similar way according to

$$RNH_2 + S=CCl_2 \longrightarrow RN=C=S + 2HCl \quad (17)$$

$$RNH_2 + S=SCl_2 \longrightarrow RN=S=S + 2HCl \quad (18)$$

$$(R = aryl—)$$

In the presence of guinoline CF$_3$SNH$_2$ condenses with S$_2$Cl$_2$, forming CF$_3$SN=S=S (19). The corresponding reaction with S=CF$_2$ in the presence of NaF leads to CF$_3$SN=C(F)SSCF$_3$ (20). The expected CF$_3$SNCS was not formed and even as an intermediate it was not detected. This is not surprising, as sulfenylisothiocyanates are almost unknown. The only characterized example is CFCl$_2$SNCS, obtained from FCl$_2$SN=CCl$_2$ and P$_4$S$_{10}$ in boiling xylol in 50% yield (22).

N-(Thiosulfinylamines) are very reactive compounds and decompose to RN=S=NR and sulfur. Some isothiocyanates behave similarly. In the presence of 2,5-dihydrophosphol-1-oxide they condense to

RN=S=NR, according to

$$2RN{=}S{=}S \longrightarrow RN{=}S{=}NR + \text{"}S{=}S{=}S\text{"}$$

[R = aryl (24); R = CF$_3$S (19)]

$$2R'{-}N{=}C{=}S \xrightarrow{\text{catalyst}} R'N{=}C{=}NR' + SCO\ (25)$$

As a result of the reaction with the catalyst, SCO, rather than S=C=S (which is similar to the unstable "S=S=S"), is formed. Equivalency is also observed in the chlorination and in the fluorination reaction, e.g.,

$$R\ddot{N}{=}\ddot{S}{=}\ddot{S} + 2Cl_2 \longrightarrow R\ddot{N}{=}\ddot{S}Cl_2 + SCl_2 \quad [R = \text{aryl}\ (24)]$$

$$CF_3\ddot{N}{=}C{=}\ddot{S} + 2Cl_2 \longrightarrow CF_3\ddot{N}{=}CCl_2 + SCl_2\ (26)$$

$$C_2H_5N{=}C{=}S + HgF_2 \longrightarrow C_2H_5N{=}CF_2 + HgS\ (73)$$

$$CF_3SN{=}S{=}S + HgF_2 \longrightarrow CF_3\ddot{S}N{=}\ddot{S}F_2\ (74) + HgS$$

The two isomeric paraelements —$\dot{N}{=}CX_2$ and $XN{=}C(X)$— resemble formally the sulfur analogs —$\dot{N}{=}\ddot{S}X_2$ and $X\ddot{N}{=}\ddot{S}(X)$.

While compounds of the type $R_f\ddot{N}{=}CX_2$, $R_f\ddot{N}{=}\ddot{S}X_2$ and $R_fC(X){=}\ddot{N}X$ are already known, hitherto no substances of the type $R_f\ddot{S}(X){=}\ddot{N}X$ with $R_f \neq X$ (X = halogen) have been described. While the fluorination of (CF$_3$CN)$_3$ at 240°C leads to CF$_3$C(F)=NF (27) this reaction cannot be applied for the synthesis of CF$_3\ddot{S}$(F)=NF, as the corresponding starting material (CF$_3$SN)$_3$ is not known. The only available compounds are (CF$_3$SN)$_4$ and (CF$_3$SN)$_x$ (28) but they do not react with halogens at ambient temperatures. The other method, addition of halogen to R$_f$CN (29), is not applicable to R$_f\ddot{S}{\equiv}$N:, as such compounds are not available. The successful preparation of CF$_3$S(X)=NX was achieved via halogenation of CF$_3$SN[Si(CH$_3$)$_3$]$_2$ according to the equations

$$CF_3SN[Si(CH_3)_3]_2 + F_2 \xrightarrow{-60°C} CF_3{-}\ddot{S}(F){=}\ddot{N}F + 2(CH_3)_3SiF$$

$$+ Cl_2 \xrightarrow{-40°C} CF_3S(Cl){=}\ddot{N}Si(CH_3)_3 \xrightarrow[-20°C]{Cl_2}$$

$$CF_3S(Cl){=}NCl + 2(CH_3)_3SiCl$$

$$+ Br_2 \xrightarrow{20°C} 1/x(CF_3SN)_x + 2(CH_3)_3SiBr$$

$$+ ClF \xrightarrow{-60°C} CF_3S(Cl){=}NCl + [CF_3SCl, CF_3Cl, (CH_3)_3SiF, (CH_3)_3SiCl]$$

No reaction took place with iodine (19).

Another paraelement pair, —N=C=N— and —N=S=N—, seem

to have a lot in common. From both a large number of organic and perfluororganoderivatives are known. The carbodiimide group exists also in an isomeric form as cyanoamide (=N—C≡N). Up to now the corresponding isomer, —N—S≡N:, is not known. The only evidence for such a group to exist is the reaction of secondary amines with $Cl_3S_3N_3$ in boiling CCl_4 in the presence of pyridin according to

$$3R_2NH + Cl_3S_3N_3 \longrightarrow 3RN{=}S{=}NR + 3HCl$$

[R = C_6F_5S (75), CF_3S (76), CF_3Se (77)]

It is assumed that at elevated temperatures $Cl_3S_3N_3$ pyrolyzes to ClS≡N which condenses with the imine to form $R_2NS≡N$ as an intermediate. This rearranges immediately to the stable RN=S=NR compound. There might be a chance to isolate a molecule with an N—S≡N group providing a cyclic imine is found with stable element–nitrogen bonds. Tetrakis(trifluoromethylthio)pyrrol might fulfil these requirements and react with $Cl_3S_3N_3$ to give the wanted substances. If rearrangement occurs, ring opening has to take place. When reacting the substituted pyrrole with $Cl_3S_3N_3$ in refluxing CCl_4 in the presence of pyridine, a light yellow substance can be isolated by sublimation *in vacuo*. On standing it changes color continuously, ending with deep purple, without changing its physical properties. On subliming aged material *in vacuo,* light yellow material is recovered almost quantitatively. Analytical and spectroscopic investigations show that the product is the suggested molecule, which is formed according to

[Reaction scheme: tetrakis(trifluoromethylthio)pyrrole (N–H) + ClS≡N + C_5H_5N → tetrakis(trifluoromethylthio)pyrrole with N–S≡N substituent + [C_5H_5NH]Cl]

According to molecular weight determinations the product is monomeric and shows no tendency to trimerize to a trithiazyl ring. Such cyclizations are observed for FSN (78) and ClSN (79), which are comparable with FCN and ClCN. This suggests that —S≡N is similar to —C≡N and not to N—O compounds. An unexpected connection can be drawn between F—S—N— and HC compounds starting from $F_4S_4N_4$. If the paranitrogen FS≡ and nitrogen are completely replaced by pseudonitrogen CH, $(FS)_4N_4$ is converted to cyclooctatetraene. Both substances have localized double bonds [d(C=C) = 1.34 Å, d(C—C) = 1.48 Å (80) compared with d(S=N) = 1.54 Å, d(S—N) = 1.66 Å]. The bond lengths are of course not equal but the differences between double and single bond ($\Delta = 0.14$, $\Delta' = 0.12$ Å) are very close. In its boot conformation $(CH)_8$ and the S_4N_4 frame have almost the same structure. The four basic atoms in $F_4S_4N_4$ do not form an ideal square but are a little eclipsed.

IV. Halogen-Like Properties of Parahalogens of Different Derivation

A. Equivalence between CF_3 and Fluorine

Impressive similarities between fluorine and a parafluorine are observed with F and CF_3. In a number of reactions this can be demonstrated in a convincing manner.

(1) The association of two SF_2 or CF_3SF molecules (30), according to

$$2F-S-F \rightarrow F_3-S-S-F$$

$$2CF_3-S-F \rightarrow CF_3SF_2SCF_3$$

(2) The acid–anhydrofluoride behavior of perfluorinated acids, not dissociating in H_2O

$$2FOH \longrightarrow \ddot{O}=\ddot{O} + 2HF \quad (31)$$

$$2F-\underset{\underset{F}{|}}{N}H \longrightarrow F\ddot{N}=\ddot{N}F + 2HF \quad (32)$$

$$2Cl-CF_2H \xrightarrow{600 \text{ to } 800°C} F_2C=CF_2 + 2HCl \quad (33)$$

$$CF_3OH \longrightarrow F_2C=O + HF \quad (34)$$

$$CF_3-\underset{\underset{CF_3}{|}}{N}-H \longrightarrow CF_3N=CF_2 + HF \quad (35)$$

$$(CF_3)_3CH \longrightarrow (CF_3)_2C=CF_2 + HF$$

The generation of CF_2 can also be achieved by hydrolysis of $ClCF_2H$ (33). Neither direct conversion of CF_3H to CF_2, and hence $CF_2=CF_2$, nor decomposition of $(CF_3)_3CH$ to $(CF_3)_2C=CF_2$, is observed. Addition of HF in the presence of F^- to $(CF_3)_2C=CF_2$, forming $(CF_3)_3CH$, takes place under normal reaction conditions (36). Some of the anhydrofluorides react with metal fluorides (bases in HF) to form saltlike compounds.

TABLE II

Physical Data of CF_3SeF_4X, F_5SeX (X = F, Cl), CF_3SeF_3, and SeF_4

Substance	D (g/cm³)	mp (°C)	bp (°C)	ΔH (kJ/mol)	ΔS (J/mol K)
CF_3SeF_5	2.0 ± 0.15	−76.5 ± 2	8.5 ± 2	26.5 ± 1.5	94.3 ± 5
SeF_6	1.93 (62)	−34.6 (63)	−45.7^a (62)	27.6 (63)	121.8 (63)
CF_3SeF_4Cl	2.5 ± 0.15	−88.0 ± 2	43.5 ± 2	32.5 ± 1.5	102.7 ± 5
SeF_5Cl (64)	—	−19.0	4.5	26.0	93.6
CF_3SeF_3	2.7 ± 0.15	26.0 ± 1	103 ± 4^b	53.5 ± 3^a	142 ± 10^a
SeF_4	2.77 (66)	− 9.5 (65) −38.87 (115)	101.0 (65)	47.0 (65)	125.4 (65)

^{a} Sublimation value.
^{b} Extrapolated from $\ln(p/760\ \text{Torr}) = 17.1 − 6435/T$.

$$MF + CF_2O \longrightarrow MOCF_3 \quad (M = K, Rb, Cs)\ (37)$$

$$HgF_2 + 2CF_3N{=}CF_2 \longrightarrow Hg[N(CF_3)_2]_2\ (38)$$

$$HgF_2 + (CF_3)_2C{=}CF_2 \longrightarrow Hg[C(CF_3)_3]_2\ (39)$$

These equations not only demonstrate the fluorine-like behavior of the parafluorine elements FO—, F_2N—, F_3C, CF_3O, $(CF_3)_2N$, and $(CF_3)_3C$, but a similarity to oxygen of NF, CF_2, CF_3N, and $(CF_3)_2C$.

(3) Although perfluoroorganosulfur(VI) compounds such as CF_3SF_5, CF_3SF_4Cl, and CF_3SO_3H are well known, corresponding selenium compounds, apart from $C_2F_5SeF_4Cl$ (fully characterized) and $C_2F_5SeF_5$ (61) (obtained only in traces), are unknown. This fact is surprising, since SeF_6 and SeF_5Cl are stable substances. One would expect that replacement of F by parafluorine, e.g., CF_3, would lead to stable compounds. Fluorination of CF_3SeF_3 with liquid F_2 at −196°C or more conveniently with AgF_2 at 65–70°C produces a complex mixture out of which CF_3SeF_5 was isolated by codistillation in 3 to 4% yield. Similarly, CF_3SeF_4Cl was made from CF_3SeF_3 and ClF (partial pressure should not exceed 200 mbar) at −105°C (12 hr) in 20% yield. Both compounds decompose at 60°C (1 hr) primarily into CF_4, CF_3Cl, and SeF_4. Physical and spectroscopic data of CF_3SeF_4X, SeF_5X (X = F, Cl), CF_3SeF_3, and SeF_4 are given in Tables II and III. Going from SeF_6 to CF_3SeF_5 or SeF_5Cl to CF_3SeF_4Cl the mp decreases by about 42 or 69°C and the bp increases by 54 or 39°C. These deviations are mainly due to change of the molecular shape differing from a regular octahedron. The other physical data are more or less comparable. The couple SeF_4/CF_3SeF_3 shows a different mp but the other figures are almost equal. Substan-

TABLE III

^{19}F-, ^{77}Se-, AND ^{13}C-NMR-Spectroscopic Data of CF$_3$SeF$_4$X, SeF$_5$X (X = F, Cl), CF$_3$SeF$_3$, AND SeF$_4$[a]

	CF$_3$SeF$_5$	CF$_3$SeF$_4$Cl (−30°C)	SeF$_6$	SeF$_5$Cl (64)	CF$_3$SeF$_3$ (−40°C)	SeF$_4$ (−140°C) (70)
$\delta_F(CF_3)$	−51.2 (quint of d)	−51.9 (quint)	—	—	−65.1 (71)	—
$\delta_F[SeF_{4(e)}]$	+29.8 (d of q)	+97.1 (q)	—	132	−59.7 (71)	−37.7
$\delta_F[SeF_{(a)}]$	+63.8 (quint of q)	—	+48 (67)	71.3	−21.9 (71)	−12.1
$\delta^{77}Se$	+626.8 (quint of d of q)	+650.1 (quint of q)	632.3 (68)	?	1072 (69)	1114.2 (68)
$\delta^{13}C$	+125.3 (q of quint)	+131.0 (q of quint)	—	—	126.6 (69)	—
$^3J_{CF_3-SeF_{4(e)}}$	23.7	25.2	—	—	—	—
$^3J_{CF_3-SeF_{(a)}}$	3.1	—	—	—	—	—
$^2J_{SeF_{4(e)}-SeF_{(a)}}$	184.1	—	—	213.6	94.4 (71)	26
$^1J_{^{77}Se-SeF_{4(e)}}$	1322	1177	—	1258	—	302
$^1J_{^{77}Se-SeF_{(a)}}$	1311	—	1420.9 (68)	1352	—	1200
$^2J_{^{77}Se-CF_3}$	381	389	—	—	41 (69)	—
$^2J_{^{13}C-SeF_{4(e)}}$	32.9	37.1	—	—	—	—
$^1J_{^{13}C-CF_3}$	352.9	353.4	—	—	346.6 (69)	—

[a] External standards: ^{19}F(CFCl$_3$); ^{77}Se(CH$_3$SeCH$_3$); ^{13}C(SiMe$_4$); δ in parts per million; J in Hertz.

tial differences are observed between Se(VI) and Se(IV) compounds. Comparable values of $\delta(^{77}Se)$ are observed for CF_3SeF_5, CF_3SeF_4Cl, and SeF_6 (that for SeF_5Cl has not been measured). They differ significantly from those of CF_3SeF_3 and SeF_4, which are similar to each other. The coupling constant $J(F_3C-^{77}Se)$ for $CF_3Se(VI)$ compounds is very high compared with values for $CF_3Se(I, II, and IV)$ molecules and therefore characteristic.

Hydrolysis of neither CF_3SeF_5 nor CF_3SeF_4Cl gave CF_3SeO_3H, but oxidation of CF_3SeO_2H with a neutral saturated $KMnO_4$ solution led to good yields of CF_3SeO_3K. Treating the salt with 70 proc. $HClO_4$ in aqueous solution gave the acid CF_3SeO_3H. Attempts to isolate the acid failed. Only concentrations up to 90% were obtained when a solution of CF_3SeO_3H in H_2O was evaporated *in vacuo*. Above this value decomposition to CF_4, COF_2, $2SeO_2$, and H_2O took place. This suggests that $[H_3O]^+[CF_3SeO_3]^-$ is formed but cannot be freed from water. Neutralization reactions of CF_3SeO_3H with Ag_2O or NH_3 resulted in the formation of the salts CF_3SeO_3M (M = Ag, NH_4) (*72*).

B. EQUIVALENCE BETWEEN $(CF_3S)_2N-$ AND FLUORINE

The bis(trifluoromethylthio)amino radical must be considered as a second-order derivative parafluorine and its high group electronegativity of 3.7 (*47*) supports this classification. The hydrogen derivative $(CF_3S)_2NH$ is a weak acid—pK_D (in dioxan/water) = 9.99 (*48*)—and forms with HgO the compound $Hg[N(SCF_3)_2]_2$ (*49*). Similarly to F_2NNF_2 (*50*), O_2NNO_2 (*51*), and $(CF_3S)_3CC(SCF_3)_3$ (*52*), its dimer $(CF_3S)_2NN(SCF_3)_2$ dissociates at 20°C like a halogen (*53*) according to

$$(CF_3S)_2NN(SCF_3)_2 \rightleftharpoons 2(CF_3S)_2N\cdot$$

In the ^{1}H-NMR spectrum of $(CF_3S)_2NCH_2N(SCF_3)_2$, prepared from $(CF_3S)_2NH$ and HC(O)H at 0°C (0.5 hr) in the presence of 100% H_2SO_4, $\delta(CH_2)$ = 5.15 ppm (*54*) and compares well with $\delta(CH_2)$ = 5.99 ppm in CH_2F_2. In boron trihalides $(CF_3S)_2N\cdot$ is able to replace halogens, forming $[(CF_3S)_2N]_nBX_{3-n}$. Only monosubstitution to $[(CF_3S)_2N]BX_2$ is observed when BCl_3 or BBr_3 reacts with $(CF_3S)_2NH$. The corresponding $(CF_3S)_2NBF_2$ can only be made by thermal decomposition of $(CF_3S)_2NB(SCF_3)_2$ at 40°C. It is not very stable and above 60°C gives off BF_3, forming $[(CF_3S)_2N]_2BF$ (*59*). Further substitution of BX_3 (X = Cl, Br) is only achievable in the reaction of stoichiometrical amounts of $(CF_3S)_2NBX_2$ and $Hg[N(SCF_3)_2]_2$, giving $[(CF_3S)_2N]_2BX$. A complete exchange takes place when excess $Hg[N(SCF_3)_2]_2$ reacts with BBr_3, giving $B[N(SCF_3)_2]_3$ in 93% yield. When BCl_3 is used instead of BBr_3 the yield drops to 55% at 80°C. Completely unexpected is the formation of the trisaminoborane from $B(SCF_3)_3$ and the mercurial according to

$$3Hg[N(SCF_3)_2]_2 + 2B(SCF_3)_3 \longrightarrow 2B[N(SCF_3)_2]_3 + 3Hg(SCF_3)_2$$

In order to understand the reactions of $(CF_3S)_2NH$ and $Hg[N(SCF_3)_2]_3$ with BX_3 (X = F, Cl, Br, SCF_3) it is necessary to consider the radicals CF_3S as parachlorine (p-Cl) and $(CF_3S)_2N$ as parafluorine (p-F) according to the element displacement principle. With BF_3 and H(p-F) [= $(CF_3S)_2NH$] and $Hg(p-F)_2$ ($Hg[N(SCF_3)_2]_2$) no reaction takes place. Similarly no metathetical exchange occurs between BF_3 and HF or HgF_2. But BCl_3 and BBr_3 react with H(p-F) to give monosubstituted boranes in agreement with HF fluorinations of BX_3 (X = Cl, Br). The only difference between (p-F)BX_2 and FBX_2 is that the former are more stable than the mixed boron halides and the reaction with H(p-F) stops after monosubstitution. When (p-F)$_2$Hg is used, then, according to reaction conditions, mono-, di-, and trisubstituted boranes are formed with BX_3. In addition, the reaction between $Hg[N(SCF_3)_2]_2$ and $B(SCF_3)_3$ is in agreement with the preparation of BF_3 from BCl_3 and HgF_2, according to

$$2B(p\text{-}Cl)_3 + 3Hg(p\text{-}F)_2 \longrightarrow 2B(p\text{-}F)_3 + 3Hg(p\text{-}Cl)_2$$

$$2BCl_3 + 3HgF_2 \longrightarrow 2BF_3 + 3HgCl_2$$

Decreasing yields going from BBr_3, $B(SCF_3)_3$, to BCl_3 at 25°C are in good agreement with known reactivities of BX_3 and also show that CF_3S (p-Cl) is more reactive than Cl in this metathesis (60).

C. EQUIVALENCE BETWEEN CF_3S AND CHLORINE

Besides the similarities between CF_3S and chlorine already given in footnote 1, new examples have been found to substantiate the parachlorine behavior of the CF_3S radical.

1. With the synthesis of $CF_3SP(O)Cl_2$ and $CF_3SP(S)Cl_2$, two new compounds are available which can be derived from the basic substances $OPCl_3$ and $SPCl_3$ by replacing chlorine by the parachlorine CF_3S. This can be confirmed by physical data given in Table IV and chemical reactions.

Partial hydrolysis of $CF_3SP(O)Cl_2$ with stoichiometric amounts of H_2O in CH_3CN leads to $CF_3SP(O)(OH)_2$, which on standing or with-

TABLE IV

Physical Data of $CF_3SP(X)Cl_2$ and $ClP(X)Cl_2$ (X = O, S)

Substance	Properties	D (g/cm³); (°C)	bp (°C)	δ(P) (ppm)	n_D (°C)	mp (°C)	$\Delta H v$ (kJ/mol)	Trouton constant (J/kmol)
$CF_3SP(O)Cl_2$	Colorless, highly refracting	1.72 (20)	122	14.27	1.4322 (20) 1.4309 (25)	−65 to −63	49.99	114.22
$ClP(O)Cl_2$ (5)	Colorless, highly refracting	1.73 (20) 1.645 (25)	105.1	2.2	1.460 (25)	1.25	33.7	80.1
$CF_3SP(S)Cl_2$	Colorless	1.933 (25)	135	33.38	1.5009 (20)	−68 to −65	40.69	99.8
$ClP(S)Cl_2$ (6)	Colorless	1.6271 (25)	125	28.8	1.563 (20)	−35	32.70	82.13

drawing of the solvent condenses to

$$CF_3SP(O)P(O)SCF_3$$
$$||$$
$$OHOH$$

and then to oligomers of the formula

$$CF_3S-\underset{\underset{OH}{|}}{\overset{\overset{O}{\|}}{P}}O-\left[-\underset{\underset{SCF_3}{|}}{\overset{\overset{O}{\|}}{P}}-O-\right]_n -\underset{\underset{OH}{|}}{\overset{\overset{O}{\|}}{P}}-SCF_3 \quad (40)$$

Hydrolysis of $OPCl_3$ with 1 mol H_2O gives $HOP(O)Cl_2$, which can be isolated in 90% purity. At 300°C/12 Torr it condenses to $Cl_2P(O)OP(O)Cl_2$, which forms at 100°C/760 Torr $P_3Cl_6O_3$ (41). Like $SPCl_3$, $CF_3SP(S)Cl_2$ is stable toward excess H_2O and does not react with it in a homogeneous phase at 20°C during 1 hr (40).

2. The hydrolysis of per(fluorohalogeno)organosulfenylchlorides is still an unresolved problem and is also a good example for an element–paraelement comparison.

With the series $CF_nCl_{3-n}SCl$ (n = 0, 1, 2, 3) there are compounds available, from which we know that, e.g., the CF_3S group behaves very similarly to chlorine (footnote 1). Its hydrolysis was studied in 1955 by Haszeldine and Kidd (42) and found to take place according to

$$CF_3SCl \xrightarrow[-HCl]{H_2O} [CF_3SOH] \longrightarrow CF_3SSCF_3 + CF_3SO_2SCF_3$$

The postulated intermediate CF_3SOH was not detected.

With the preparation of $CF_3S(O)SCF_3$ by controlled hydrolysis of $CF_3SF_2SCF_3$ and its characterization, it was possible to detect $CF_3S(O)SCF_3$ as an intermediate in the reaction of CF_3SCl and water (43). So the reaction scheme is now more complete:

$$CF_3SCl \xrightarrow[-HCl]{H_2O} [CF_3SOH] \xrightarrow[-HCl]{CF_3SCl} CF_3S(O)SCF_3 \longrightarrow CF_3SSCF_3 + CF_3SO_2SCF_3$$

The other product of the series, CCl_3SCl, hydrolyzes in a completely different manner, forming $ClC(O)SCl$ according to

$$CCl_3SCl \xrightarrow[-HCl]{H_2O} [Cl_3CSOH] \xrightarrow[-HCl]{} Cl_2CSO \longrightarrow \left[Cl_2C\begin{smallmatrix}S\\||\\O\end{smallmatrix}\right] \longrightarrow ClC(O)SCl$$

The relatively stable intermediate $Cl_2C{=}S{=}O$ was isolated (44). The main difference between these two reactions is the HCl elimination from the postulated X_3CSOH. If $X = F$ an intermolecular reaction takes place and for $X = Cl$ an intramolecular elimination occurs.

In this connection it was interesting to study the hydrolysis of CF_2ClSCl and $CFCl_2SCl$. Since both sulfenylchlorides are insoluble in water, hydrolysis was always a two-phase reaction. The following substances are obtained at 20°C:

$$F_2ClCSCl \xrightarrow{H_2O} F_2ClCS(O)SCF_2Cl + F_2ClCSO_2SCF_2Cl$$

$$FCl_2CSCl \xrightarrow{H_2O} FCl_2CS(O)SCFCl_2 + FCl_2CSO_2SCFCl_2$$

The volatile products consisted of HCl, H_2S, CO_2, and COS and the aqueous phase contained HCl, HF, and H_2SO_4. No $CF_nCl_{3-n}SSCF_nCl_{3-n}$ ($n = 1, 2$) is formed in the first 15 hr; but later formation of some disulfide is observed. The products of the hydrolysis were identified by ^{19}F- and ^{13}C-NMR-spectroscopy (see Table V). Neither F_2CSO nor FClCSO appeared, so intramolecular HCl elimination can be excluded. This proves that $F_nCl_{3-n}CSCl$ ($n = 1, 2$) resembles in this respect CF_3SCl and not CCl_3SCl. Two surprising observations can be made here. They are (1) the relative stability of $CF_nCl_{3-n}S(O)SCF_nCl_{3-n}$ ($n = 1, 2$) and (2) the absence of the corresponding disulfides.

For a better understanding of the reaction paths the observed products $CF_nCl_{3-n}S(O)SCF_nCl_{3-n}$ and $CF_nCl_{3-n}SO_2SCF_nCl_{3-n}$ ($n = 1, 2$)

TABLE V

^{19}F- AND ^{13}C-NMR DATA[a]

(a)	(b)	F(a)	F(b)	C(a)	C(b)	$^5J_{(F-F)}$	$^1J(^{13}C-^{19}F)$	
							(a)	(b)
CFCl$_2$S–	–S(O)CFCl$_2$	–13.6	–54.9	118.4	125.5	—	336.3	344.0
CF$_2$ClS–	–S(O)CF$_2$Cl	–20.1[b]	–56.0[b]	129.4	131.1	—	327.9	345.5
CF$_3$S–	–S(O)CF$_3$	–34.2	–69.5	129.7	125.6	1.5	311.3	336.9 (43)
CFCl$_2$S–	–S(O)$_2$CFCl$_2$	–21.4	–58.9	117.3	123.8	17.8	346.1	343.5
CF$_2$ClS–	–S(O)$_2$CF$_2$Cl	–24.9	–61.0	127.3	126.7	9.8	335.9	339.1
CF$_3$S–	–S(O)$_2$CF$_3$	–36.3	–76.8 (55)	122.9	116.0 (43)	5.0 (55)	314.9	327.1 (43)
CFCl$_2$–	–SO$_2$H	–65.3						
CF$_2$Cl–	–SO$_2$H	–66.4						

[a] δ in parts per million, J in Hertz.
[b] ABXY spectrum; coupling constants in the range of 1 Hz.

were synthesized by special methods on a preparative scale and used for the following reactions:

$$CF_nCl_{3-n}\overset{\overset{O}{\|}}{S}-SCF_nCl_{3-n} \xrightarrow{HCl/H_2O} CF_nCl_{3-n}SCl + CF_nCl_{3-n}SO_2H$$

$$\xrightarrow{HCl\ (gas)} CF_nCl_{3-n}SCl + H_2O$$

$$CF_nCl_{3-n}\overset{\overset{O}{\|}}{\underset{\underset{O}{\|}}{S}}-SCF_nCl_{3-n} \xrightarrow{H_2O} CF_nCl_{3-n}SO_2H + [CF_nCl_{3-n}SOH]$$

The conditions are the same as those of the hydrolysis reactions. The sulfinic acids were identified in the aqueous layer by ^{19}F-NMR-spectroscopy (see Table V).

With this additional information, the picture of the hydrolysis of $CF_nCl_{3-n}SCl$ (n = 1, 2, 3) becomes more complete and a qualitative description of the reaction paths can be provided.

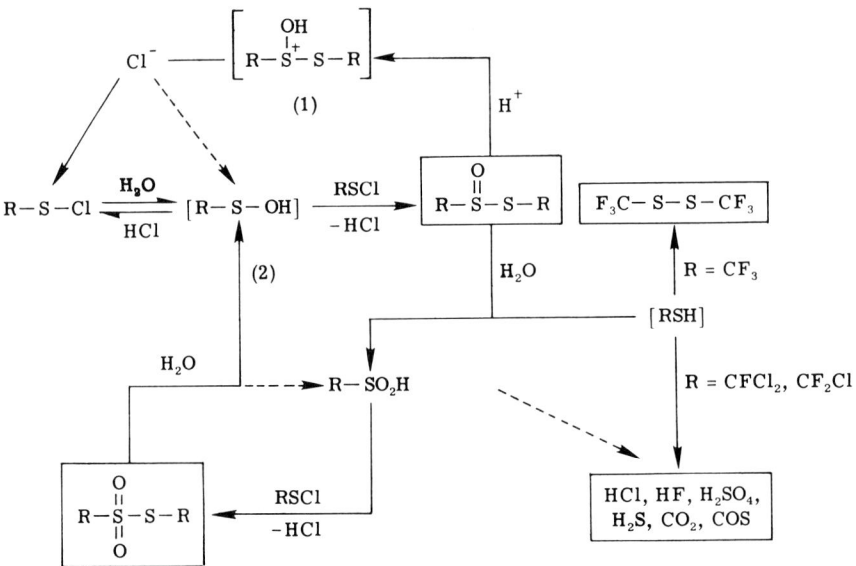

From this scheme it can be deduced that RSSR is not a primary product, as the disulfides are stable to water and HCl under the conditions studied. The precursor needed is RSH, which only with R = CF_3 reacts with CF_3SCl or CF_3SOH to form CF_3SSCF_3. For R′ = CF_2Cl and

$CFCl_2$, decomposition of R'SH is faster than condensation with R'SCl or R'SOH. Therefore no R'SSR' is observed. Two reaction cycles, (1) and (2), maintain the equilibria found. The scheme below provides possible intermediates and their reactions.

$$RSCl \underset{}{\overset{H_2O}{\rightleftharpoons}} [RSOH] \xrightarrow{RSCl} \begin{bmatrix} R-S\!-\!\overset{\frown}{O-H} \\ \downarrow \\ R-S\!-\!Cl \end{bmatrix}_A \xrightarrow{-HCl} R-\underset{}{\overset{O}{\underset{\|}{S}}}\!-\!S\!-\!R \xrightarrow{HOH}$$

$$R-\underset{O}{\overset{O}{\underset{\|}{S}}}\!-\!S-R \xleftarrow{-HCl} \begin{bmatrix} R-\overset{O}{\underset{\|}{S}}-\overset{\frown}{O-H} \\ \downarrow \\ R-S-Cl \end{bmatrix}_C \xleftarrow{RSCl} R-\overset{O}{\underset{\|}{S}}-OH \xleftarrow{-RSH} \begin{bmatrix} R-\overset{O}{\underset{\|}{S}}\!\!\!\overset{S-R}{\underset{\uparrow}{}} \\ H\overset{O}{\diagup}\!\diagdown H \end{bmatrix}_B$$

The first step is the formation of RSOH, which is attacked electrophilically by excess RSCl, forming an intermediate A. Stabilization takes place by giving off HCl and producing RS(O)SR. The electrophilic center in the thiosulfinate is S(IV), which on hydrolysis forms RS(O)OH and RSH via intermediate B. Condensation of RS(O)OH with RSCl gives, via C, RSO_2SR. This proposed mechanism is in good agreement with reactions of organic sulfenic acids and organosulfenylhalides (46, 58). The system RSCl and H_2O is sensitive toward reactions conditions, e.g., the amount of H_2O (concentration of H^+), temperature, and reaction time (57). According to the element displacement principle RS is a first-derivative parachlorine. Therefore the hydrolysis model mentioned can be transferred—with reservation—to the reaction of Cl_2 plus H_2O. An overall picture is gained by looking at the following scheme.

$$Cl-Cl \underset{HCl}{\overset{H_2O}{\rightleftharpoons}} ClOH \xrightarrow{Cl-Cl} \begin{bmatrix} Cl-\overset{\frown}{O-H} \\ \downarrow \\ Cl-Cl \end{bmatrix}_{A'} \xrightarrow{-HCl} \begin{bmatrix} \overset{O}{\underset{\|}{Cl}}-Cl \end{bmatrix} \xrightarrow{HOH} \begin{bmatrix} \overset{O}{\underset{\|}{Cl}}\!\!\!\overset{Cl}{\underset{\uparrow}{}} \\ H\overset{O}{\diagup}\!\diagdown H \end{bmatrix}_{B'}$$

$$\downarrow -HCl$$

$$\underset{O}{\overset{O}{\underset{\|}{Cl}}}-OH \xleftarrow[-HCl]{H_2O} \begin{bmatrix} \overset{O}{\underset{\|}{Cl}}-Cl \\ \overset{\|}{O} \end{bmatrix} \xleftarrow{-HCl} \begin{bmatrix} \overset{O}{\underset{\|}{Cl}}-\overset{\frown}{O-H} \\ \downarrow \\ Cl-Cl \end{bmatrix}_{C'} \xleftarrow{Cl-Cl} HOClO$$

The reaction steps are obtained by replacing RS by Cl. The hydrolysis of chlorine is only complete in basic media. In H_2O the equilibrium $Cl_2 + H_2O \rightleftharpoons HCl + HOCl$ is almost completely on the left side. The first step is the formation of HOCl, which is attacked by the electrophile Cl_2 to form A'. By elimination of HCl an unstable intermediate [Cl(O)Cl] is formed, which reacts with H_2O via B' to HOClO. This mechanism is repeated and via C' [Cl(O$_2$)Cl] and its hydrolysate $HOClO_2$ is formed. Unfortunately, little is known about the hydrolysis of chlorine. Only the formation of HOCl has been well investigated and an equilibrium constant of 6×10^{-4} was measured at 20°C (45). The disproportionation of HOCl to $HClO_3$ and HCl is not significant at room temperature but its rate increases with higher temperature. For the reaction $3H_2O + 3Cl_2 \rightleftharpoons HClO_3 + 5HCl$, $K = [H^+]^6[Cl^-]^5[ClO_3^-][Cl_2]^{-3} = 4.3 \times 10^{-7}$ at 91°C (56).

Transferring the results, obtained from the reactions of RSCl with H_2O, from parachlorine to chlorine we get a suggestive mechanism which remains speculative till further results on this subject are available. One piece of support for this mechanism can already be provided. Kinetic studies on the reaction $Cl_2 + H_2O \rightarrow H^+ + Cl^- + HClO$ show that the most probable first step is the nucleophilic addition of H_2O at the electrophile atom in Cl_2, forming the intermediate

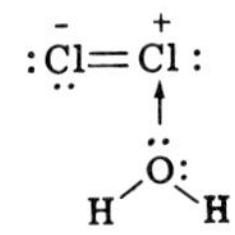

$$:\overset{-}{\underset{..}{Cl}}=\overset{+}{Cl}:$$
$$\uparrow$$
$$\underset{H \quad H}{\overset{..}{O}:}$$

which on losing H^+ forms Cl_2OH^-, which stabilizes to HOCl and Cl^- (45) according to

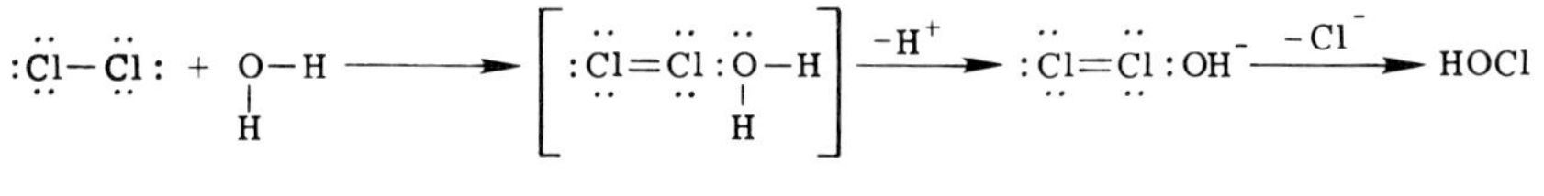

$$:\overset{..}{\underset{..}{Cl}}-\overset{..}{\underset{..}{Cl}}: + \underset{H}{\overset{|}{O}}-H \longrightarrow \left[:\overset{..}{\underset{..}{Cl}}=\overset{..}{\underset{..}{Cl}}:\overset{..}{\underset{H}{O}}-H\right] \xrightarrow{-H^+} :\overset{..}{\underset{..}{Cl}}=\overset{..}{\underset{..}{Cl}}:OH^- \xrightarrow{-Cl^-} HOCl$$

Although this step has not been discussed in the hydrolysis of RSCl, the first intermediates $[:\overset{..}{\underset{..}{Cl}}=\overset{..}{\underset{..}{Cl}}:\overset{..}{O}:H_2]$ and $[:\overset{..}{\underset{..}{Cl}}=\overset{..}{\underset{..}{Cl}}OH]^-$ are similar to B and B' and also their decompositions are comparable with the mechanism given above.

D. EQUIVALENCE BETWEEN TETRAKIS(TRIFLUOROMETHYLTHIO)PYRROLYL— AND CHLORINE

Pseudohalogens according to Grimm (*1*) (OH, NH$_2$, CH$_3$), to Birkenbach and Kellermann (*2*) (CN, SCN, NCS, OCN, NCO, N$_3$ . . .), and to Lagowski (*3*) (CF$_3$, CF$_3$S) can be considered with the aid of the element displacement principle as a unique, causative, and continuous class of molecule radicals that originate in the periodic system. The tetrakis(trifluoromethylthio)pyrrolle radical reacts in spite of its size, complexity, and derivation as a parahalogen. It dimerizes according to Eq. (1) at 20°C reversibly to 2,2′,3,3′,4,4′,5,5′-octakis(trifluoro-

$$\text{(1)}$$

methylthio)-2,2′-bi-2*H*-pyrrolle (*81*). The hydrogen derivative 2,3,4,5-tetrakis(trifluoromethylthio)pyrrole is an acid (pK_D = 9.2 in water/dioxane) and its proton can be replaced by various cations according to Eq. (2) (*82*). The radical is made by reacting the silver salt

$$\text{(2)}$$

[M = Ag (*83*), K, Na, NH$_4$, C$_6$H$_5$NH, (CH$_3$)$_3$NH (*82*), Zn, CuI, (C$_6$H$_5$)$_3$PC$_2$H$_5$, (C$_6$H$_5$)$_4$As, (C$_6$H$_5$)$_4$Sb, (CH$_3$)$_3$S, (C$_2$H$_5$)$_3$Se, and (C$_6$H$_5$)$_2$J (*84*)]

with iodine in pentane [Eq. (1)] or by oxidizing the tetrasubstituted pyrrolle by PbO$_2$ or nickel peroxide in C$_6$H$_6$ or C$_6$F$_6$ (*81*). Among the many salts synthesized so far the silver and potassium salts are very good starting materials for N-substitutions. They react with S$_x$Cl$_2$ or RSCl, forming the corresponding derivatives according to Eq. (3). Among these, pentakis(trifluoromethylthio)pyrrole was used to study its sulfenylating properties and to compare them with CF$_3$SCl. Compounds containing an acidic hydrogen such as primary, secondary, and

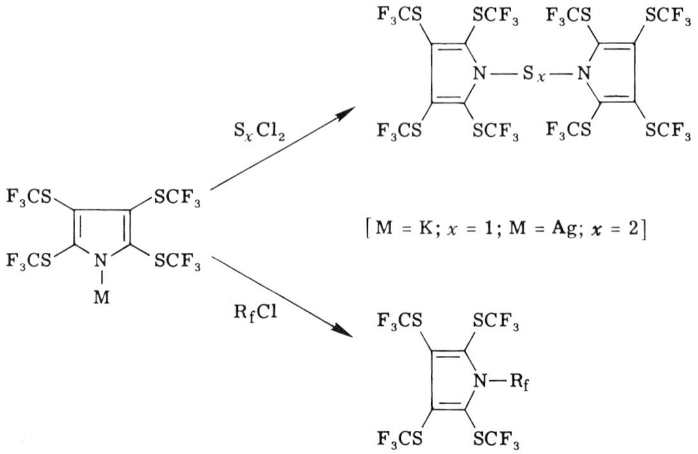

<img placeholder for reaction scheme 3>

[M = K; x = 1; M = Ag; x = 2]

(3)

[M = Ag, R_f = CF_3S, CF_2ClS, $ClC(O)S$, CF_3Se;
M = Ag or K, R_f = $CFCl_2S$, CCl_3S; M = K,
R_f = $(CF_3S)_2CClS$]

tertiary alcohols, thiols, and secondary amines react with $(CF_3SC)_4NSCF_3$, substituting H for CF_3S according to Eq. (4). It is

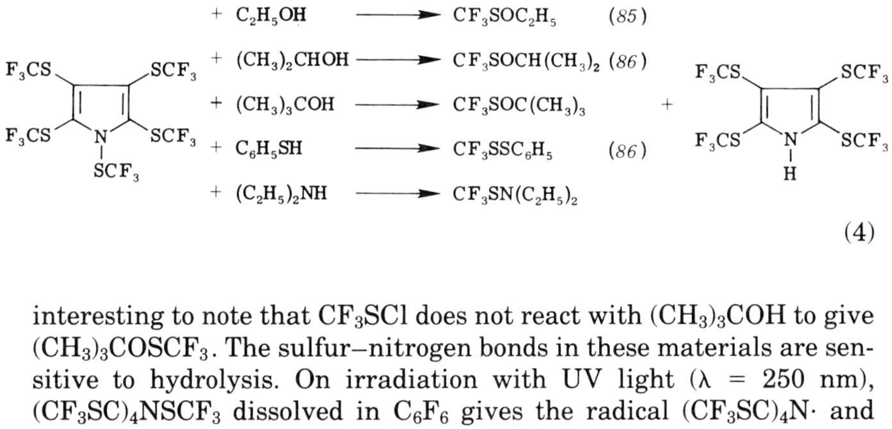

$+ C_2H_5OH \longrightarrow CF_3SOC_2H_5$ (85)

$+ (CH_3)_2CHOH \longrightarrow CF_3SOCH(CH_3)_2$ (86)

$+ (CH_3)_3COH \longrightarrow CF_3SOC(CH_3)_3$

$+ C_6H_5SH \longrightarrow CF_3SSC_6H_5$ (86)

$+ (C_2H_5)_2NH \longrightarrow CF_3SN(C_2H_5)_2$

(4)

interesting to note that CF_3SCl does not react with $(CH_3)_3COH$ to give $(CH_3)_3COSCF_3$. The sulfur–nitrogen bonds in these materials are sensitive to hydrolysis. On irradiation with UV light (λ = 250 nm), $(CF_3SC)_4NSCF_3$ dissolved in C_6F_6 gives the radical $(CF_3SC)_4N\cdot$ and CF_3SSCF_3 (82). When the radical dimerizes, surprisingly not the N—N but the C—C linked dimer is obtained. This could be proved by an X-ray structure analysis of both isomers. On heating the C—C dimer to 120–130°C, rearrangement to the N—N dimer takes place according to Eq. (5). The structure of both compounds is shown in Figs. 1 and 2. Surprisingly, the N—N dimer is rather stable and melts at 122–123°C without dissociation (81). Methylation of $(CF_3SC)_4NH$ succeeds with

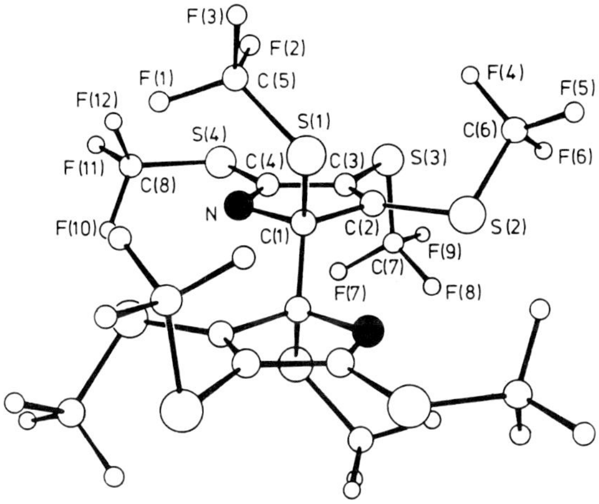

FIG. 1. A C—C linked dimer.

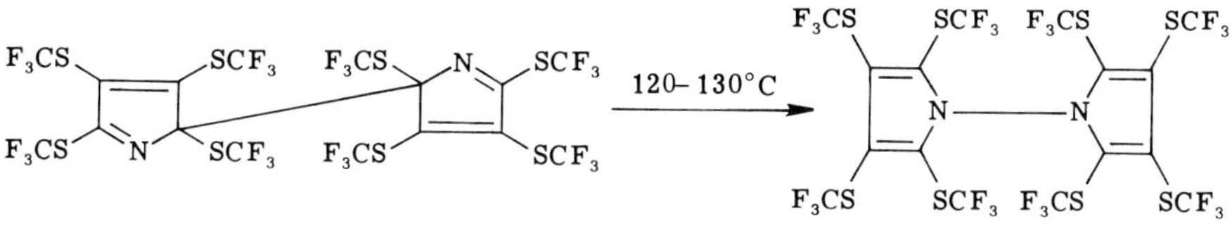

CH$_3$I in a two-phase reaction catalyzed by strong bases, e.g., KOH (87). After many unsuccessful attempts to chlorinate tetrakis(trifluoromethylthio)pyrrolle or its salts with Cl$_2$ or other chlorinating reagents, it was finally possible to synthesize the wanted material. When (CF$_3$S—C)$_4$NH is treated with NaOCl or NaOBr the corresponding N—Cl or N—Br substituted compounds are obtained in good yield. By metathetical reaction (CF$_3$SC)$_4$NCl is converted by NaF in the presence of Sb$_2$O$_3$ to (CF$_3$SC)$_4$NF. Similarly, by using KCN or NaSCN the desired N-substitution takes place. The reactions can be illustrated by Eq. (6).

Unexpectedly, (CF$_3$SC)$_4$NCl, in the few cases studied so far, does not act as a chlorinating agent. Under the influence of UV light (24 hr) with toluene, (CF$_3$SC)$_4$N · CH$_2$ · C$_6$H$_5$ and HCl are formed. In boiling CCl$_4$ it reacts with C$_6$H$_5$COOH analogously to form (CF$_3$SC)$_4$-NC(O)C$_6$H$_5$, which can also be made from the potassium salt and C$_6$H$_5$C(O)Cl. These reactions show that the N—Cl bond is unexpectedly polarized to $\overset{\delta-}{N}$—$\overset{\delta+}{Cl}$ and therefore it acts as a pyrrolating agent.

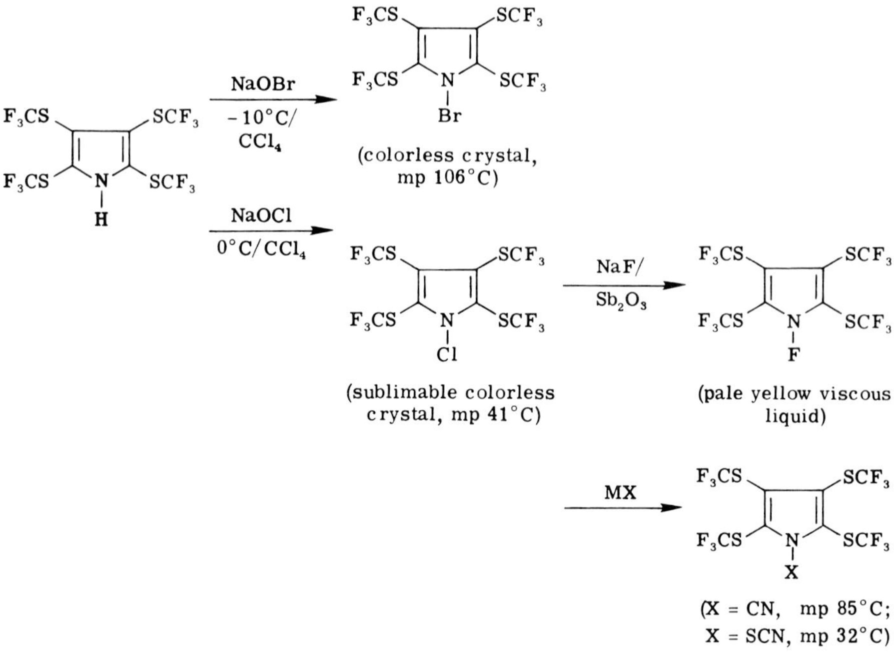
NaOBr
-10°C/
CCl4
NaOCl
0°C/CCl4
NaF/
Sb2O3
MX
(colorless crystal,
mp 106°C)
(sublimable colorless
crystal, mp 41°C)
(pale yellow viscous
liquid)
(X = CN, mp 85°C;
X = SCN, mp 32°C)

(6)

The N—Br derivative is less stable and decomposes slowly at 20°C in Br_2 and the C—C dimer (*88*). In order to understand this unexpected bond polarization it was necessary to determine the group electronegativity of the radical. This can be done by applying the Kagarise (*112*) equation $\nu(C{=}O)cm^{-1} = 1536.5 + 48.85\ (\kappa_x + \kappa_y)$ for molecules of the type XC(O)Y 89, provided a compound like $(CF_3SC)_4NC(O)F$ can be prepared. It was made by reacting the potassium salt with FC(O)Cl dissolved in ether at 20°C (3 hr) in good yields, according to Eq. (7), and showed $\nu(C{=}O)$ in the IR spectrum at 1872 cm^{-1}. The group electronegativity for the tetrakis(trifluoromethylthio)pyrrolyl radical was estimated to 2.90, which lies between chlorine (3.0) and the CF_3S group (2.7). This value accounts very well for the observed polarization and reactivity of $(CF_3SC)_4NCl$. The close relation between the radical and chlorine can also be demonstrated in an additional case. In an aqueous two-phase system $(CF_3SC)_4NH$ reacts with CH_2I_2 in the presence of concentrated NaOH to 1,1′-bis[tetrakis(trifluoromethylthio)pyrrolyl]-methylene according to Eq. (7). It shows in the ^1H-NMR-spectrum

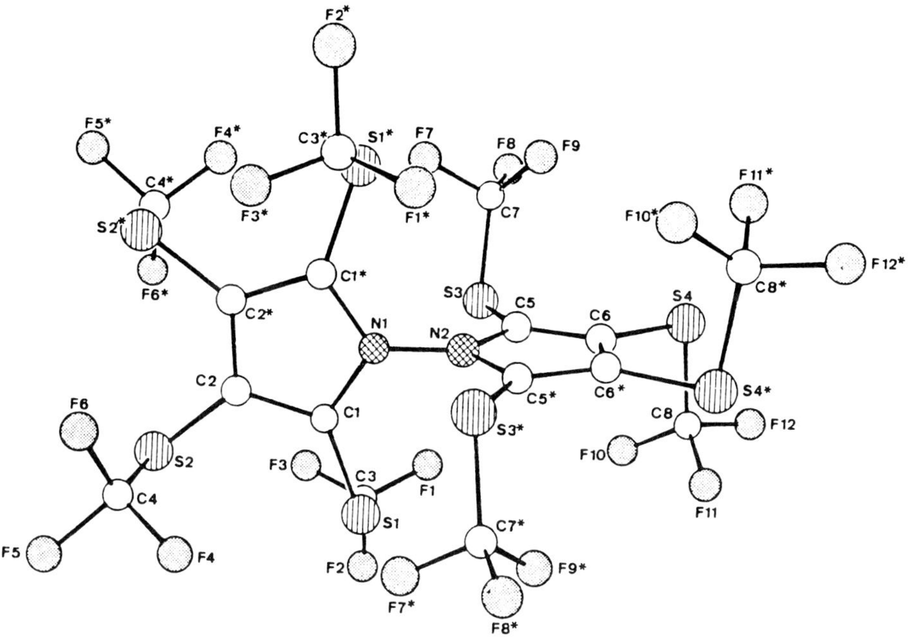

THE ELEMENT DISPLACEMENT PRINCIPLE 191

FIG. 2. An N—N linked dimer.

$\delta(CH_2) = 6.9$ ppm (84). Such unusually high chemical shifts for methylene protons are also observed in dihalogenomethanes, e.g., $\delta(CH_2)$ in $CH_2Cl_2 = 5.3$ ppm.

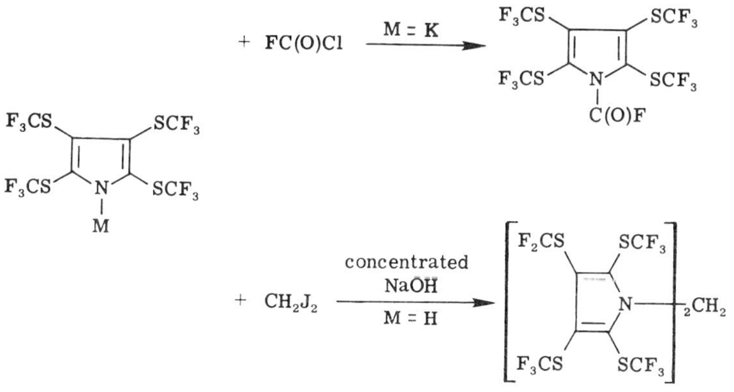

The tetrakis(trifluoromethylthio)pyrrolyl radical replaces halogen in PX_3 (X = Cl, Br, I). Monosubstitution is observed when stoichiometric amounts of the pyrrolle or its salts are reacted with PX_3.

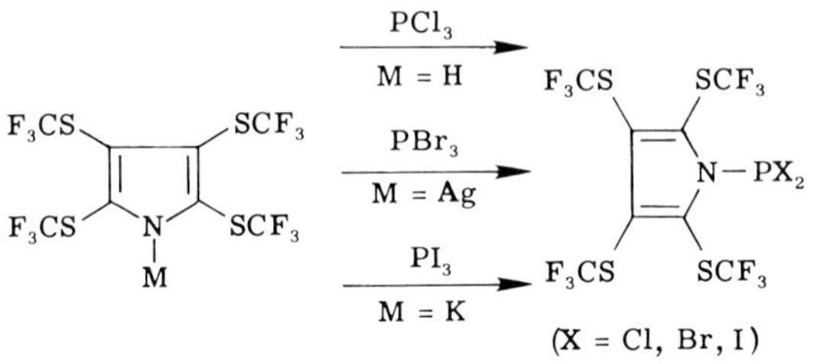

When the dichlorophosphane is treated with the potassium salt in a 1:1 molar reaction disubstitution and with an excess a complete substitution is observed. It was, however, not possible to convert PCl_3 either by the silver or by the potassium salt to the trisubstituted phosphane at once. But the reaction between the potassium salt and $AsCl_3$ or $SbCl_3$ goes straight through to a complete replacement. Formation of intermediates is not observed. Complete substitution cannot be achieved with phosphorus(V) halides. If $P(E)Cl_3$ or $C_6H_5PF_4$ is treated with the silver salt only one halogen is replaced. These reactions are shown below

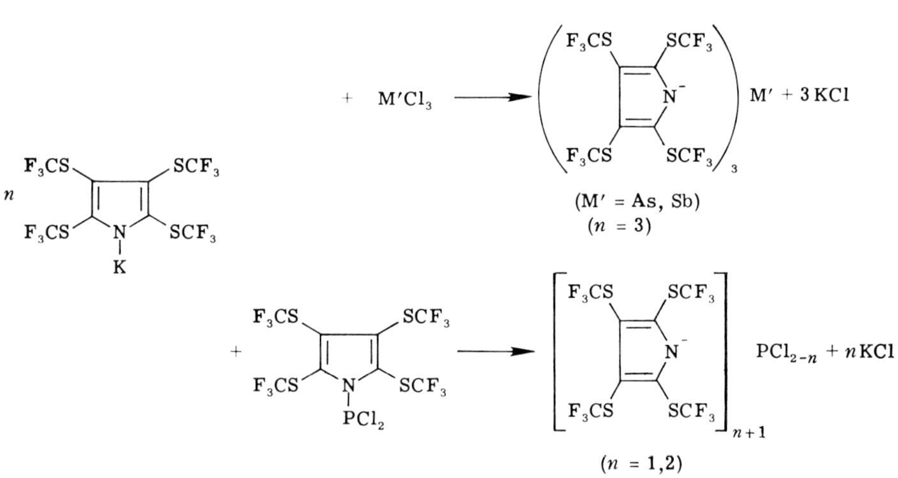

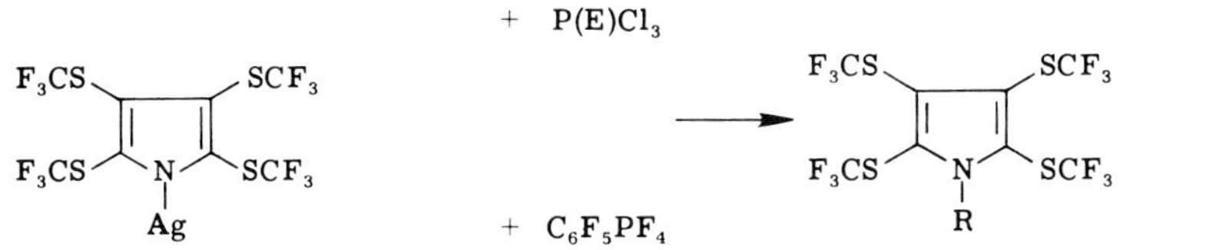

[Scheme: reaction of silver pyrrole derivative with P(E)Cl$_3$ giving N-substituted pyrrole + C$_6$F$_5$PF$_4$]

[R = P(O)Cl$_2$, P(S)Cl$_2$, P(C$_6$H$_5$)F$_3$]

All these substances are very sensitive to hydrolysis (*89*). Functionalization at the nitrogen also takes place in the reaction of the silver salt with ClSO$_2$NCO. The sulfonyl isocyanate obtained does not show typical isocyanate reactions. It solvolyzes with H$_2$O or CH$_3$OH without forming intermediates to (CF$_3$SC)$_4$NH. However, with moist air partial hydrolysis leads to the corresponding sulfonamide according to Eq. (8). A number of acyl derivatives were synthesized via Eq. (9). The oxalyl

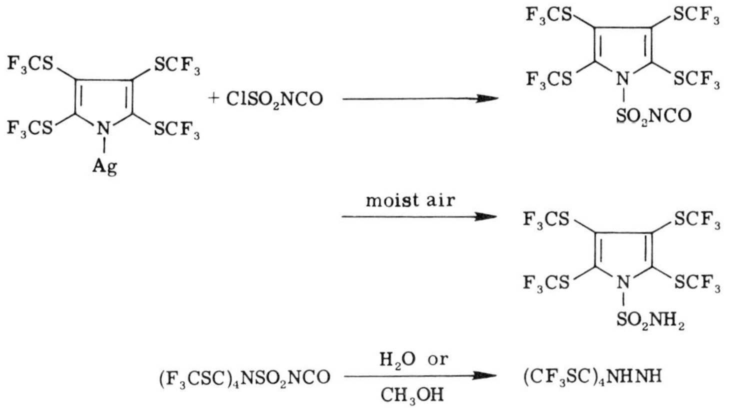

[Reaction (8): silver pyrrole + ClSO$_2$NCO → N-SO$_2$NCO pyrrole; moist air → N-SO$_2$NH$_2$ pyrrole]

(8)

$$(F_3CSC)_4NSO_2NCO \xrightarrow{H_2O \text{ or } CH_3OH} (CF_3SC)_4NHNH$$

[Reaction (9): M-substituted pyrrole + RC(O)X → N-C(O)R pyrrole + MX]

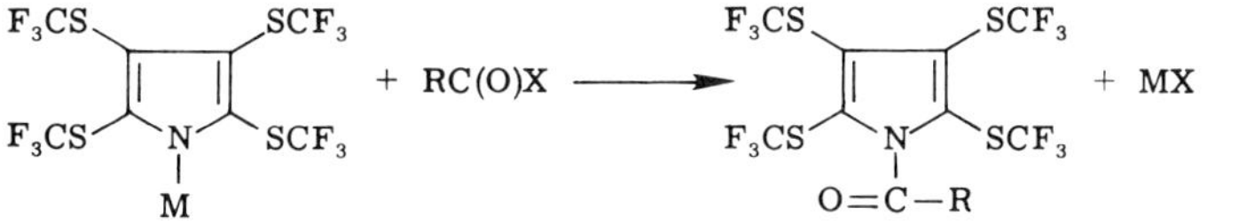

(9)

M = Ag, X = Cl, R = C$_6$H$_5$
M = K, X = Cl, R = CH$_3$, CCl$_3$,
M = K, X = OC(O)CF$_3$, R = CF$_3$

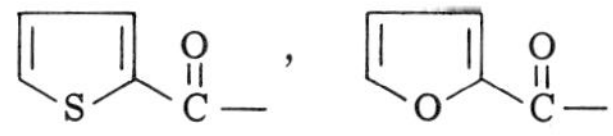

[structures of thiophene-2-carbonyl and furan-2-carbonyl groups]

$$2\,(CF_3SC)_4NK + ClC(O)C(O)Cl \longrightarrow (CF_3SC)_4N-\underset{\underset{O}{\|}}{C}-\underset{\underset{O}{\|}}{C}-N(CSCF_3)_4 + 2\,KCl$$

derivative $(CF_3SC)_4NC(O)C(O)N(CSCF_3)_4$ is very stable toward UV light (λ = 254 nm). It remains unaffected even after several days of irradiation (91). The manifold reactivities of this halogen-like parachlorine and its influence as a radical on other functional groups are so far not fully elucidated. Additional interesting results can be expected in this field of chemistry.

V. Interchangeability of Elements and Paraelements in Known Structures

The structures of known compounds made up entirely of elements remain unaltered if the elements are substituted partially or totally by pseudo- or paraelements.

A. Transition from Hexazabenzene (—N=N—)$_3$ to Borazole ($-\overset{\delta+}{N}H-\overset{\delta-}{B}H-$)$_3$

The planar aromatic benzene molecule may be thought of as a resonance-stabilized trimeric pseudonitrogen molecule (HC—CH)$_3$. Hexazabenzene (—N=N—)$_3$, which is the associated analogous parent compound containing a single element, is stable only under extreme conditions (110). If, however, alternate N atoms or all the N atoms in (—N=N—)$_3$ are replaced by a paranitrogen such as, for example, FC, ClC, BrC, IC, CF$_3$C, etc. (= X), the following series is obtained.

$$(-N=N-)_3 \longrightarrow \begin{array}{c} N=N \\ XC \quad CX \\ N-N \end{array} \longrightarrow (X-C=N)_3 \begin{array}{c} \nearrow (-N=N-)(-\underset{X}{C}=\underset{X}{C}-)_2 \quad \text{1,2-Diazine} \\ \searrow (-\underset{X}{C}=N-)_2(-\underset{X}{C}=\underset{X}{C}-) \quad \begin{array}{c}\text{1,3-Diazine}\\ \text{1,4-Diazine}\end{array} \end{array}$$

Hexazabenzene → Tetrazine (90) → Triazine

$(XC=CX)_3 \longleftarrow (-\underset{X}{C}=\underset{X}{C}-)_2(-\underset{X}{C}=N)$

Benzene ← Pyridine

By partial substitution of C—X by BH_2 (pseudonitrogen) in triazine $(\cdot \underset{\cdot\cdot}{N}=BH_2)_3$ is first produced and this is stabilized by resonance through rearrangement to borazol $(\cdot \overset{\delta+}{NH}-\overset{\delta-}{BH})_3$, which has polarized B—N bonds.

B. EXCHANGEABILITY WITHIN ADAMANTANE STRUCTURES

Phosphorus, arsenic, and antimony form oxides with the formula E_4O_6 (E = P, As, Sb) which have the adamantane structure. No nitrogen compound N_4O_6 with a corresponding structure is known other than N_2O_3 and this is stable only at low temperatures. If, though, the oxygen is replaced by a pseudooxygen unit such as, e.g., CH_2, hexamethylenetetramine $N_4(CH_2)_6$ is obtained.

It has been established quite generally that compounds with the adamantane structure are formally made up of elements of Group V and VI in the ratio 4:6. These may be replaced either completely or in part by the corresponding pseudo- or paraelements. Accordingly, adamantane, $(CH)_4(CH_2)_6$, consists of four pseudonitrogen and six pseudooxygen atoms. All substances with an adamantane structure so far synthesized correspond to this composition (92, 93). Adamantanes containing paraelements are, for example, $(BrC)_4(CH_2)_6$, $(CH_3C)_4S_6$, $(ClSi)_4(CH_2)_6$, $P_4(NCH_3)_6$, and $(RSi)_4O_6$ (93).

Very recently it was possible to synthesize two basic adamantanes of the formula $(HSi)_4S_6$ and $(HSi)_4Se_6$. They were made by reacting $HSiCl_3$ with $(H_3Si)_2E$ (E = S, Se). An X-ray structure determination proved the expected adamantane structure for $(HSi)_4S_6$ and spectroscopic investigations left no doubt that $(HSi)_4Se_6$ has the same structure (94).

Among the elements able to form adamantane structures boron and its higher analogs should also be considered. Like the elements of Group V they have as sp^2 hybrids three unpaired electrons at their disposal. Compounds of the formula MX_3 (M = B, Al, Ga ...) are planar and are not equivalent to $M'X_3$ (M' = N, P, As ...). A formal equivalence between the elements of Groups III and V can be arrived at through adduct formation with Lewis bases, e.g., $R_3N:BX_3$. Here boron is tetrahedrally coordinated and is electronically isovalent with, e.g., nitrogen. This element–paraelement relationship should make possible the substitution of, e.g., N, CH ... for $R_3N:B\equiv$, and in fact it does. B. M. Mikkailov (95) has provided methods of preparation for 1-boradamantane. The following synthesis is chosen for illustration:

The relatively stable etherate of 1-boradamantane dissociates partially in its components even at 20°C. It can be used for synthesizing complexes with other ligands just by mixing the etherate with stronger Lewis bases. 1-Boradamantane forms colorless, well-formed prisms which on heating to 190–200°C lead to a liquid mass. The boron atom is sp^3 hybridized and the CBC angle is almost tetrahedral. Complex formation with Lewis bases take place through the vacant sp^3-type orbital (95). The preparation of $(HC)_4(BCH_3)_6$ (96) is an example of an adamantane with CH_3B as a chalkogen type of ligand. The crystal structure determined proves the adamantane shape and gives angles at boron at 117.2° and those at carbon in the range of 103.1 to 106.4° (97). Oxidation of P_4O_6 to P_4O_{10} depends on coordination bonding of the electrophilic oxygen atom to the free electron pair of a phosphorus atom without change in the adamantane structure. Since the pseudonitrogen atom CH possesses no free electron pair, oxidation products of the composition $(OCH)_4(CH_2)_6$, for example, are unknown. The electrophile O in P_4O_{10} may be replaced by others, whereby compounds of the general formula $Y_nP_4O_6$ are obtained. A number of derivatives have been prepared (see Table VI).

C. Interchangeability within the P_4S_3 Structure

It is very difficult at first sight to understand why substances like P_7R_3, $P_4[Si(CH_3)_2]_3$ and also Sb_7^{3-} should have the same structure as P_4S_3. With the aid of the element displacement principle this question can be answered without difficulty. In the cases mentioned the molecules contain a P_4 and an Sb_4 unit and the other parts can be written as

THE ELEMENT DISPLACEMENT PRINCIPLE

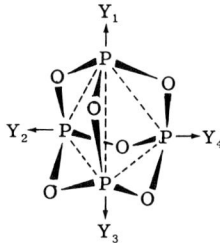

TABLE VI

	x	Y_1	Y_2	Y_3	Y_4
$S_xP_4O_6$	1	S	—	—	—
(98)	2	S	S	—	—
	3	S	S	S	—
	4	S	S	S	S
$Se_xP_4O_6$	1	Se	—	—	—
(98)	2	Se	Se	—	—
	3	Se	Se	Se	—
$(H_3B)_xP_4O_6$	1	BH_3	—	—	—
(99)	2	BH_3	BH_3	—	—
	3	BH_3	BH_3	BH_3	—
$[(OC)_3Ni]_xP_4O_6$	1	$Ni(CO)_3$	—	—	—
(100)	2	$Ni(CO)_3$	$Ni(CO)_3$	—	—
	3	$Ni(CO)_3$	$Ni(CO)_3$	$Ni(CO)_3$	—
	4	$Ni(CO)_3$	$Ni(CO)_3$	$Ni(CO)_3$	$Ni(CO)_3$
$[(OC)_4Fe]_xP_4O_6$	1	$Fe(CO)_4$			
(101)					

PR, $Si(CH_3)_2$, or Sb^-. The first two radicals are first-order derivative paraelements equivalent to sulfur. The formula P_7R_3 can be written as $P_4(PR)_3$ with R = H, Li (102), $(CH_3)_3Si$ (103), or CH_3 (104), thus accounting for the observed similarity to P_4S_3 (105). Analogously, it is found that $Sb_7^{3-} = Sb_4(Sb^-)_3$ has a P_4S_3 structure (106, 107). In $P_4[Si(CH_3)_2]_3$ the parasulfur $Si(CH_3)_2$ replaces PR without changing the P_4S_3 cage (108, 109).

VI. Electronegativities

Paraelements, derived from element displacement, exhibit chemical properties which are largely determined by the coordinated central element and which resemble those of neighbors on the right. Thus, e.g.,

CF_3 resembles fluorine and CF_3S resembles chlorine. Such an assignment is very difficult for higher order derivative paraelements, e.g., $(CF_3S)_2N$ and $(CF_3SC)_4N\cdot$ [tetrakis(trifluoromethylthio)pyrrolyl]. Group electronegativities help to solve such problems. According to the values estimated, $(CF_3S)_2N$ is like fluorine and $(CF_3SC)_4N\cdot$ is like chlorine. The chemical properties of these paraelements confirm these assignments. The presumption that Grimm's pseudoelements are, as far as chemical properties are concerned, superior to paraelements, cannot be maintained. The members of the series OH, NH_2, and CH_3 have chemically much less in common with fluorine than OF, NF_2, and CF_3, although the former are isoelectronic and isoprotonic with fluorine and the others only electronically isovalent. The group electronegativities of the two series show this quite convincingly.

OH 2.78–3.51   OF 4.70

NH_2 2.47–2.61 NF_2 3.60–3.64

CH_3 2.27–2.34 CF_3 3.46–3.55

J. E. Huheey (111) has calculated group electronegatives for a substantial number of radicals. The values obtained are, despite the general irregularities, in good agreement with those calculated from carbonyl frequency data (112) and from phosphoryl frequency data (113). They might be a good guide in assigning paraelements.

Group electronegativities of fluorinated paraelements which have been determined experimentally may be placed between those of fluorine and bromine, giving the following sequence: F (4.0); OF (3.8) (114); OOF (3.8) (114); $(CF_3S)_2N$ (3.7) (47); CF_3OO (3.7) (114); CF_3 (3.3) (3); CF_3SNH- (3.2) (47); Cl (3.0); $(CF_3SC)_4N\cdot$ (2.9); CF_3S (2.7) (3); Br (2.8). This shows that incorporation of paraelements in the periodic system smoothes out sudden changes in the properties of elements in going over from one period to the next.

The concept put forward in this article is the outcome of collecting relevant results from the literature, completing them with the aid of new ideas, and arranging them according to the periodic system of functional groups.

In other words, it can be said that available mosaic stones were complemented by new ones and arranged to form a mosaic picture. The overall picture is now understood but the boundaries are diffuse and should be made clearer by additional work.

Acknowledgments

Thanks are expressed to my co-workers for their enthusiastic experimental work and valuable discussions. Reading of the manuscript by Dr. M. Lieb and help in working on the literature by Dr. R. Walz are gratefully acknowledged.

References

1. Grimm, H. G., *Z. Elektrochem.* **31,** 474 (1925).
2. Birkenbach, L., and Kellermann, K., *Ber. Dtsch. Chem. Ges.* **58,** 786 (1925).
3. Lagowski, J. J., *Q. Rev. Chem. Soc.* **13,** 233 (1959).
4. Birkenbach, L., and Huttner, K., *Z. Anorg. Allg. Chem.* **190,** 1 (1930).
5. Madelung, W., and Kern, E., *Liebigs Ann.* **427,** 1 (1922).
6. Zintl, E., *Angew. Chem.* **52,** 1 (1939).
7. Klemm, W., and Bussmann, E., *Z. Anorg. Allg. Chem.* **319,** 297 (1963).
8. Morse, S. D., and Shreeve, J. M., *J. Chem. Soc. Chem. Commun.* 360 (1976).
9. Morse, S. D., and Shreeve, J. M., *Inorg. Chem.* **17,** 2169 (1978).
10. Kumar, R. C., and Shreeve, J. M., *J. Am. Chem. Soc.* **103,** 1951 (1981).
11. Swindell, R. F., Zaborowski, L. M., and Shreeve, J. M., *Inorg. Chem.* **10,** 1635 (1971).
12. Morse, S. D., and Shreeve, J. M., *Inorg. Chem.* **16,** 33 (1977).
13. Lustig, M., *Inorg. Chem.* **5,** 1317 (1966).
14. Young, J. A., Durell, W. S., and Dresdner, R. D., *J. Am. Chem. Soc.* **81,** 1587 (1959).
15. Haas, A., and Schott, P., *Chem. Ber.* **101,** 3407 (1968).
16. Kühle, E., *Ger. Pat.* 1149712 (1963); *C. A.* **59,** 12704 (1963).
17. Bögemann, M., Petersen, S., Schultz, D. E., and Söll, H., "Methoden der Organischen Chemie (Houben–Weyl–Müller)," 4. Infl., Vol. IX, p. 790. Thieme, Stuttgart, 1955.
18. Barton, D. H. R., and Robson, M. J., *J. Chem. Soc. Perkin I* **1,** 1245 (1974).
19. Haas, A., and Walz, R., *J. Fluorine Chem.* **23,** 472 (1983); and unpublished results.
20. Gielow, P., and Haas, A., *Z. Anorg. Allg. Chem.* **394,** 53 (1972).
21. Middleton, W. J., and Krespan, C. G., *J. Org. Chem.* **30,** 1398 (1965).
22. Kühle, E., Hagemann, H., and Oehlmann, L., *Angew. Chem.* **87,** 707 (1975); *Angew. Chem. Int. Ed.* **14,** 698 (1975).
23. Swindell, R. F., Babb, D. P., Quelette, P. J., and Shreeve, J. M., *Inorg. Chem.* **11,** 242 (1972).
24. Shermolovich, Y. G., Vasilev, V. V., and Markovskii, L. N., *J. Org. Chem. USSR* **13,** 664 (1977).
25. Campbell, T. W., Monagle, J. J., and Foldi, V. S., *J. Am. Chem. Soc.* **84,** 3673 (1962).
26. Dahms, G., Haas, A., and Klug, W., *Chem. Ber.* **104,** 2732 (1973).
27. Hynes, J. B., Bishop, B. C., Bandyopadhyay, P., and Bigelow, L. A., *J. Am. Chem. Soc.* **85,** 83 (1963).
28. Bielefeld, D., and Haas, A., *Chem. Ber.* **116,** 1257 (1983).
29. Bishop, B. C., Hynes, J. B., and Bigelow, L. A., *J. Am. Chem. Soc.* **86,** 1827 (1964); Chembers, W. J., Tullock, C. W., and Coffman, D. D., *J. Am. Chem. Soc.* **84,** 2337 (1962); Stacey, F. W., E. I. duPont de Nemours, U.S.P. 3076843, 2655 (1963); Tullock, C. W., E. I. duPont de Nemours, B. P. 870328 (1960); *C. A.* **56,** 8561 (1962).

30. Gombler, W., Haas, A., and Willner, H., *Z. Anorg. Allg. Chem.* **469**, 135 (1980).
31. Appleman, E. A., *Acc. Chem. Res.* **6**, 113 (1973).
32. Lawton, E. A., Pilipovich, D., and Wilson, R. D., *Inorg. Chem.* **4**, 118 (1965).
33. Burton, D. J., and Hahnfeld, J. L., *Fluorine Chem. Rev.* **8**, 119 (1977).
34. Seppelt, K., *Angew. Chem.* **89**, 325 (1977); *Angew. Chem. Int. Ed.* **16**, 322 (1977).
35. Barr, D. A., and Haszeldine, R. N., *J. Chem. Soc.* 2532 (1955); Young, J. H., Tsoukalas, S. N., and Dresdner, R. D., *J. Am. Chem. Soc.* **80**, 3604 (1958).
36. Knunyants, I. L., Krasuskaya, M. P., and Gambaryan, N. P., *Izv. Akad. Nauk SSSR Ser. Khim.* 723 (1965); *Bull. Acad. Sci. USSR Chem. Div.* 702 (1965).
37. Redwood, M. E., and Willis, C. J., *Can. J. Chem.* **43**, 1893 (1965).
38. Young, J. H., Tsoukalas, S. N., and Dresdner, R. D., *J. Am. Chem. Soc.* **80**, 3604 (1958).
39. Aldrich, P. E., Howard, E. G., Linn, W. J., Middleton, W. J., and Sharkey, W. H., *J. Org. Chem.* **28**, 184 (1963).
40. Haas, A., and Kortmann, W., *Z. Anorg. Allg. Chem.* **501**, 79 (1983).
41. Grunze, H., *Z. Anorg. Allg. Chem.* **313**, 316 (1962); **298**, 152 (1959).
42. Haszeldine, R. N., and Kidd, J. M., *J. Chem. Soc.* 2901 (1955).
43. Gombler, W., *Angew. Chem.* **89**, 740 (1977); *Angew. Chem. Int. Ed. Engl.* **16**, 723 (1977).
44. Silhanek, J., and Zbirovsky, M., *Chem. Commun.* 878 (1969).
45. Eigen, M., and Kustin, K., *J. Am. Chem. Soc.* **84**, 1355 (1962).
46. Kice, J. L., *Adv. Phys. Org. Chem.* **17**, 65 (1980).
47. Borowski, H. E., and Haas, A., *Chem. Ber.* **115**, 533 (1982).
48. Lorenz, R., Dissertation, Bochum, 1972.
49. Flegler, K. H., and Haas, A., *Chem. Ztg.* **100**, 339 (1976).
50. Freeman, J. P., *Inorg. Chim. Acta Rev.* **1**, 65 (1967).
51. Vosper, A. J., *J. Chem. Soc. A* 625 (1970).
52. Haas, A., Schlosser, K., and Steenken, S., *J. Am. Chem. Soc.* **101**, 6282 (1979).
53. Schlosser, K., and Steenken, S., *J. Am. Chem. Soc.* **105**, 1504 (1983).
54. Borowski, H. E., and Haas, A., *Chem. Ber.* **115**, 523 (1982).
55. De Marco, R. A., and Shreeve, J. M., *Inorg. Chem.* **12**, 1896 (1973).
56. Olson, A. R., *J. Am. Chem. Soc.* **42**, 896 (1920).
57. Haas, A., Wanzke, W., and Welcman, N., *J. Fluorine Chem.* **23**, 471 (1983); and unpublished results.
58. Kice, J. L., and Cleveland, J. P., *J. Am. Chem. Soc.* **95**, 104 (1973).
59. Haas, A., Häberlein, M., and Krüger, C., *Chem. Ber.* **109**, 1769 (1976).
60. Haas, A., and Willert-Porada, M., *J. Fluorine Chem.* **23**, 427 (1983); and *Chem. Ber.*, in press (1984).
61. Lau, C., and Passmore, J., *J. Fluorine Chem.* **7**, 261 (1976).
62. Klemm, W., and Henkel, P., *Z. Anorg. Allg. Chem.* **207**, 73 (1932).
63. Yost, D. M., and Claussen, W. H., *J. Am. Chem. Soc.* **55**, 887 (1933).
64. Schack, C. J., Wilson, R. D., and Hon, J. F., *Inorg. Chem.* **11**, 208 (1972).
65. Peacock, R. D., *J. Chem. Soc.* 3617 (1953).
66. Prideaux, E. B. R., and Cox, C. B., *J. Chem. Soc.* 1606 (1928).
67. Muetterties, E. L., and Phillips, W. D., *J. Am. Chem. Soc.* **81**, 1084 (1959).
68. Birchall, P., Gillespie, R. J., and Vekris, S. L., *Can. J. Chem.* **43**, 1672 (1965).
69. Gombler, W., Habilitation, Bochum, 1982.
70. Seppelt, K., *Z. Anorg. Allg. Chem.* **416**, 12 (1975).
71. Lehmann, E., *J. Chem. Res. Suppl.* 42 (1978).

72. Haas, A., and Weiler, H. U., *J. Fluorine Chem.* **23,** 467 (1983); *Chem. Ber.,* in press (1984).
73. Sheppard, W. A., *J. Am. Chem. Soc.* **87,** 4338 (1965).
74. Haas, A., and Schott, P., *Chem. Ber.* **101,** 3407 (1968).
75. Golloch, A., and Kuss, M., *Z. Naturforsch.* **27b,** 280 (1972).
76. Haas, A., and Westebbe, U., unpublished results.
77. Haas, A., and Tebbe, K., *Z. Naturforsch.* **39b,** 897 (1984).
78. Glemser, O., Meyer, H., and Haas, A., *Chem. Ber.* **97,** 1704 (1964).
79. Glemser, O., and Richert, H., *Z. Anorg. Allg. Chem.* **307,** 313 (1961).
80. Traetteberg, M., *Acta Chem. Scand.* **20,** 1724 (1966).
81. Gerstenberger, M. R. C., Haas, A., Kirste, B., Krüger, C., and Kurrek, H., *Chem. Ber.* **115,** 2540 (1982).
82. Ceacareanu, D. M., Gerstenberger, M. R. C., and Haas, A., *Chem. Ber.* **116,** 3325 (1983).
83. Dorn, S., Eggenberg, P., Gerstenberger, M. R. C., Haas, A., Niemann, U., and Zobrist, P., *Helv. Chim. Acta* **62,** 1442 (1979).
84. Haas, A., and Maciej, T., *J. Fluorine Chem.* **23,** 480 (1983); and *Z. Anorg. Allg. Chem.,* in press (1984).
85. Andreades, S., U.S. Patent 3081350 (1961/63); *Chem. Abstr.* **59,** 5024 (1963).
86. Andreades, S., Harris, J. F., and Sheppard, W. A., *J. Org. Chem.* **29,** 898 (1964).
87. Gerstenberger, M. R. C., Haas, A., and Liebig, F., *J. Fluorine Chem.* **19,** 461 (1982).
88. Haas, A., and Klare, Ch., *J. Fluorine Chem.* **23,** 490 (1983); and unpublished results.
89. Haas, A., and Weiss, F., *Z. Anorg. Allg. Chem.,* in press (1984).
90. Brown, H. C., Gisler, H. J., and Cheng, M. P., *J. Org. Chem.* **31,** 781 (1966).
91. Haas, A., and Kolasa, A., *Chem. Ber.* **117,** 1896 (1984).
92. Stetter, H., *Angew. Chem.* **74,** 361 (1962).
93. Averina, N. V., and Zefirov, N. S., *Usp. Khim.* **45,** 1077 (1976); *Russ. Chem. Rev.* **45,** 544 (1976).
94. Haas, A., Hitze, R., Krüger, C., and Angermund, K., *Z. Naturforsch.* **39b,** 890 (1984).
95. Mikhailov, B. M., *Pure Appl. Chem.* **52,** 691 (1980).
96. Brown, M. P., Holliday, A. K., and Way, G. M., *J. Chem. Soc. Dalton Trans.* 148 (1975).
97. Rayment, I., and Shearer, H. M. M., *J. Chem. Soc. Dalton Trans.* 136 (1977).
98. Walker, M., Peckenpaugh, D., and Mills, J. L., *Inorg. Chem.* **18,** 2792 (1979).
99. Riess, J. G., and Van Wazer, J. R., *J. Am. Chem. Soc.* **88,** 2339 (1966).
100. Riess, J. G., and Van Wazer, J. R., *J. Am. Chem. Soc.* **87,** 5506 (1965); **88,** 2166 (1966).
101. Walker, M. L., and Mills, J. L., *Inorg. Chem.* **14,** 2438 (1975).
102. Baudler, M., Ternberger, H., Faber, W., and Hahn, J., *Z. Naturforsch.* **34b,** 1690 (1979).
103. Fritz, G., and Holderich, W., *Naturwissenschaften* **62,** 573 (1975).
104. Baudler, M., Faber, W., and Hahn, J., *Z. Anorg. Allg. Chem.* **469,** 15 (1980).
105. Hänle, W., and v. Schnering, H. G., *Z. Anorg Allg. Chem.* **440,** 171 (1978).
106. Diehl, L., Khodadadch, K., Kummer, D., and Strähle, J., *Chem. Ber.* **109,** 3404 (1976).
107. Corbett, J. D., Adolphson, D. G., Merryman, D. J., Edwards, P. A., and Armatis, F. J., *J. Am. Chem. Soc.* **97,** 6267 (1975).

108. Fritz, G., and Uhlmann, R., *Z. Anorg. Allg. Chem.* **440,** 168 (1978).
109. Hänle, W., and v. Schnering, H. G., *Z. Anorg. Allg. Chem.* **440,** 171 (1978).
110. Vogler, A., Wright, R. E., and Kunkely, H., *Angew. Chem.* **92,** 745 (1980); *Angew. Chem. Int. Ed.* **9,** 717 (1980).
111. Huheey, J. E., *J. Phys. Chem.* **69,** 3284 (1965).
112. Kagarise, R. E., *J. Am. Chem. Soc.* **77,** 1377 (1955).
113. Bell, J. V., Heisler, J., Tannenbaum, H., and Goldenson, J., *J. Am. Chem. Soc.* **76,** 5185 (1954).
114. Hargittai, I., *Z. Naturforsch.* **34a,** 755 (1979).
115. "Selected Values of Chemical Thermodynamic Properties." U.S. Natl. Bur. Stands.; *Circulation* **500** (1961).

COMPOUNDS OF PENTACOORDINATED ARSENIC(V)

R. BOHRA* and H. W. ROESKY**

*Department of Chemistry, University of Rajasthan, Jaipur, India, and
**Institute of Inorganic Chemistry, University of Göttingen, Göttingen, Federal Republic of Germany

I. Introduction 203
II. As—C-Containing Compounds 204
III. As—N-Containing Compounds 216
IV. As—O- and As—S-Containing Compounds 223
V. As—Halogen-Containing Compounds 238
VI. Conclusion 248
References 249

I. Introduction

Since about 1969 there has been increased interest in the pentacoordination of main-group elements and also in the role of pentacoordination in reaction mechanisms (*101–106, 112, 118, 124, 136, 151, 163, 186, 201*). Structural studies have shown that compounds that contain pentacoordinated elements form a continuous range of conformations extending from the ideal trigonal bipyramid to the square or rectangular pyramid. In this respect, the compounds of pentacoordinated phosphorus (*105, 106, 124*), arsenic, and antimony (*112*) are unique and interesting. During these years a considerable number of papers have been published dealing with the chemistry of compounds of pentacoordinated arsenic(V) (arsorane) and organoarsenic(V) (organoarsorane), but have been mentioned in a scattered fashion in some review articles, books, and treatises (*41, 49, 50, 111, 130, 153, 178*). Although since 1967, the work published on the chemistry of organoarsoranes has been abstracted in *Annual Surveys of Organometallic Chemistry*, there is as such no review article dealing exclusively with compounds of pentacoordinated arsenic(V). This account has been constructed mainly to focus attention on the dynamic stereochemistry, structure, and bonding aspects of these compounds. An attempt has been made to cover almost all of the pertinent literature through the end of 1982.

II. As—C-Containing Compounds

Many compounds of pentacoordinated arsenic(V) containing only As—C bonds are reported in the literature and are summarized in Table I.

Although more than 100 years ago Cahours (35) claimed to have obtained pentamethylarsorane by the reaction of dimethylzinc with tetramethylarsonium iodide, subsequent attempts (55, 195) to synthesize pentaalkylarsoranes by the reaction of tetraalkylarsonium salts with organometallic compounds were unsuccessful. However, Mitschke and Schmidbaur (137) have obtained pentamethylarsorane in 80% yield by the following reaction:

$$Me_3AsCl_2 + 2MeLi \xrightarrow[-60°C]{Me_2O} Me_5As + 2LiCl$$

On the basis of spectral studies a trigonal–bipyramidal structure has been proposed for this compound. Low-temperature ^{1}H-NMR spectra in toluene indicate that all methyl protons are magnetically equivalent down to $-95°C$ and that the exchange of methyl groups at the equatorial and axial positions in Me_5As is faster than the NMR scale (137).

As compared to pentaalkylarsorane, a considerable number of pentaarylarsoranes are known (Table I). Pentaphenylarsorane has been prepared in 65% yield by the reaction of phenyllithium and tetraphenylarsonium bromide (194). Reaction of triphenyldichloroarsorane and phenyllithium also gives the same compound, but in poor yield. Formation of this compound has also been reported by the reaction of phenyllithium with triphenylarsenic oxide or with the imine p-$CH_3C_6H_4N{=}AsPh_3$ (198). The reactions of pentaphenylarsorane have been investigated (194, 198). It reacts with excess halogens to give the corresponding tetraphenylarsonium halides and halobenzenes. It can also be cleaved by acids. It decomposes on heating to give mainly triphenylarsane with smaller amounts of biphenyl and benzene.

The unit cell of Ph_5As has been reported by Wheatley and Wittig (187). From these data it has been concluded that the molecule adopts the expected trigonal–bipyramidal conformation, very similar to that found for pentaphenylphosphorane (189). It is interesting to mention here that the unsolvated Ph_5Sb crystallizes with square–pyramidal geometry (15, 188) in a triclinic cell rather than in the monoclinic Cc

TABLE I

COMPOUNDS OF PENTACOORDINATED ARSENIC(V) CONTAINING As—C BONDS

Compound	Melting point (°C)	Physical measurements	Reference
Me_5As	−6 to −7 (bp −10/0.1)	Vibrational spectra, ^1H NMR, DTA, mass spectra	137, 195
Ph_5As	Colorless crystals 139.5 (decomp.)	Dipole moment, vibrational spectra, mass spectra, X-ray diffraction	14, 31, 96, 126, 187, 194, 196, 198
$Ph_5As \cdot 0.5C_6H_{12}$	149–150	Lattice energy calculations, X ray	30, 32
$(p\text{-}MeC_6H_4)_5As$	139–140	^1H NMR, ^{13}C NMR	84, 87, 121
$(p\text{-}ClC_6H_4)_5As$	149	^1H NMR	87
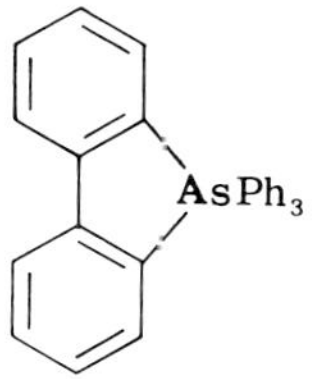 ![structure: dibenzarsole with AsPh3]	188–189	Thermal decomposition	87, 196, 198
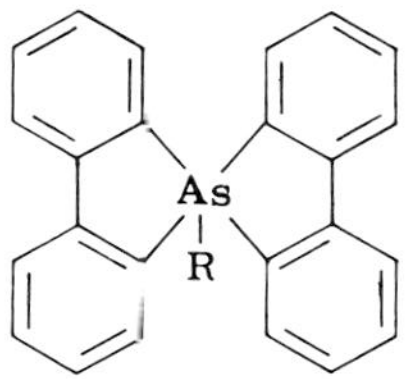 ![structure: spirobi-dibenzarsole with As—R], R = Me	Colorless crystals 215–216	IR, ^1H NMR, mass spectra, thermal decomposition	87, 96

(continued)

TABLE I (*continued*)

Compound	Melting point (°C)	Physical measurements	Reference
R = Et	173–174 (decomp.)	IR, mass spectra, thermal decomposition	87
R = *i*-Pr	148–149 (decomp.)	IR, ^{1}H NMR, mass spectra, thermal decomposition	94
R = Bu	166–166.5 (decomp.)	IR, mass spectra, thermal decomposition	87, 199
R = *t*-Bu	102–103 (decomp.)	IR, ^{1}H NMR, mass spectra, thermal decomposition	94
R = Vinyl	176–177 (decomp.)	IR, ^{1}H NMR, mass spectra, thermal decomposition	94
R = Cyclopentyl	114–115 (decomp.)	IR, ^{1}H NMR, mass spectra, thermal decomposition	94
R = Ph	233–235	IR, mass spectra, thermal decomposition	87, 96, 196, 198, 199
R = *cis*-Styryl	124–126	IR, ^{1}H NMR, thermal decomposition	94
R = *trans*-Styryl	115–117	IR, ^{1}H NMR, thermal decomposition	94
R = *p*-ClC$_6$H$_4$	196–197		87
R = *p*-CH$_3$C$_6$H$_4$	211–212		87
R = *p*-(CH$_3$)$_2$NC$_6$H$_4$	234–236	Thermal decomposition	87, 198

R =			
PhCH₂	142–144	IR, ¹H NMR, mass spectra	88, 95, 97
2-biphenylyl	141–143	IR, ¹H NMR, mass spectra	88, 95, 97
1-naphthyl	223–224	IR	97
8-fluoro-1-naphthyl	211–213	IR, ¹H NMR	97
8-chloro-1-naphthyl	220–221	IR, ¹H NMR	97
8-methyl-1-naphthyl	215–217	IR, ¹H NMR	97
8-amino-1-naphthyl	182–184	IR, ¹H NMR	97

(*continued*)

TABLE I (continued)

Compound	Melting point (°C)	Physical measurements	Reference
(structure: As with dibenzofuran-like rings and biphenyl substituent)	230–232	Optical activity	82, 126
(structure: dimethyl-substituted As with Ph)	191–193	Optical activity	198
(structure: As with naphthyl fused ring and R substituent) R = p-$(CH_3)_2NC_6H_4$	190		198

	198–200	¹H NMR	98
R = Ph	224–227	¹H NMR	98
R =	221–224	¹H NMR, mass spectra	89, 93

(continued)

TABLE I (continued)

Compound	Melting point (°C)	Physical measurements	Reference
	185–188	¹H NMR	93

cell observed for the arsenic and phosphorus molecules. It has been suggested that crystal-packing forces in the solid state cause Ph_5Sb to exhibit square–pyramidal geometry (15, 188). Later, a detailed study has been reported of the low-frequency (below 400 cm^{-1}) solid state and CH_2Cl_2 or CH_2Br_2 solution vibrational spectra of both Ph_5As and Ph_5Sb (14). From this study, it has been suggested that both compounds retain their solid-state structures in solution and that the structure of Ph_5Sb in the solid state might not be due to packing effects. The role of coulombic interactions in explaining the anomalous structure of Ph_5Sb has been advanced (31).

The structure of the cyclohexane solvate of pentaphenylarsorane, $Ph_5As \cdot \frac{1}{2}C_6H_{12}$, has been determined by X-ray crystallographic analysis (30). The substance crystallizes in space group $P\bar{1}$, cell dimensions $a = 10.448(9)$, $b = 10.566(21)$, $c = 14.905(25)$ Å, $\alpha = 121.09(5)$, $\beta = 106.38(4)$, $\gamma = 92.44(5)°$, with one Ph_5As and one-half C_6H_{12} molecule in the asymmetric unit. This molecule is an undistorted trigonal bipyramid with average axial and equatorial As—C bond lengths of 2.105(7) and 1.964(11) Å, respectively. The molecular conformation, as measured by phenyl ring rotations, of Ph_5As has been found to be very similar to that of the analogous phosphorus compound (189) and to that of $Ph_5Sb \cdot \frac{1}{2}C_6H_{12}$ (27). ^{13}C-NMR solution spectral study indicates that α-carbons of both pentaphenylarsorane and pentaphenylstiborane are magnetically equivalent down to 173 K. This suggests a rapid interconversion between axial and equatorial sites, presumably through an intermediate square–pyramidal geometry (171). Semiempirical calculations have been made for crystals of $Ph_5M \cdot \frac{1}{2}C_6H_{12}$ (M = P, As, or Sb) (32). From this study, it has been suggested that the molecular packing is not considerably affected by a change in the central group VA element.

The ^{1}H-NMR spectrum of penta-p-tolylarsorane exhibits a single signal of the methyl protons (84). The magnetic equivalence of the five methyl groups is not consistent with either a trigonal–bipyramidal or a square–pyramidal structure. It has been assumed that rapid pseudorotation averages the environment of the five groups attached to the arsenic atom. ^{1}H- and ^{13}C-NMR investigations of this compound as well as of pentaaryl group VA compounds have also been reported (87, 121). The question of the stereochemistry of these compounds in solution is not readily answered. There is no static solution structure, except at low temperatures. The low-energy barrier indicates that ligand size is not a dominant factor in limiting the exchange process.

Reaction of the imine p-$CH_3C_6H_4N$=$AsPh_3$ with 2,2'-biphenylenedilithium yields the heterocyclic compound **I** (198). This compound can also be obtained by the reaction of a dilithium compound with an imine of type **II** or by the reaction of spirocyclic arsonium halides with either

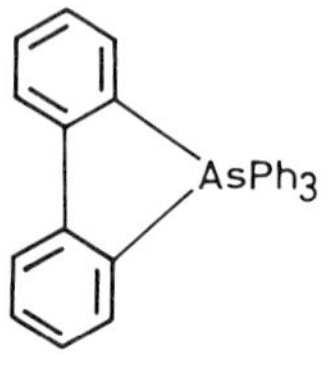

I

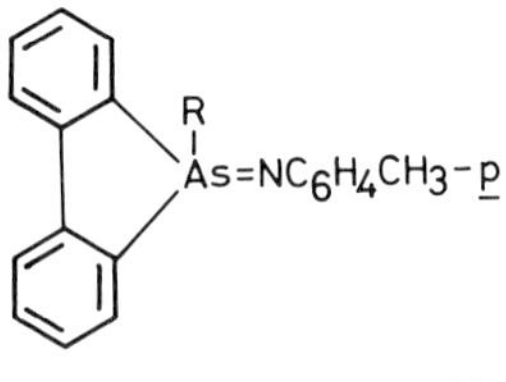

II (R = alkyl or aryl)

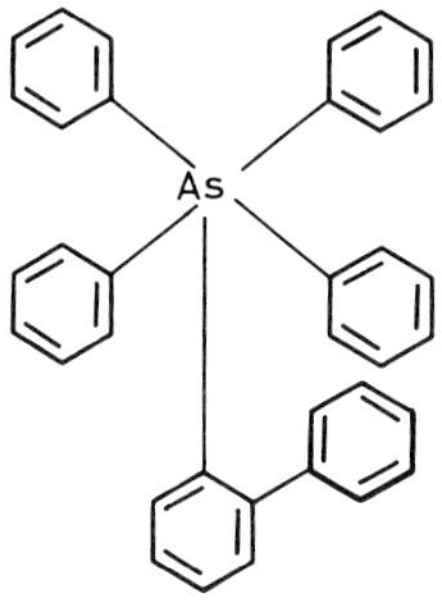

III

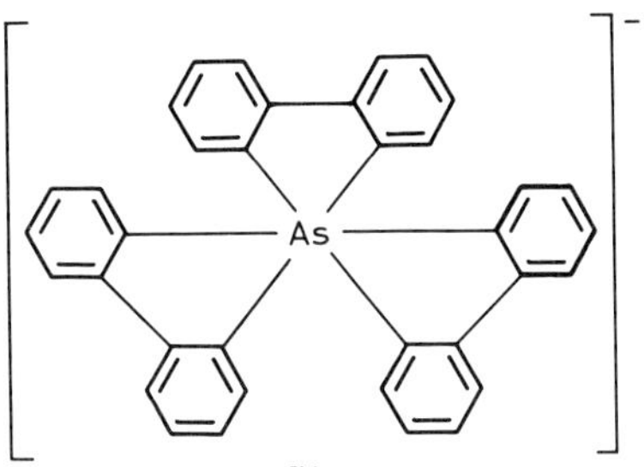

IV

lithium (*87, 198*) or Grignard (*198*) reagents. Compound **III** may be obtained by the acid cleavage of the lithium salts of **IV** or by the following reaction (*86*):

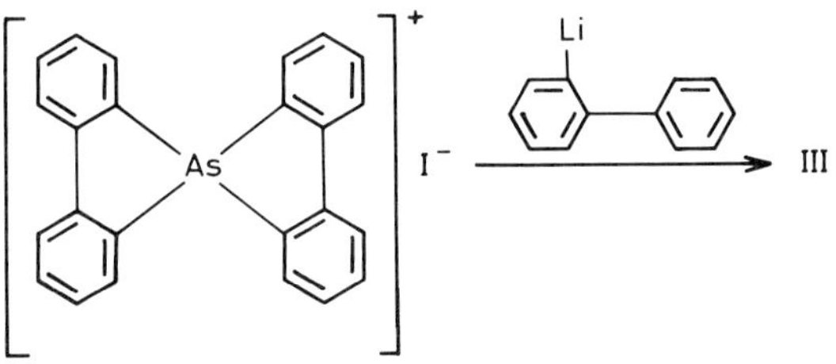

Compound **III** has been found to be optically inactive, suggesting pseudorotation between trigonal–bipyramidal and tetragonal conformations (*86*).

Compounds of type **V** have been prepared by the following reaction (*86*):

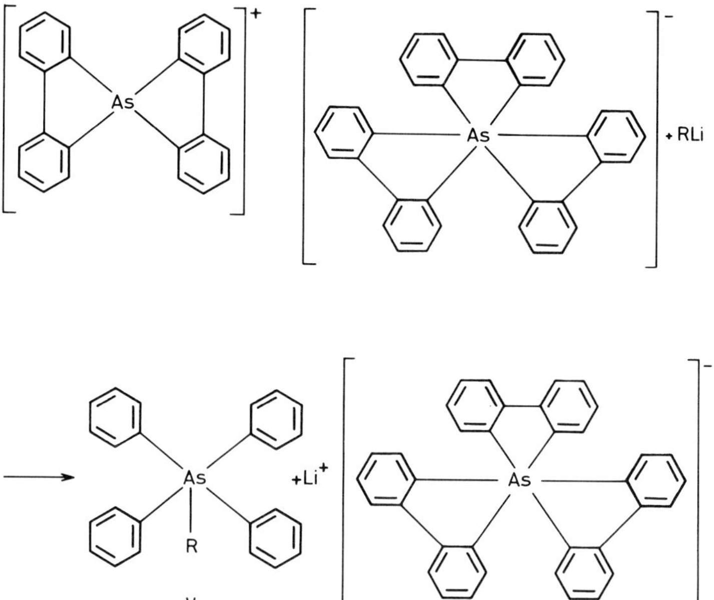

Alternatively, use may be made of exchange reactions of the following type:

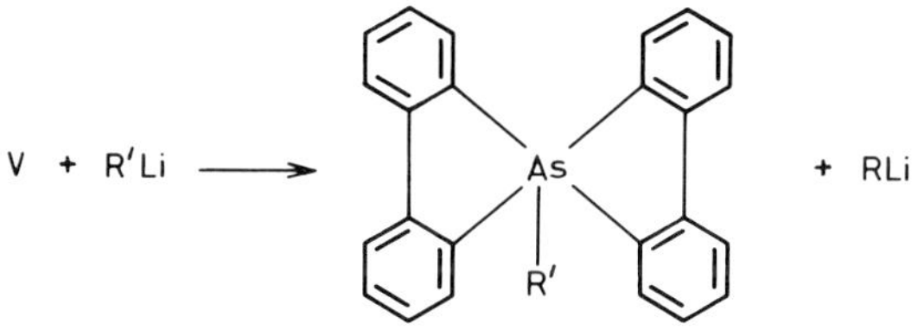

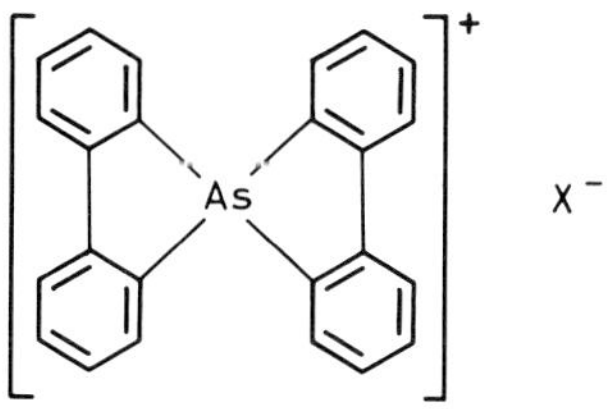

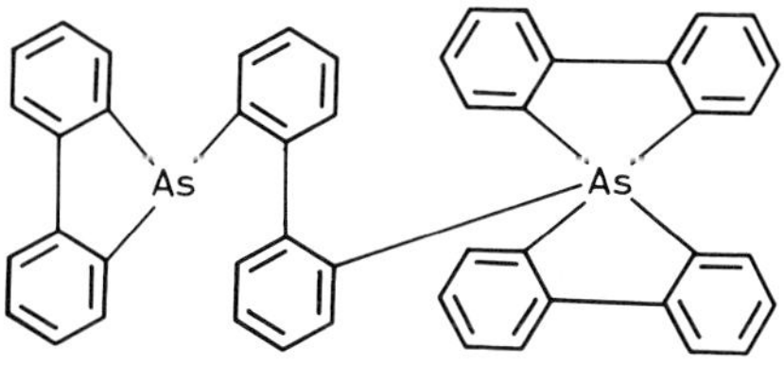

The isopropyl compound **V**, (R = Me₂CH) has readily been obtained from **VI** and isopropylmagnesium chloride in ether solution (*94*); but when R = cyclopentyl, THF has been found to be necessary for effecting the reaction. When R = *tert*-butyl the Grignard reaction yielded the dimer **VII**, but the desired **V** (R = *tert*-butyl) has been obtained from *tert*-butyllithium and the arsonium salt **VI**. Compound **V** (R = *tert*-butyl) has been found to be unstable at room temperature. It decomposes to the tertiary arsane **VIII** and isobutylene. The reactions of

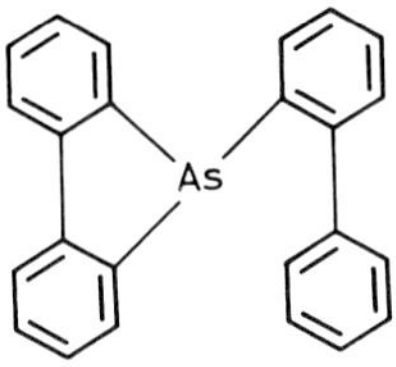

VIII

the compounds of type **V** have been found to be quite similar to those of the pentaphenylarsorane (*87, 94*). For example, **V** (R = Ph) reacts with several electrophilic reagents with cleavage of one of the heterocyclic rings and the formation of an arsonium compound. The cleavage of the compounds **V** (R = Me, Et, or PhCH$_2$) with boiling alcohol has also been investigated. The tertiary arsane **VIII** has been obtained in each case. The reaction has been followed by means of deuterium-labeled ethanol.

When compounds of type **V** are subjected to strong nucleophiles such as lithium organyls, the spiro skeleton remains unaffected and only the single ligand R is exchanged (*85, 87*). It has been shown that the exchangeability of R increases along the following sequence (*87*):

$$n\text{-}C_4H_9 \ll p\text{-}Me_2NC_6H_4 \sim CH_3 \sim p\text{-}MeC_6H_4 < p\text{-}ClC_6H_4 < C_6H_5$$

When compounds **V** (R = ethyl, isopropyl, cyclopentyl, *n*-butyl, or *tert*-butyl) have been heated 30–50°C above their melting points, **VIII** and the corresponding olefins have been obtained by β elimination (*87, 94*). The course of the reaction has been controlled by the use of the deuterium-labeled compound **V** (R = CD$_2$Me). It has been shown that all of the deuterium occurred in the ethylene formed by thermolysis (*94*). When R is methyl or phenyl, the thermolysis of **V** results in formation of **IX**. The styryl derivatives of **V** (R = *cis*-CH=CHPh and *trans*-CH=CHPh) have also been synthesized from **VI** and the corresponding Grignard reagents (*94*). Interestingly, thermolysis of these deriva-

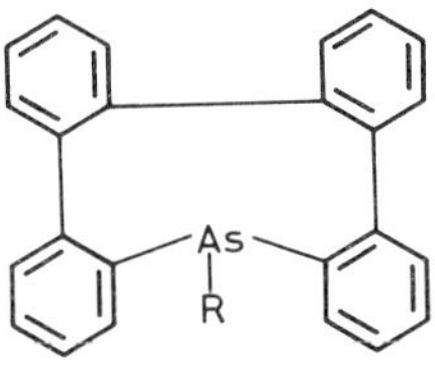

IX

tives did not produce phenylacetylene by β elimination as expected, but rather **X** (R = *cis*-styryl and *trans*-styryl, respectively). Thermolysis of the vinyl compound **V** (R = CH=CH$_2$) gives a mixture of acetylene, **VIII**, and the rearranged product **X**.

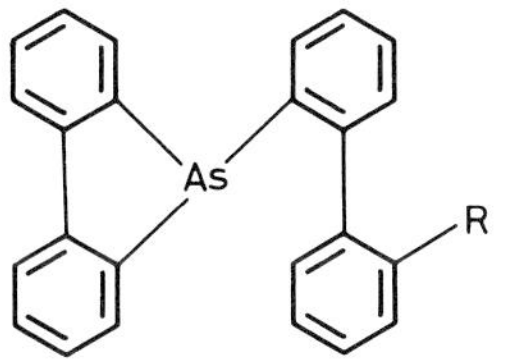

X

Reaction of **VI** with neopentylmagnesium bromide yields a product of molecular weight, as found by mass spectroscopy, corresponding to **V** (R = CH$_2$CMe$_3$). However, this compound exhibits all the properties of the tertiary arsane **X** (R = CH$_2$CMe$_3$) (*94*).

Mass spectra of pentaphenylarsorane and some compounds of type **V** have been reported and compared with the corresponding derivatives of group VA elements (*92, 94, 96*). The mass spectra of some of the phosphorus- and arsenic-substituted biphenyl systems show doubly charged parent ions of higher abundance than the singly charged molecular ions (*92*). Characteristic IR bands of compounds of type **V** have been summarized in Table II.

Dynamic NMR spectroscopy of compounds of types **V** and **XI** (R = organyl) indicates that these, like analogous phosphorus compounds, possess trigonal bipyramidal ground states that show intramolecular ligand equilibrations even at low temperatures (*95*). Activation energies of the order of 12–19 kcal/mol have been observed, depending on the bulkiness of the organyl group R (*84, 90, 94, 95, 97*).

Hellwinkel *et al.* (*98*) have also carried out an ^{1}H-NMR investigation of some overcrowded asymmetric phosphoranes, arsoranes, and stiboranes. In **XII** (R = phenyl or biphenyl), the two different positions are reversibly equilibrated at elevated temperatures.

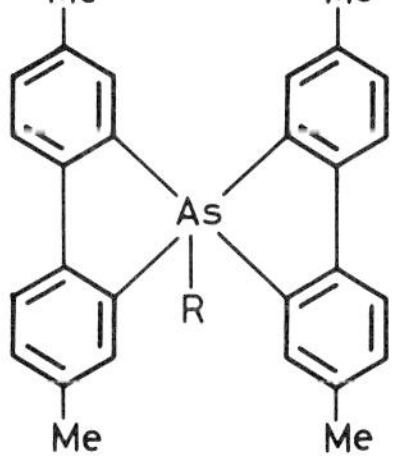

XI 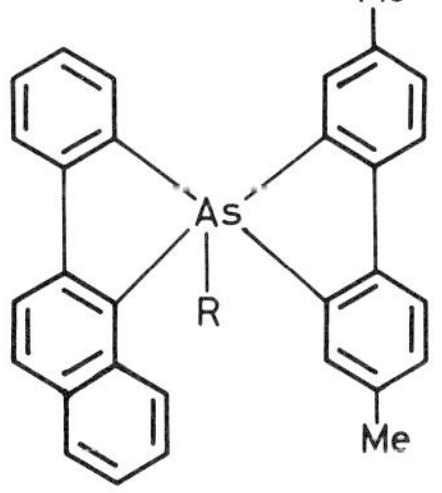   XII

TABLE II

CHARACTERISTIC INFRARED BANDS OF

R	Bands (cm^{-1})	Reference
Methyl	652, 667	87
Ethyl	652, 667	87
Isopropyl	643, 665	94
Butyl	650, 667	87
tert-Butyl	645, 664	94
Cyclopentyl	643, 665	94
Vinyl	650, 662	94
cis-Styryl	653, 669	94
trans-Styryl	655, 666	94
Benzyl	645, 665	94
Phenyl	653, 667	87
p-Tolyl	653, 670	87
p-Chlorophenyl	654, 670	87
p-Dimethylaminophenyl	653, 667	87
2-Biphenylyl	655, 668	86
2'-(2,2'-Biphenylylenearsino)-2-biphenylyl	653, 667	94

The free enthalpies of activation have been measured and an interpretation is offered starting with a trigonal–bipyramidal ground state conformation. The ligand exchange phenomenon is discussed in terms of a pseudorotation process and trigonal–bipyramidal transition states with diequatorial biarylylene groups.

III. As—N-Containing Compounds

Compounds of pentacoordinated arsenic(V) containing at least one As—N bond are summarized in Table III. Only some of these need discussion. Cycloarsa(V)azanes, which are apparently five-coordinated, have been prepared by aminolysis of pentamethoxyarsorane (75, 80). Various products have been reported, depending on the nature

TABLE III

COMPOUNDS OF PENTACOORDINATED ARSENIC(V) CONTAINING As—N BONDS

Compound	Melting point (°C)	Physical measurements	Reference
[(MeO)$_3$AsNPh]$_2$	120–121	Molecular weight	80
[(MeO)$_2$(n-PrNH)AsNPr]$_2$	107–108	Molecular weight	80
[(MeO)$_2$(n-BuNH)AsNPr]$_2$	54–55	Molecular weight	80
[(MeO)$_2$(PhCH$_2$NH)AsNCH$_2$Ph]$_2$	80–81	Molecular weight	80
[(MeO)(PhCH$_2$NH)$_2$AsNCH$_2$Ph]$_2$	72–74	Molecular weight	80
[(MeO)$_2$AsFNPr]$_2$	—	Molecular weight	82
(MeO)$_3$As$\begin{smallmatrix}R\\|\\N\\ \end{smallmatrix}$As(OMe)$_3$	—	^{1}H NMR, mass spectra	81, 148
R = n-Pr, n-Bu			
(MeO)$_8$As$_4$(NPh)$_6$	180 (decomp.)	Molecular weight	80
[(RNH)$_2$AsFNF]$_2$ R = benzyl	175	Molecular weight	82
(PrNH)$_8$As$_4$(NPh)$_6$	155 (decomp.)	IR, ^{1}H NMR, ^{19}F NMR	156

3-CF$_3$C$_6$F$_4$ Me C$_6$H$_4$CF$_3$-3
\\ | /
O=C—N—As—N—C=O
     |
     Cl
     |
     N
     /
    Me
	144–146	IR, ^{1}H NMR, ^{19}F NMR	156

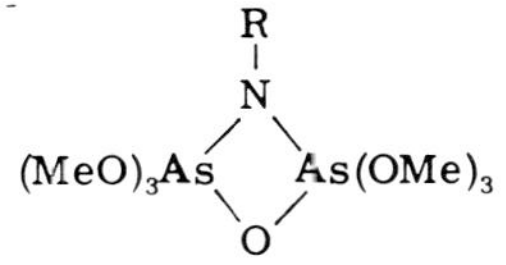

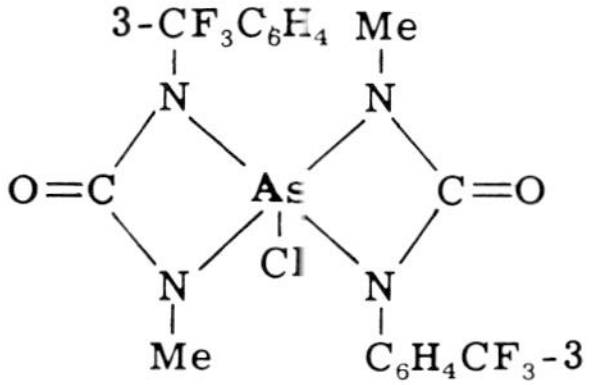

(continued)

TABLE III (continued)

Compound	Melting point (°C)	Physical measurements	Reference
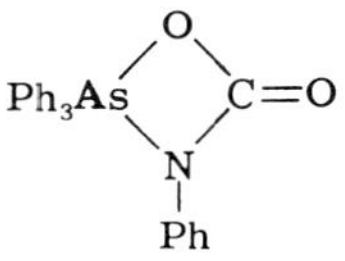 Ph₃As–O–C(=O)–N(Ph) (ring)	—	—	29, 56, 57, 138
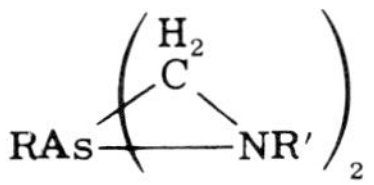 (RAs(CH₂–NR')–)₂	—	—	4
[(CF₃)₂AsClNSiMe₃]₂	116	IR, ¹⁹F NMR, mass spectra, X ray	23, 24, 157
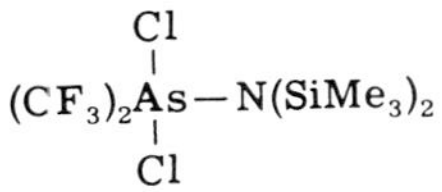 (CF₃)₂As(Cl)–N(SiMe₃)₂ with Cl	90	IR, ¹H NMR, ¹⁹F NMR, mass spectra, X ray	23, 25
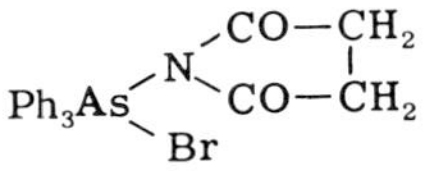 Ph₃As(–N(CO–CH₂)₂–)Br	132–134	IR, ¹H NMR	43
Ph₃As(NCS)₂	105; mp 104	IR	150, 21, 200
(p-Tolyl)₃As(NCS)₂	158–160	IR	20, 21
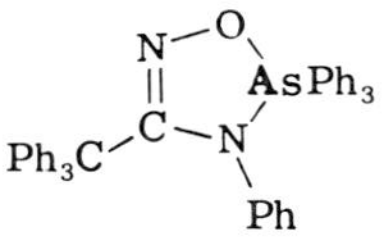 Ph₃C–C(=N–O–AsPh₃)–N–Ph (ring)	182	IR, mass spectra	58

218

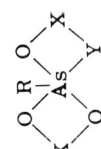

Compound	mp (°C)	Methods	Ref.
R = CH$_3$, X = CH$_2$CH$_2$, Y = NCH$_3$	40–42	^{1}H NMR, mass spectra	128
R = CH$_3$, X = (o-xylylene), Y = NH	178–180	^{1}H NMR, mass spectra, X ray	128, 204
R = CH$_3$, X = (2,3-naphthylene), Y = NH	230–232	^{1}H NMR, mass spectra	128
R = C$_6$H$_5$, X = CH$_2$CH$_2$, Y = NCH$_3$	100	^{1}H NMR, mass spectra	128
R = C$_6$H$_5$, X = (o-xylylene), Y = NH	178–180 (decomp.)	^{1}H NMR, mass spectra	128
R = C$_6$H$_5$, X = (2,3-naphthylene), Y = NH	277–279 (decomp.)	^{1}H NMR, mass spectra	128
$\left[\begin{array}{c} \text{R} \\ -\text{As}-\text{NH}-(\text{R}'\text{H})-\text{NH}\cdot \\ \text{R} \end{array} \right]_n$	—	Molecular weight	36

$n = 4000;\ 10^5$

of the amine and the molar ratio of the reagents, as shown below (**XIII**):

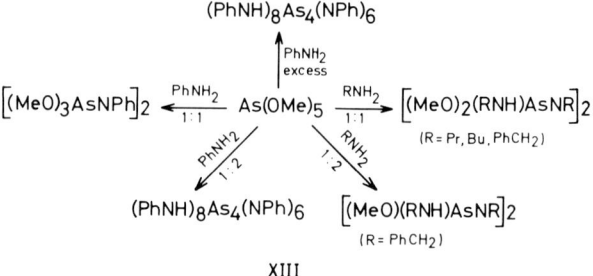

XIII

Aminolysis of other arsenic compounds also results in the formation of cyclodiars(V)azanes (82):

$$As(OMe)_4F + RNH_2 \longrightarrow As(OMe)_3(NHR)F + MeOH$$

$$2As(OMe)_3(NHR)F \xrightarrow{vacuum} [As(OMe)_2(NR)F]_2 + 2MeOH$$

$$\left.\begin{array}{l} As(OMe)_4F + RNH_2 \\ \quad (excess) \\ As(OMe)_3F_2 + RNH_2 \end{array}\right\} \longrightarrow [(RNH)_2AsFNR]_2$$
$$(R = PhCH_2)$$

It has been suggested that dimers are probably four-membered As—N rings, with five-coordinated arsenic. However, no physical evidence is available except molecular weight measurements. Polymeric structures (**XIV**) may be suggested for $X_8As(NR)_6$; but again, any physical evidence is so far lacking:

XIV

The synthesis, IR, ^{19}F-NMR, and X-ray crystal and molecular structures of **XV** have been reported. It is the first well-characterized four-membered As—N ring compound with arsenic atoms of coordination number five. It has been prepared by the following reaction:

$$2(CF_3)_2AsN(SiMe_3)_2 + 2Cl_2 \longrightarrow [(CF_3)_2AsClN(SiMe_3)]_2 + 2Me_3SiCl$$
(5–10% excess)

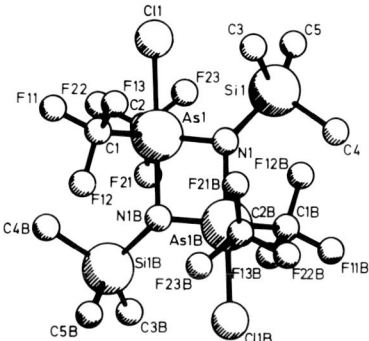

(XV) Molecular structure of $[(CF_3)_2 AsClN(SiMe_3)]_2$

The ^{19}F-NMR spectrum of **XV** indicates only one signal at -55.1 ppm at room temperature as well as at 153 K. Compound **XV** crystallizes in the orthorhombic space group $Pbca$ with $a = 11.979(5)$, $b = 15.451(4)$, $c = 13.166(7)$ Å, $Z = 4$. The four-membered As—N ring is planar and the compound has a trigonal–bipyramidal geometry with axial and equatorial As—N bond distances of $1.933(7)$ and $1.768(7)$ Å. The difference of 0.165 Å between the two As—N bond lengths is somewhat greater than that in the corresponding cyclodiphos(V)azane (0.12–0.14 Å) (*170*). The axial As—Cl bond distance is $2.296(4)$ Å. Thermal decomposition of **XV** gives six- and eight-membered As—N rings (*23, 24*):

$$[(CF_3)_2AsClN(SiMe_3)]_2 \longrightarrow \underset{\displaystyle \substack{F_3C\quad CF_3 \\ N=As\\ CF_3\,|\quad\;\,|\,CF_3 \\ \;\;As\!-\!As \\ CF_3\;F_3C\;\;CF_3}}{} + \underset{\displaystyle \substack{F_3C\;\;\;\;\;\;\;\;\;\;CF_3 \\ F_3C\!-\!As\!=\!N\!-\!As\!-\!CF_3 \\ F_3C\;\;N\quad\;\;\;N\;\;CF_3 \\ F_3C\;\;As\!-\!N\!=\!As\;\;CF_3 \\ F_3C\;\;\;\;\;\;\;\;\;\;CF_3}}{}$$

When the reaction of $(CF_3)_2AsN(SiMe_3)_2$ with Cl_2 is carried out in exactly 1:1 molar ratio, **XVI** is obtained (*25*). Compound **XVI** also possesses trigonal–bipyramidal geometry with chlorine atoms occupying axial positions. The crystals of **XVI** are monoclinic $P2_{1/m}$, $a = 9.298(4)$, $b = 12.841(3)$, $c = 15.090(2)$ Å, $\beta = 95.96(5)°$, $Z = 4$. Both ^{1}H- and ^{19}F-NMR spectra of **XVI** in CH_2Cl_2 show one signal relatively unchanged from 193 K to room temperature.

$$\underset{\displaystyle \substack{F_3C\;\;Cl \\ \;\;\;\,|\\ As\!-\!N(SiMe_3)_2 \\ F_3C\;\;Cl}}{}$$

XVI

Although compounds of the type $As_2O(OMe)_6NR$ ($R = C_3H_7, C_4H_9$) have been characterized by ^{1}H-NMR and mass spectra, they could not be obtained in a pure state (81, 148).

Compound **XVII** is formed by the reaction between $Ph_3As=NPh$ and triphenylacetonitrile N-oxide (58). The structure **XVII** has been assigned on the basis of that of the corresponding phosphorus compound (109):

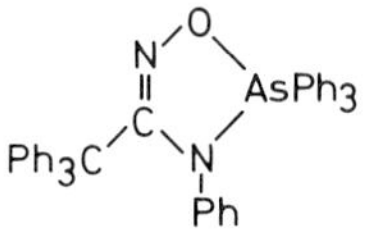

Ph₃C–C(=N–O–AsPh₃)–N(Ph)

XVII

Compound **XVIII** has been synthesized by the following reaction (156):

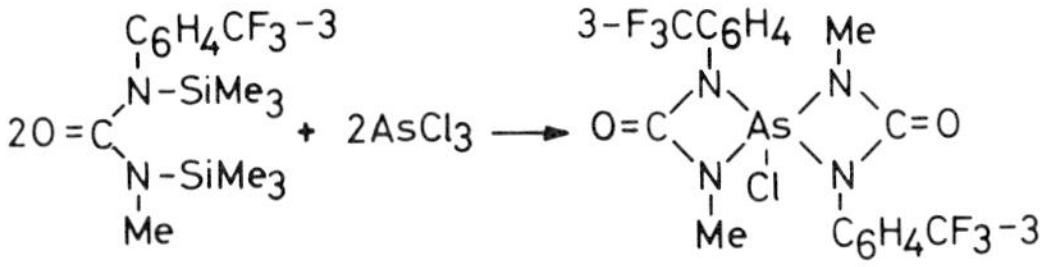

$C_6H_4CF_3$-3, N-SiMe₃, 20=C, N-SiMe₃, Me + 2AsCl₃ → 3-F₃CC₆H₄, Me, N, N, O=C, As, C=O, N, Cl, N, Me, $C_6H_4CF_3$-3

XVIII

^{1}H- and ^{19}F-NMR data indicate the presence of only one isomer in solution.

The compounds of type **XIX** have been prepared by the condensation reaction of phenylarsonic acid or by ester interchange of methyltetramethoxyarsorane with the corresponding amine (128, 204). The crystal

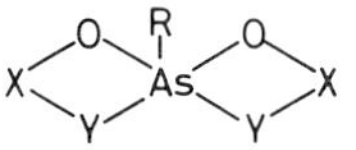

X–O–As(R)(=O)–O–X, Y–Y

R = Me, Ph
X = CH_2CH_2, (benzene), (naphthalene)
Y = NCH_3, NH

XIX

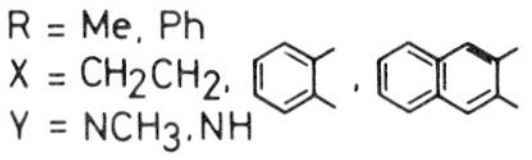

structure of one of the compounds **XIX** (R = Me, X = (methylbenzene), Y = NH) has been reported (204). The crystals of $MeAs(C_6H_4ONH)_2$ (**XX**) are monoclinic $P2_{1/c}$, $a = 12.285(5)$, $b = 9.508(3)$, $c = 10.848(2)$ Å, $\beta = 104.66(2)°$, $Z = 4$. The geometry of the pentacoordinated arsenic atom is closer to that of a trigonal bipyramid (72%) than of a rectangular pyramid. The two O atoms occupy axial positions [As—O = 1.893(5) and 1.860(5) Å]. The axial O—As—O angle is 169.6° and the equatorial N—As—N and two N—As—C angles are 125.9, 118.9, and 115.2°, respectively (204). It has been pointed out that the structural distor-

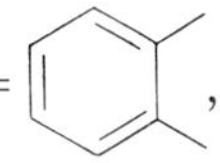

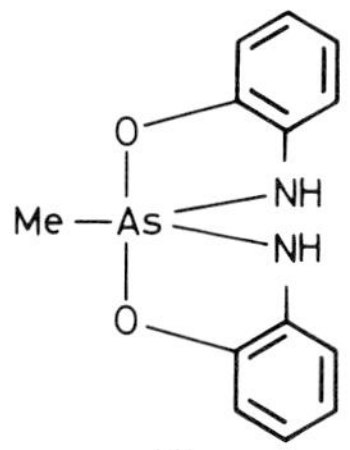

XX

tions from the idealized geometries deviate significantly from the Berry exchange coordinate.

Single-crystal X-ray analysis of the phenyl derivative of (**XX**) [PhAs-$(C_6H_4ONH)_2$] reveals a trigonal–bipyramidal structure. The structure is displaced 22.9% from the trigonal bipyramid toward the rectangular pyramid (*46a*).

IV. As—O- and As—S-Containing Compounds

Compounds of pentacoordinated arsenic(V) containing As—O and As—S bonds are summarized in Table IV. Acyclic compounds of type $As(OR)_5$ (R = Me, Et), $R_2As(OMe)_3$ (R = Me, Ph), and $R_3As(OMe)_2$ (R = Ph) have been prepared according to the following general reactions (*45, 59, 149*):

$$R_nAsX_{3-n} + Br_2 \longrightarrow R_nAsX_{3-n}Br_2$$

$$R_nAsX_{3-n}Br_2 + (5-n)NaOR' \longrightarrow R_nAs(OR')_{5-n} + (5-n)Na^+ + (3-n)X^- + 2Br^-$$

(X = halide)

Mass spectra of these compounds have been examined, and their fragmentation behavior has been compared to that of the corresponding phosphorus compounds. The variable temperature ^{1}H-NMR spectra of molecules $R_2As(OMe)_3$ and $Ph_3As(OMe)_2$ have been examined by Dale and Froyen (*45*) down to 173 K, and the results have been interpreted in terms of a pseudorotation process among structures with trigonal-bipyramidal geometry. According to these workers, only $Ph_2As(OMe)_3$ shows qualitative differences in the NMR spectrum at different temperatures. The low-temperature spectrum has been found to be in agreement with the predicted structure that the phenyl groups occupy equatorial positions. This geometry leads to greater shielding of the equatorial methoxy group. From the coalescence temperature and the chemical shift difference between the two types of methoxy

TABLE IV

Compounds of Pentacoordinated Arsenic(V) Containing As—O and As—S Bonds

Compound	Properties (°C)	Physical measurements	Reference
As(OMe)$_5$	bp 39/2[a]	Mass spectra	59, 149
As(OEt)$_5$	bp 47/0.3	Mass spectra	59, 149
As(OMe)$_3$(n-OBu)$_2$		Mass spectra	149
As(OMe)$_4$(n-OBu)		Mass spectra	149
MeAs(OMe)$_4$	bp 30/0.3	^{1}H NMR	45
Me$_2$As(OMe)$_3$	bp 32/0.2	^{1}H NMR, ^{13}C NMR	45, 47
Me$_3$As(OMe)$_2$		IR, Raman spectra	131, 147
Me$_4$As(OMe)	bp 38/17	IR, Raman spectra, ^{1}H NMR, ^{13}C NMR	137, 147, 164
Me$_4$As(OEt)		IR, Raman spectra, ^{1}H NMR	52
PhAs(OMe)$_4$	bp 70/0.1	^{1}H NMR	45
Ph$_2$As(OMe)$_3$	bp 163/1	^{1}H NMR	45
Ph$_3$As(OMe)$_2$		^{1}H NMR	45
Me$_4$AsONH$_2$	bp 61–63/5	IR, Raman spectra	147
Me$_4$AsONHCH$_3$	bp 70–71/10	IR, Raman spectra, ^{1}H NMR	147
Me$_4$AsON(CH$_3$)$_2$	bp 60–62/10	IR, Raman spectra, ^{1}H NMR	147
Me$_4$AsON=CHCH$_3$	mp 86–90	IR, Raman spectra, ^{1}H NMR	147
Me$_4$AsON=C(CH$_3$)$_2$	mp 76–78	IR, Raman spectra, ^{1}H NMR	147
Me$_3$As[ON(CF$_3$)$_2$]$_2$	White crystalline solid	IR	7
Me$_2$CF$_3$As[ON(CF$_3$)$_2$]$_2$	Colorless liquid, mp 10–20	IR	7
Me(CF$_3$)$_2$As[ON(CF$_3$)$_2$]$_2$	mp 22 and 33 (two isomers)	IR	6, 7
Me$_2$As(OSiMe$_3$)$_3$	mp 53–56	IR, ^{1}H NMR	100
Ph$_3$As(OAc)$_2$	mp 210	IR	34, 71
(p-Tolyl)$_3$As(OAc)$_2$	mp 248	IR	21
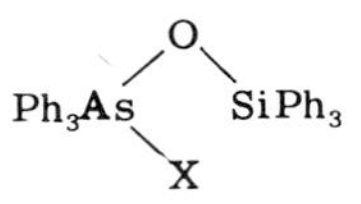 Ph$_3$**As**⟨O–SiPh$_3$ / X⟩		IR	21

224

Compound	mp/property	Characterization	Ref.
X = Cl	mp 106–109 (decomp.)	¹H NMR	*42*
X = Br	mp 150 (decomp.)	¹H NMR	*42*
$(C_6F_5)_3AsCl(OR)$			
R = Me	mp 215	IR	*146*
R = Et	mp 210	IR	*146*
R = Ph	mp 220	IR	*146*
Me₃As⟨O–CH₂–CH₂⟩ (5-ring with H₂C, CH₂)		¹H NMR, ¹³C NMR	*76, 165*
Ph₃As⟨O–CH₂–CH₂⟩ (5-ring)	mp 116–117	¹H NMR, thermal decomposition	*76*
(MeO)₃As(O–CH₂, O–CH₂)	mp 75/1	¹H NMR	*44*
R¹As(O–CR²₂–CR²₂–O)₂			
R¹ = OH, R² = H	mp 120	¹H NMR, ¹³C NMR, X ray	*54*
R¹ = OH, R² = Me	mp 118	¹H NMR	*69*
R¹ = OMe, R² = H		¹H NMR	*54, 67, 161*
			54, 67, 161

(*continued*)

TABLE IV (continued)

Compound	Properties (°C)	Physical measurements	Reference
R^1 = OMe, R^2 = Me		^{1}H NMR	54, 67, 161
Ph$_2$As–O–CH$_2$–CH$_2$–O–MeO		^{1}H NMR	44
Ph$_3$As(O–CH$_2$–CH$_2$–O)	mp 93–96	^{1}H NMR	44, 61
Ph$_3$As(O–CMe$_2$–CMe$_2$–O)	mp 92–95 mp 107–108	^{1}H NMR, ^{13}C NMR, mass spectra	44, 61 13
Ph$_2$As(O–CH$_2$–C(H$_2$)–C(H$_2$)–CH$_2$–O)MeO		^{1}H NMR	44
PhAs(O–CMe$_2$–CMe$_2$–O)$_2$	mp 103 mp 108	^{13}C NMR, X ray	161 68, 69
PhAs(O–C$_6$H$_4$–S)$_2$			5

	X	Y		
	OCH$_3$	OCH$_3$	mp 190 (decomp.)	72
	OC$_2$H$_5$	OC$_2$H$_5$	mp 192–193 (decomp.)	72
		Cl	mp 158 (decomp.)	72
	N(C$_2$H$_5$)$_2$	Cl	mp 167 (decomp.)	72
	N(CH$_3$)C$_6$H$_5$	Cl	mp 147 (decomp.)	72
	SC$_2$H$_5$	SC$_2$H$_5$		72

R = CMe$_2$CMe$_2$	mp 63–65 bp 131–132/3	^{1}H NMR — 161, 192
R = CHMeCHMe	bp 53–56/0.01	^{1}H NMR — 192
R = CH$_2$CF$_2$	mp 52–54 bp 110–111/4	^{1}H NMR — 161, 192

(*continued*)

TABLE IV (continued)

Compound	Properties (°C)	Physical measurements	Reference
R = o-C$_6$H$_4$, R' = CMe$_2$CMe$_2$	mp 73–75	^{1}H NMR	192
R = CMe$_2$CMe$_2$, R' = CH$_2$CH$_2$	bp 119–122/0.1 bp 53–56/0.1	^{1}H NMR	192
R = Me	Canary yellow crystals mp 150–152	^{1}H NMR, X ray	161, 192, 203
R = Bu	Canary yellow crystals mp 83		161
R = PhCH$_2$	Canary yellow crystals mp 147		161
R = CMe$_2$CMe$_2$	mp 176–178	^{1}H NMR	192
R = o-C$_6$H$_4$	mp 203–205	^{1}H NMR	192
R = CH$_2$CH$_2$		^{1}H NMR, mass spectra	73

R = o-C₆H₄ ... ¹H NMR, mass spectra ... 73

Yellow crystals mp 251 ... ¹H NMR ... 73

R	R'	X	Y		
Me	H₂CCH₂	O	S	mp 65–67	¹H NMR, mass spectra	128
Me	H₂CCH₂	O	NCH₃	mp 40–42	¹H NMR, mass spectra	128
Me	C₆H₄	O	NH	mp 178–180	¹H NMR, mass spectra	128
Me	C₁₀H₆	O	NH	mp 230–232	¹H NMR, mass spectra	128
Ph	H₂CCH₂	O	S	mp 99	¹H NMR, mass spectra	128
Ph	H₂CCH₂	O	NCH₃	mp 100	¹H NMR, mass spectra	128
Ph	C₆H₄	O	NH	mp 178–180	¹H NMR, mass spectra	128
Ph	C₁₀H₆	O	NH	mp 277–279	¹H NMR, mass spectra	128

R = Me; R¹, R², R³, R⁴ = H ... bp 110–111 ... ¹H NMR ... 37, 44, 60, 161, 162
R, R¹ = Me; R², R³, R⁴ = H ... bp 115.5–116 ... 161
R, R¹, R², R³, R⁴ = Me ... bp 131–132 ... 161

(*continued*)

TABLE IV (continued)

Compound	Properties (°C)	Physical measurements	Reference
R = Bu; R¹, R², R³, R⁴ = H	mp 20		161
	bp 140.5–141.5		161
R = Bu; R¹ = Me; R², R³, R⁴ = H	bp 142.6–143.4		161
R = Bu; R¹, R², R³, R⁴ = Me	bp 169–170		37, 44, 161
R = Ph; R¹, R², R³, R⁴ = H	mp 105.5	¹H NMR	54, 67, 161
R = Ph; R¹, R², R³, R⁴ = Me	mp 176	¹H NMR	37
R = PhCH₂; R¹, R², R³, R⁴ = H		¹H NMR	37
R = p-MeOC₆H₄; R¹, R², R³, R⁴ = H		¹H NMR	44
R = OMe; R¹, R², R³, R⁴ = H			

[structure: arsenic heterocycle with R group]

| R = Ph | mp 235 | IR, ¹H NMR | 9, 28, 37 |
| R = p-ClC₆H₄ | mp 223 | IR | 28 |

[structure: bis-arsenic bicyclic compound]

| | | | 192 |

R	X	Y		
H$_2$CCH$_2$	S	S	mp 175–180	193
H$_2$CCH$_2$	O	S	mp 232	193
Me$_2$CCMe$_2$	O	O	mp 158	193

Colorless crystals, IR, X ray — 26

— 64

[a] Divided values indicate °C/mm.

signals (58 Hz), the energy of activation has been estimated as 14.4 kcal/mol.

The variable-temperature ^{1}H- and ^{13}C-NMR spectra of Me$_2$As(OMe)$_3$ have been reinvestigated by Denney et al. (47). These findings suggest that it exists as a trigonal bipyramidal structure with one equatorial and two axial methoxy groups. The coalescence temperature for the process that renders the methoxy protons equivalent has been found to be 273 K, and the activation energy, 13.6 kcal/mol. The coalescence temperature for the process that renders the carbon atoms of the methoxy groups equivalent has been found to be 294 K, and the activation energy, 14.1 kcal/mol. It has been suggested that a slow intermolecular process cannot be entirely ruled out.

Reactions of bis(trifluoromethyl)nitroxyl with a number of methyl- and trifluoromethyl-substituted arsanes at room temperature give compounds of type Me$_{3-n}$(CF$_3$)$_n$M[ON(CF$_3$)$_2$]$_2$ (n = 0, 1, 2) (6, 7). A free radical mechanism has been proposed for these oxidative addition reactions.

Trialkyl- and triarylperoxyarsoranes have been obtained by the reaction of triorganyl dihaloarsoranes with either an alkylhydroperoxide in the presence of a tertiary amine or with the sodium salt of alkylhydroperoxides. These can also be prepared by the reaction between amino halides R$_3$As(NH$_2$)X and an alkylhydroperoxide or by the following exchange reactions:

$$R_3As(OOR'')_2 + 2R'OOH \longrightarrow R_3As(OOR')_2 + 2R''OOH$$

These moisture-sensitive diperoxides are stable at room temperature, but explode when heated in a flame. No further investigations have been made on these compounds.

Reaction of triphenylarsane with lead tetraacetate lead to the formation of Ph$_3$As(OAc)$_2$ (21, 34, 71). This compound reacts with amines to give Ph$_3$As=NR:

$$Ph_3As + Pb(OAc)_4 \longrightarrow Ph_3As(OAc)_2 \xrightarrow{+RNH_2} Ph_3As=NR$$

A number of crystalline arsenic compounds of the type R$_3$As(X)OSiPh$_3$ have been obtained by the reaction of triphenylarsane with silyl hypohalites, Ph$_3$SiOX (X = Cl, Br) (42).

Monomeric covalent compounds of type Me$_4$AsX (X = OR, ONH$_2$, ONR$_2$, or ON=CR$_2$) have been prepared by the reaction of pentamethylarsorane with equimolar amounts of alcohols, hydroxylamines, or oximes (52, 147, 164). According to vibrational spectra and low-temperature ^{1}H-NMR spectra, these compounds possess trigonal–bipyramidal structures.

Cyclization of $Me_3As=CH_2$ with ethylene oxide gives **XXI**:

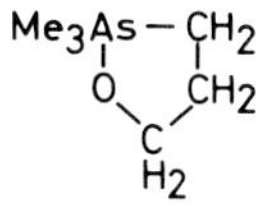

$$Me_3As-CH_2 \atop OCH_2 \atop C \atop H_2$$

XXI

Variable-temperature ^{1}H- and ^{13}C-NMR data suggest a trigonal–bipyramidal structure for **XXI**, with a methyl group and an oxygen atom in axial positions. The phenyl derivative has been obtained by the following reaction (76):

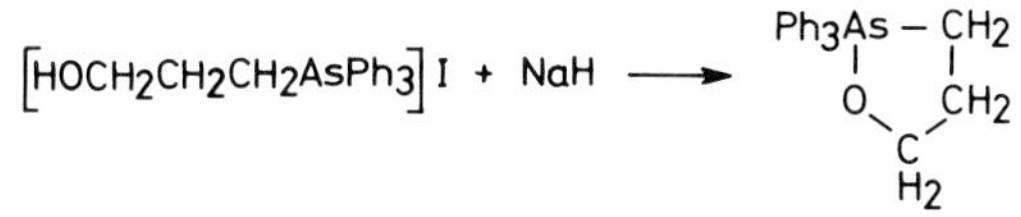

$$[HOCH_2CH_2CH_2AsPh_3]I + NaH \longrightarrow \begin{array}{c} Ph_3As-CH_2 \\ OCH_2 \\ C \\ H_2 \end{array}$$

The very stable spirocyclic compounds of types **XXII** and **XXIII** have been known for a long time (10, 54, 60, 161). These have been prepared by allowing 1 mol of the appropriate arsonic acid to react with 2 mol of the 1,2-dihydroxy compound, the water formed being removed either by performing the reaction in acetic anhydride or by azeotropic distillation from a suitable solvent, e.g., benzene (161).

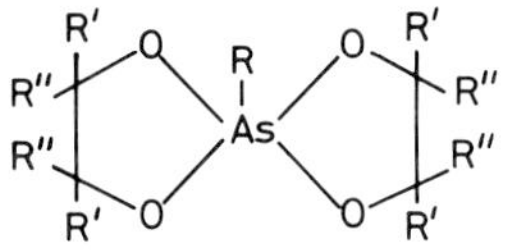

XXII 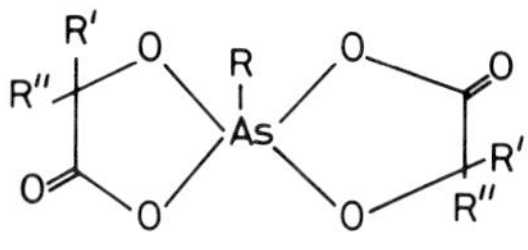   **XXIII**

Mixing of the arsorane $RAs(OMe)_4$ (R = OMe, Me, or Ph) with the diol compound in a 1:1 molar ratio leads to the formation of monocyclic arsoranes quantitatively only in the case where the diol is pinacol and R = OMe:

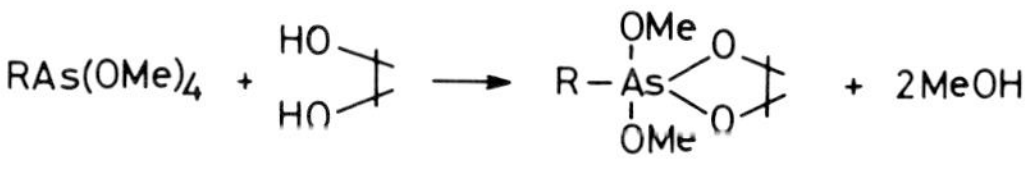

$$RAs(OMe)_4 + \begin{array}{c}HO \\ HO\end{array}\!\!\!\!\!\!>\!\!\!\!< \longrightarrow R-As\!\!\begin{array}{c}OMe \\ OMe\end{array}\!\!\begin{array}{c}O \\ O\end{array}\!\!\!\!\!\!>\!\!\!\!< + 2MeOH$$

When R = Me or Ph, a mixture of products is obtained (44). The reaction of $RAs(OMe)_4$ with pinacol or neopentyl glycol in a 1:2 molar ratio leads to the formation of five- and six-membered cyclic spiroarsoranes, respectively:

$$RAs(OMe)_4 + 2\,HO{-}\!\!\!\diagup\!\!\!{-}HO \longrightarrow \text{[spiro diester]} + 4\,MeOH$$

$$RAs(OMe)_4 + 2\,HO{-}\!\!\!\diagup\!\!\!{-}HO \longrightarrow \text{XXIV} + 4\,MeOH$$

Compounds of type **XXV** have been obtained by the reaction of $Ph_nAs(OMe)_{5-n}$ ($n = 2$ or 3) with a diol. It has been observed that compounds containing two rings are thermodynamically more stable than compounds with one ring. This is independent of ring size.

XXV (R = OMe, Ph)

The role of the pseudorotation process in compounds of types **XXII**, **XXIII**, **XXIV** (*37, 44, 67, 164*), and analogous phosphorus compounds (*108, 124*) has been studied by variable-temperature NMR spectroscopy. Goldwhite (*67*) reported the NMR spectral studies of a series of spirocyclic compounds of type **XXII** (R′ = R″ = H, Me; R = OH, OMe, Me, Ph) and concluded that these compounds exist as trigonal bipyramids with the two rings spanning axial–equatorial positions. The NMR results suggested rapid pseudorotation involving the two rings, even at temperatures as low as 173 K (**XXVI**). It has been suggested that the compounds of pentacoordinated arsenic(V) undergo pseudorotation more rapidly than the corresponding phosphorus compounds.

XXVI (Pseudorotation process)

This latter conclusion has been challenged by Casey and Mislow (*37*). These workers studied the barriers of pseudorotation in spiroarsoranes **XXII** (R = Me, $PhCH_2$, $p\text{-MeOC}_6H_4$, $p\text{-NO}_2C_6H_4$, Ph; R′ = R″ = Me). The ^{1}H-NMR spectra of each compound showed that two methyl signals, associated with the ring-methyl protons, coalesced to a single

TABLE V

Crystal Data for PhAs[O$_2$C$_2$(CH$_3$)$_4$]$_2$ and HOAs(O$_2$C$_2$H$_4$)$_4$

Cell constants	PhAs[O$_2$C$_2$(CH$_3$)$_4$]$_2$	HOAs(O$_2$C$_2$H$_4$)$_4$
a	9.150(5) Å	9.415(4) Å
b	12.699(8) Å	6.791(2) Å
c	17.386(7) Å	12.426(5) Å
β	103.73(5)°	119.11(6)°
Space group	$P2_{1/c}$, $Z = 4$	$P2_{1/c}$, $Z = 4$

signal at elevated temperatures. They suggested that this process can only be explained by a Berry pseudorotation in which one of the rings spans two equatorial positions. A process of this type requries a high energy of activation. For the various compounds studied, the $\Delta G^\ddagger$ values lie in the range 20–23 kcal/mol. These values are slightly higher than values obtained for somewhat analogous compounds of phosphorus (108). These results suggest that the barriers of pseudorotation in arsenic and phosphorus compounds are comparable.

The dynamic stereochemistry of spiroarsoranes containing five- and six-membered ring systems has also been studied by Dale and Froyen (44). The variable-temperature NMR results have been interpreted in terms of pseudorotation processes. It has been concluded that the observed spectra do not allow deduction as to whether trigonal–bipyramidal, rectangular–pyramidal, or any other intermediate structure is the most stable configuration in solution (44, 45).

The crystal and molecular structures of two spiroarsoranes of type **XXI** (R = Ph, R' = R'' = CH$_3$ and R = OH, R' = R'' = H) have been determined by single-crystal X-ray diffraction analyses (68, 69). The crystal data for these compounds are summarized in Table V. Both compounds have a geometry at the arsenic atom that lies on the Berry coordinate between rectangular–pyramidal and trigonal–pyramidal. These structures show close parallels between the structures of related arsenic and phosphorus systems. It has been concluded that, since the solid-state structures of these compounds lie close to the Berry coordinate, the dynamic process in solution is distortion along that coordinate (68, 69).

Although the oxidation of RAs(OR)$_2$ with selenium dioxide leads to the esters of the corresponding arsenic(V) acids, Wieber et al. (192) have shown that oxidation of the cyclic esters with SeO$_2$ in the presence of a diol results in compounds of type **XXVII**:

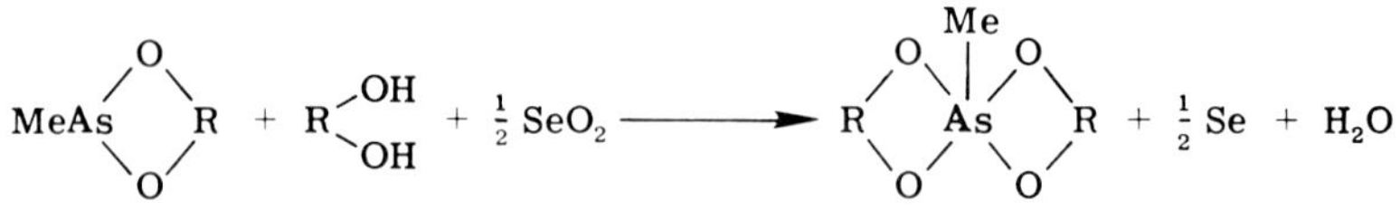

$$\text{MeAs}\begin{array}{c}\diagup\text{O}\diagdown\\ \diagdown\text{O}\diagup\end{array}\text{R} + \text{R}\diagup\begin{array}{c}\text{OH}\\ \text{OH}\end{array} + \tfrac{1}{2}\text{SeO}_2 \longrightarrow \text{R}\begin{array}{c}\diagup\text{O}\diagdown\overset{\overset{\text{Me}}{|}}{\text{As}}\diagup\text{O}\diagdown\\ \diagdown\text{O}\diagup\diagdown\text{O}\diagup\end{array}\text{R} + \tfrac{1}{2}\text{Se} + \text{H}_2\text{O}$$

XXVII

Unsymmetrical spirocyclic compounds have also been prepared:

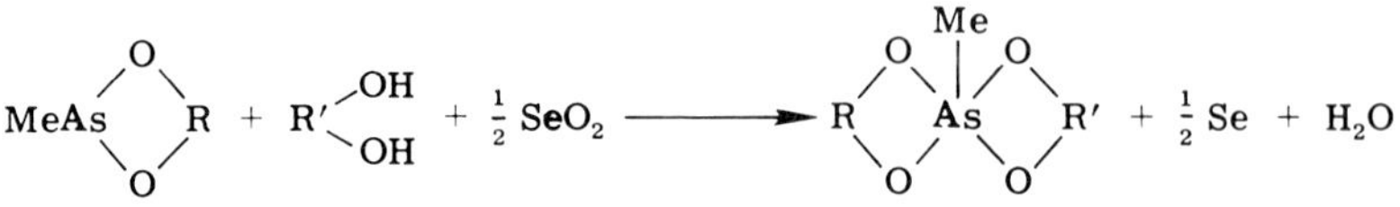

$$\text{MeAs}\begin{array}{c}\diagup\text{O}\diagdown\\ \diagdown\text{O}\diagup\end{array}\text{R} + \text{R}'\diagup\begin{array}{c}\text{OH}\\ \text{OH}\end{array} + \tfrac{1}{2}\text{SeO}_2 \longrightarrow \text{R}\begin{array}{c}\diagup\text{O}\diagdown\overset{\overset{\text{Me}}{|}}{\text{As}}\diagup\text{O}\diagdown\\ \diagdown\text{O}\diagup\diagdown\text{O}\diagup\end{array}\text{R}' + \tfrac{1}{2}\text{Se} + \text{H}_2\text{O}$$

The reaction of cyclic esters with quinones give either symmetrical or unsymmetrical spirocyclic compounds. For example, symmetrical spirocyclic compounds have been synthesized as follows (*192*):

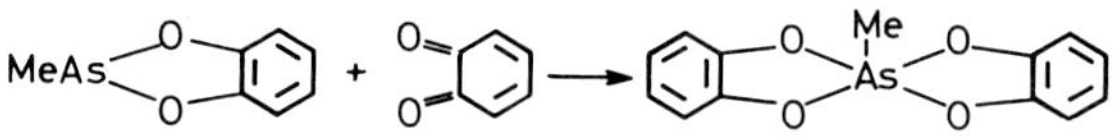

[Reaction scheme showing MeAs cyclic ester + benzoquinone → spirocyclic product with Me-As center]

The oxidation of 2-iodo-1,3,2-diheteroarsolanes by tetrachloro-*o*-benzoquinone has been studied by Wieber and Götz (*193*). In this reaction the cleavage of the As—I bond occurred with the formation of spirocyclic arsoranes:

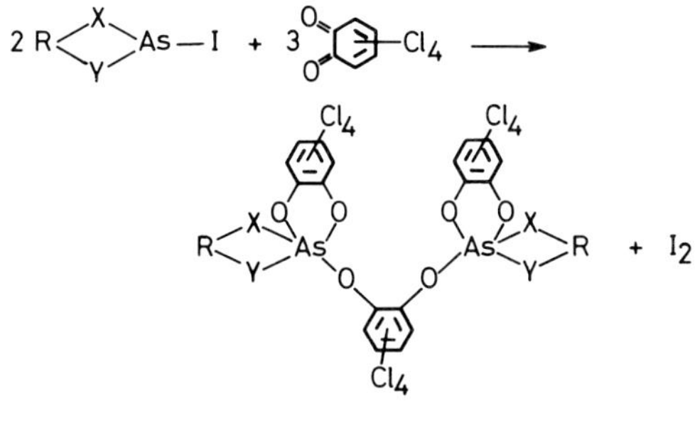

[Reaction scheme: 2 R(X,Y)As—I + 3 tetrachlorobenzoquinone → bis-spirocyclic arsorane product + I₂]

(X, Y = O or S; and R = CH₂CH₂ or Me₂CCMe₂)

It has additionally been observed that the same oxidizing agent can also cleave both As—I bonds in methyl diiodoarsane as depicted below:

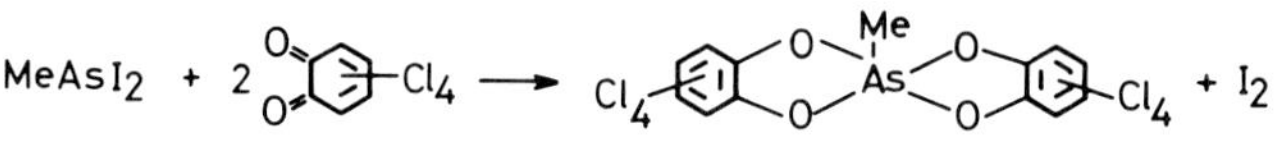

$$\text{MeAsI}_2 + 2\,\text{O=C}_6\text{Cl}_4\text{=O} \longrightarrow \text{Cl}_4\text{-spirocyclic-Me-As-spirocyclic-Cl}_4 + \text{I}_2$$

XXVIII

The same authors (*72*) have also reported that several substituents other than iodine do not undergo oxidative cleavage. Instead, these substituents are retained, and the oxidation by tetrachloro-o-benzoquinone proceeds in the following manner:

$$\text{Me-As}\genfrac{}{}{0pt}{}{X}{Y} + \text{O}\!\!=\!\!\bigcirc\!\!=\!\!\text{O-Cl}_4 \longrightarrow \text{Me - As}\genfrac{}{}{0pt}{}{X}{Y}\!\!<\!\!\genfrac{}{}{0pt}{}{O}{O}\!\!\bigcirc\!\!\text{-Cl}_4$$

(X, Y = methoxy, ethoxy, chlorine, piperdine, diethylamine, etc.)

Thioesters of methylarsenic(III) acid and an equimolar amount of tetrachloro-o-benzoquinone give unsymmetrical compounds (**XXIX**) (73):

$$R\!\!<\!\!\genfrac{}{}{0pt}{}{S}{S}\!\!>\!\!\text{As}\!\!<\!\!\genfrac{}{}{0pt}{}{O}{O}\!\!\bigcirc\!\!\text{-Cl}_4$$

XXIX

Compounds of type **XXIX** are stable at room temperature but on heating rearrange to symmetrical compounds of type **XXIX**. When a second equivalent of tetrachloro-o-benzoquinone is added, the As—S bonds are oxidized, yielding symmetrical compounds of type **XXVIII** and disulfides.

Mallon and Weiber (128) have prepared spirocyclic arsoranes **XXX** by the reaction between PhAs(O)(OH)$_2$ or MeAs(OMe)$_4$ with HOXYH (where R = Me, Ph; X = CH$_2$CH$_2$, o-C$_6$H$_4$, 2,3-C$_{10}$H$_6$; Y = S, NH, NMe). Wunderlich (203) determined the crystal structure of

$$X\!\!<\!\!\genfrac{}{}{0pt}{}{O}{Y}\!\!>\!\!\text{As}\!\!<\!\!\genfrac{}{}{0pt}{}{O}{Y}\!\!>\!\!X$$

XXX

MeAs(O$_2$C$_6$H$_4$)$_2$. The compound crystallizes in the orthorhombic space group $Pca2_1$, with cell constants a = 18.086(2), b = 8.294(1), c = 8.229(1) Å, Z = 4. The geometry of the pentacoordinated arsenic atom is described as a 74% rectangular pyramid with the methyl group in the apical position. The molecule contains trans basal angles O—As—O of 158.6 and 143.1° and apical basal angles C—As—O of 108.6, 108.3, 100.2, and 100.1°.

The crystal structure of PhAs(O$_2$C$_6$H$_4$)$_2$ reveals the first truly rectangular–pyramidal structure (46a). The compound crystallizes in the monoclinic space group, $P2_{1/c}$, with a = 16.787(5), b = 6.767(3), c = 27.374(6) Å, β = 90.37(2)°, and Z = 8. The structure was refined to R = 0.041, Rw = 0.060, and showed two independent molecules per asymmetric unit. The displacement along the Berry coordinate for two molecules, based on unit bond distances, is 99.2 and 94.8% from the trigonal bipyramid toward the rectangular pyramid. These studies suggest that structural principles found for phosphoranes apply thus far to arsenic(V) [and apparently to antimony(V)] derivatives. Molecular mechanics calculations on related spirocyclic phosphoranes and arsoranes support this conclusion (*46a*).

When a mixture of $(CF_3)_2AsI$ and CF_3AsI_2 in a 1:1 molar ratio is oxidized by H_2O_2, compound **XXXI** is obtained (*26*):

$$\begin{array}{c} HO \quad CF_3 \\ O-As-O \\ F_3C \quad | \quad O \quad | \quad CF_3 \\ As-O-As \\ F_3C \quad | \quad O \quad | \quad CF_3 \\ O-As-O \\ HO \quad CF_3 \end{array}$$

XXXI

Compound **XXXI** reacts with Me_3SiCl to give **XXXII**:

$$\begin{array}{c} F_3C \quad O \quad OH \\ | \quad / \quad \backslash \quad | \\ Cl-As \quad\quad As-Cl \\ | \quad \backslash \quad / \quad | \\ HO \quad O \quad CF_3 \end{array}$$

XXXII

The crystal and molecular structures of **XXXI** and **XXXII** have been determined by single-crystal X-ray diffraction analysis (*26*). Crystals of **XXXII** are monoclinic, space group $P2_{1n}$, $a = 5.543(1)$, $b = 9.345(2)$, $c = 11.496(2)$ Å, $\beta = 92.70(1)°$, $Z = 2$. It has a distorted trigonal–bipyramidal geometry with OH and CF_3 groups occupying axial positions.

V. As—Halogen-Containing Compounds

Compounds of pentacoordinated arsenic(V) containing As—halogen bonds are summarized in Table VI. Compounds containing As—Br and As—I bonds are generally ionic and hence are not included in the table. For the same reason, compounds of type R_4AsX have been excluded.

Arsenic pentafluoride is made by the reaction between AsF_3 or the oxide and elemental fluorine (*166*). The greater volatility of AsF_5 compared with the trihalide is associated with the zero dipole moment of AsF_5. Vibrational spectra of AsF_5 have been interpreted in terms of D_{3h} symmetry (*12*). ^{19}F-NMR studies indicate that all five nuclei are equivalent even at the lowest temperature observable. This suggests

TABLE VI

COMPOUNDS OF PENTACOORDINATED ARSENIC(V) CONTAINING ARSENIC–HALOGEN BONDS

Compound	Properties (°C)	Physical measurements	Reference
AsX$_5$			
X = F	mp −79.8	Vibrational spectra, ^{19}F NMR, ^{75}As NMR, mass spectra	39, 107, 114, 117, 158, 166, 181
X = Cl	bp −52.8	Vibrational spectra, mass spectra	142, 167, 168
RAsX$_4$			
R = Ph, X = F	bp 52–53/2^a	IR, NMR	173
R = Ph, X = Cl		IR, ^{35}Cl NQR	22, 38, 48, 154
R$_2$AsX$_3$			
R = Me, X = F	mp 85	^{19}F NMR	140
R = Me, X = Cl		IR	11, 154
R = CF$_3$, X = Cl	bp 93–95/722	IR	53
R = Ph, X = F	mp 94–96	IR, ^{19}F NMR	123, 140, 159, 160, 177
R = Ph, X = Cl		Vibrational spectra, ^{35}Cl NQR	46, 63, 115, 154, 190
R = PhCH$_2$, X = Cl			133, 134
R = Cyclohexyl, X = Cl			175
R = o-Biphenylyl, X = Cl		Thermal decomposition	197, 198
Ph$_2$As(CN)Cl$_2$			132
R$_3$AsX$_2$			
R = Me, X = F	mp 69–70 bp 54/12	Vibrational spectra, ^{1}H NMR, ^{19}F NMR, mass spectra	144, 183, 202
R = Me, X = Cl	mp 156–157	Vibrational spectra, ^{1}H NMR, ^{35}Cl NQR, mass spectra, X ray	48, 110, 144, 180, 183, 202
R = CF$_3$, X = Cl	bp 98.5	IR	53

(continued)

TABLE VI (continued)

Compound	Properties (°C)	Physical measurements	Reference
R = Et, X = F		Vibrational spectra, ^{1}H NMR, ^{19}F NMR, mass spectra	183
R = Et, X = Cl		Vibrational spectra, ^{1}H NMR, mass spectra	180, 183
R = n-Pr, X = Cl		Thermal decomposition	180
R = i-Pr, X = F		Vibrational spectra, ^{1}H NMR, mass spectra	184
R = i-Pr, X = Cl		Vibrational spectra, ^{1}H NMR, mass spectra	184
R = Bu, X = Cl		Thermal decomposition	180
R = Ph, X = F	mp 135–137	Vibrational spectra, ^{1}H NMR, ^{13}C NMR, ^{19}F NMR, X ray	12, 48, 110, 144, 180, 183, 202
	mp 139–140		144
R = Ph, X = Cl	mp 205	Vibrational spectra, ^{35}Cl NQR	19, 48, 79, 120, 126, 127, 179, 180
	mp 214–215		113
R = PhCH$_2$, X = F		Vibrational spectra, ^{1}H NMR, ^{19}F NMR, mass spectra	139, 182, 184
R = PhCH$_2$, X = Cl		Vibrational spectra, ^{1}H NMR, mass spectra	182, 184
R = C$_6$F$_5$, X = Cl	mp 190	IR	146
	mp 214–216		78
R = 2-Thienyl, X = F			145
R = p-Me$_2$NC$_6$H$_4$, X = Cl		^{1}H NMR	116
![arsenic ring structure with H$_2$C–CH$_2$–CH$_2$–CH$_2$ ring, As with Cl, Cl, CH$_3$]		Thermal decomposition	70

Compound		Property	Method	Ref.
Ph₃AsCl₂ (structure)			Thermal decomposition	197, 198
R₂As(CH₂)ₙAsR₂ (with F substituents)				
	$n = 1$, R = Me	mp 166	¹H NMR, ¹³C NMR, ¹⁹F NMR	159
	$n = 1$, R = Et	bp 106/0.001	¹H NMR, ¹³C NMR, ¹⁹F NMR	159
	$n = 1$, R = i-Pr	bp 118/0.001	¹H NMR, ¹³C NMR, ¹⁹F NMR	159
	$n = 1$, R = Ph	mp 134	¹H NMR, ¹³C NMR, ¹⁹F NMR	159
	$n = 2$, R = Ph	mp 144	¹H NMR, ¹³C NMR, ¹⁹F NMR	159
	$n = 3$, R = Ph	mp 153	¹H NMR, ¹³C NMR, ¹⁹F NMR	159
RR'₂AsX₂				
	R = Me, R' = Ph, X = F	mp 96	¹H NMR, ¹³C NMR, ¹⁹F NMR	159
	R = Me, R' = Ph, X = Cl			33
	R = Et, R' = Ph, X = Cl		Thermal decomposition	1, 3
	R = n-Pr, R' = Ph, X = Cl		Thermal decomposition	1, 3
	R = Ph, R' = Me, X = Cl		Thermal decomposition	1, 3
	R = Ph, R' = Et, X = Cl		¹H NMR	2
	R = Styryl, R' = Me, X = Cl		Thermal decomposition	46
R(p-MeC₆H₄)₂AsCl₂ (R = Et, Pr Me₂CH, Bu, Me₂CHCH₂, pentyl)			Thermal decomposition	65
RR'R''AsX₂				
	R = Me, R' = Et, R'' = Ph, X = Cl			33
	R = Me, R' = α-naphthyl, R'' = Ph, X = Cl			119

a Divided values indicate °C/mm.

that the barrier to pseudorotation, which interchanges axial and equatorial halogens, is small. ^{75}As-NMR spectra also support this observation (*114*). The molecular structure of AsF$_5$ has been determined by electron diffraction studies (*39*). The molecule has been found to be a trigonal bipyramid with axial bonds 0.055 ± 0.010 Å longer than equatorial bonds and an average arsenic fluorine bond length of 1.678 ± 0.002 Å.

Arsenic pentachloride has been prepared by irradiating a solution of AsCl$_3$ in chlorine with UV light at 168 K. It has been characterized by chemical analysis and by comparison of its Raman spectrum with those of PCl$_5$ and SbCl$_5$ (*167, 168*). Arsenic pentachloride is a soft yellow solid. It melts with partial decomposition at ~223 K. The Raman spectrum also suggests that it has a trigonal–bipyramidal structure in both the liquid and solid states. The difference in stability of PCl$_5$ and AsCl$_5$ has been attributed not to the As—Cl and P—Cl bond strengths but to the difference in the ionization energies in the first step of the chlorination reaction, which suffices to make the AsCl$_3$ reaction endothermic. It is interesting to mention here that the ionization energy of AsCl$_3$ is about 28 kcal/mol higher than that of PCl$_3$. Thus the large ionization energy in an excited state will be correspondingly increased. It has been suggested that this unexpected order follows the transition metal contraction. The small screening of the large nuclear charge of arsenic by the 3d electrons causes a lowering of the energy of the 4s orbitals, giving the nonbonding electron pair in AsCl$_3$ higher s character. Nonempirical valence shell SCFMO calculations suggest that the dissociation energy for the process AsCl$_5$ → AsCl$_3$ + 2Cl is negative.

Compounds of the type RAsX$_4$ are rare. Baeyer (*11*) reported that methyl tetrachloroarsorane is an unstable crystalline compound formed by the action of chlorine on methyldichloroarsane. Attempts to prepare MeAsCl$_4$ by this method led to explosions at low temperatures. It has been concluded that this compound is extremely unstable, if it exists at all (*48*). Phenyltetrachloroarsorane and the tolyl compounds have been synthesized by the reaction of chlorine with dichloroarsanes (*33*). However, the para-substituted isomer could not be obtained by this method. Many organyl tetrafluoroarsoranes have been prepared by the following route (*172–174*):

$$\text{RAsO}_3\text{H}_2 + 3\text{SF}_4 \xrightarrow{70°\text{C}} \text{RAsF}_4 + 3\text{SOF}_2 + 2\text{HF}$$

If this reaction is run at low temperatures with less sulfur tetrafluoride, RAsOF$_2$ is formed (*174*).

All of the organyl tetrahaloarsoranes are readily hydrolyzed by water to the arsenic(V) acids. When heated in air, phenyltetrachloroarsorane produces the dichloroarsane and chlorine, but when heated to 150°C in a sealed tube, it gives chlorobenzene and trichloroarsane.

A trigonal–bipyramidal structure has generally been assumed for compounds of the type $RAsX_4$ (X = F, Cl). Smith (*173*) reported the IR and NMR spectra of $PhAsF_4$ but was unable to draw any conclusions regarding its structure. Muetterties *et al.* (*140*) observed only one single peak in the ^{19}F-NMR spectrum for this compound and suggested that a fast intramolecular fluorine exchange might be occurring.

Vibrational spectra of $PhAsCl_4$ indicate that it has a trigonal–bipyramidal structure with an equatorial phenyl group (*154*). This conclusion gets support from a ^{35}Cl-NQR study of $PhAsCl_4$ (*48*). In trigonal–bipyramidal molecules of type $RAsX_4$, an atom in an axial position has a much lower NQR frequency than that of a similar atom in an equatorial site. This is reasonable since the axial bonds are longer and therefore more ionic in character. The ^{35}Cl-NQR spectrum of $RAsCl_4$ is very similar to that of the analogous phosphorane (*48, 125, 176*). It indicates four independent chlorine sites, two of which appear from the frequencies to be considerably more ionic than the others. This observation is entirely consistent with a trigonal–bipyramidal structure for the molecule, with the phenyl group occupying an equatorial site **XXXIII**:

$$R-\underset{\underset{X}{|}}{\overset{\overset{X}{|}}{As}}\begin{matrix}X\\X\end{matrix}$$

XXXIII

Compounds of type R_2AsX_3 are also rare, but are relatively more stable than $RAsX_4$. They are generally prepared by the reaction of dry halogens on dialkyl- or diarylhaloarsanes. For example, dimethyltrichloroarsorane has been synthesized by the reaction of chlorine with dimethylchloroarsane in carbon disulfide solution or by the reaction of PCl_5 on $[(CH_3)_2As]_2O$ (*11*). Diphenyltrichloroarsorane has been prepared by the reaction of diphenylchloroarsane and chlorine. It can also be prepared by the action of thionyl chloride on diphenylarsenic(V) acid (*63*).

Bis(trifluoromethyl)trichloroarsorane has been obtained by allowing tris(trifluoromethyl)arsane to react with chlorine in a sealed tube for 1 month (*53*). Dibenzyltrichloroarsorane has been synthesized by heating tribenzylarsane with an excess of benzyl chloride at 200°C

(*134*). It can also be obtained when benzyl chloride and trichloroarsane are condensed with sodium (*133*).

Ph$_2$AsF$_3$ has been prepared by the following reactions (*177*):

$$Ph_2AsO_2H + 2SF_4 \longrightarrow Ph_2AsF_3 + 2SOF_2 + HF$$

$$2AsF_5 + 2C_6H_6 \longrightarrow Ph_2AsF_2^+AsF_6^- + 2HF$$

$$Ph_2AsF_2^+AsF^- + CsF \longrightarrow Ph_2AsF_3 + CsAsF_6$$

The diorganyl trihaloarsoranes are also moisture-sensitive compounds and are readily hydrolyzed to arsenic(V) acid. They are decomposed at relatively low temperatures. Dicyclohexyltrichloroarsorane, when warmed to 80–90°C, loses chlorocyclohexane to produce cyclohexyldichloroarsane (*175*). Diphenyltrichloroarsorane, when heated in a sealed tube to 200°C, gives chlorobenzene and phenyldichloroarsane (*122*). Di-*o*-biphenyltrichloroarsorane (*198*) loses hydrogen chloride at 265°C to yield **XXXIV**:

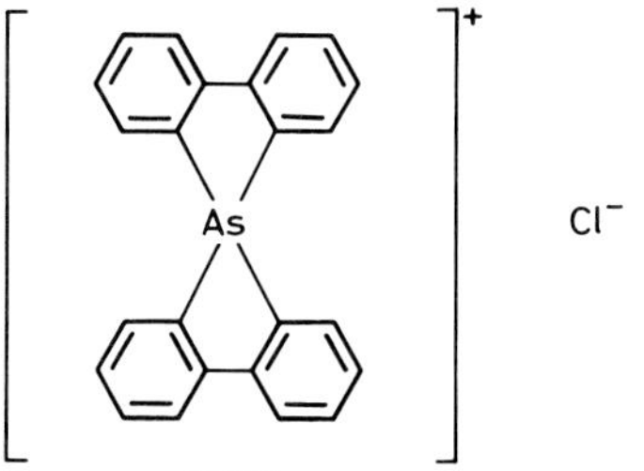

XXXIV

Vibrational spectral studies have suggested that Me$_2$AsCl$_3$ and Ph$_2$AsCl$_3$ have a trigonal–bipyramidal structure with methyl or phenyl groups in equatorial positions (**XXXV**) (*154*). These results are in complete agreement with the results of ^{35}Cl-NQR studies (*48*).

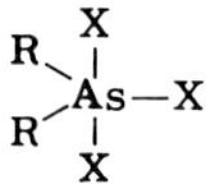

XXXV

Muetterties et al. (*140*) examined the ^{19}F-NMR spectra of diphenyltrifluoroarsorane and dimethyltrifluoroarsorane. The spectrum of Ph$_2$AsF$_3$ was found to consist of a doublet and triplet of relative intensity 2:1 with a common coupling constant of 67 Hz. It was suggested that this compound has a slightly distorted trigonal–bipyramidal structure in which two fluorine atoms occupy axial positions and the other fluorine atom occupies an equatorial position (**XXXV**). Littlefield and Doak (*123*) questioned the published ^{19}F-NMR spectrum. These workers fluorinated Ph$_2$AsH or PhAsCl with SF$_4$ to obtain a crystalline solid, Ph$_2$AsF$_3$. The ^{19}F-NMR spectrum of this compound consisted of a singlet relatively unchanged from 183 K to room temperature. These findings have been attributed to pseudorotation.

In order to settle this controversy, Tanzella and Bartlett (*177*) studied the ^{19}F-NMR spectrum of Ph$_2$AsF$_3$ again. These workers found the same ^{19}F-NMR spectrum as that reported by Muetterties *et al.* (*140*) and concluded that Ph$_2$AsF$_3$ is a rigid trigonal bipyramid, with phenyl groups in equatorial positions.

The structure of dimethyltrifluoroarsorane appears to be markedly different from that of the analogous alkyl and aryl derivatives of Group VA elements (*140*). Thus unlike Ph$_2$AsF$_3$, the dimethyl compound indicates evidence of association and is not very soluble in aromatic solvents. ^{19}F-NMR spectra in acetone or acetonitrile show a singlet at 25°C and doublet–triplet resonances at low temperatures. It has been suggested that it is probably behaving as a Lewis acid and forming an octahedral species in solution. However, more studies are required to establish the structures of these compounds with certainty.

As compared to RAsX$_4$ and R$_2$AsX$_3$, compounds of the type R$_3$AsX$_2$ have been studied in much greater detail. The usual procedure for the synthesis of R$_3$AsX$_2$ is the treatment of a trialkyl- or triarylarsane in a nonpolar solvent with the halogens in the same solvent (*19, 33, 38, 45, 51, 119, 135, 169, 191*). TlCl$_3$, PbCl$_4$, AsCl$_3$, TeCl$_4$, CuCl$_2$, or HgCl$_2$ have been used instead of Cl$_2$ as chlorinating agents (*18, 21, 33, 62, 127, 146*). The reaction of triphenylarsane oxide with 2 mol of acetyl chloride gives Ph$_3$AsCl$_2$ (*179*):

$$Ph_3AsO + 2MeC(O)Cl \longrightarrow Ph_3AsCl_2 + (MeCO)_2O$$

Fluorination of Ph$_3$AsO by aqueous HF (1–40%) gives Ph$_3$AsF$_2$ (*66*). Reaction of thionyl chloride or sulfur dichloride with triphenylarsane results in Ph$_3$AsCl$_2$ (*120*):

$$Ph_3As + SOCl_2 \longrightarrow Ph_3AsCl_2 + \tfrac{1}{2}S + \tfrac{1}{2}SO_2$$

A number of mixed aliphatic–aromatic dihaloarsoranes (RR$'_2$AsX$_2$ or RR$'$R$''$AsX$_2$) have been prepared from the arsanes and the appropriate halogens (*33, 49, 119*). The reaction of RR$'_2$AsS with PCl$_3$ or acetylchloride yields RR'_2AsCl_2 (*1, 180*). Similarily, reaction of RR'_2AsO with excess of HX gives RR'_2AsX_2 (*3, 40, 53, 134*).

Trialkyl- or triaryldifluoroarsoranes have been obtained by the metathetical reaction between trialkyl- or triaryldichloroarsorane and silver fluoride (*53, 144, 145*).

Fluorination of Ph_2AsR with fluorine in $CHCl_3$ yields Ph_2AsF_2R (R = Me, Ph). Similarly, $R_2As(CH_2)_nAsR_2$ produces $R_2AsF_2(CH_2)_nAsF_2R_2$ (R = Me, Et, Me_2CH, Ph). Ph_2AsH and fluorine results in the formation of $Ph_2AsAsPh_2$, further fluorination of which gives Ph_2AsF_3 (*159*).

Compounds of type R_3AsX_2 are generally low-melting, crystalline solids, soluble in alcohol but only slightly soluble or insoluble in nonpolar solvents. These can readily be reduced to tertiary arsanes. They are hydrolyzed by water or alkali. Compounds containing at least one alkyl group yield on pyrolysis an alkyl halide and a haloarsane (*49*):

$$R_3AsX_2 \longrightarrow R_2AsX + RX$$

Initially, three possible structures have been suggested for R_3AsX_2 (X = halogen), namely, $(R_3As)^{2+}2X^-$, $(R_3AsX)^+X^-$, and R_3AsX_2, in which the arsenic atom is surrounded by six, eight, or ten valence electrons, respectively. The structure $(R_3As)^{2+}2X^-$ was discarded on the basis that even in dilute solution these compounds do not ionize completely (*143*). For the remaining two structural possibilities, conflicting results are reported in the literature (*129, 155*). It is interesting to point out here that vibrational studies indicated that R_3PX_2 has the $(R_3PX)^+X^-$ ionic structure (*74*), while X-ray data show that R_3SbX_2 exists as a trigonal bipyramid with the two halogens at the axial positions (*185*). The Sb—X distances, however, are longer than the sum of the covalent radii and suggest that the Sb—X bond is partially ionic. Conductivity and IR data (*19, 77, 126, 144, 145*) indicate that compounds of type R_3AsX_2 (R = Me, Et, $PhCH_2$, 2-thienyl, Ph; X = F, Cl) also have a trigonal–bipyramidal structure. The conductivities of R_3AsBr_2 (R = Me, Et, Ph) in acetonitrile are larger than those of the corresponding R_3AsCl_2 compounds but are low compared to the values observed for strong 1:1 electrolytes. Infrared spectra indicate that Et_3AsBr_2 has a covalent, but Me_3AsBr_2 has an ionic structure in the solid state (*144, 145*). R_3AsI_2 forms highly conducting solutions in acetonitrile.

A ^{35}Cl-NQR study (*48*) indicates that R_3AsCl_2 (R = Me or Ph), unlike the phosphorus analogs, possesses a trigonal–bipyramidal structure in which all R groups occupy equatorial sites and the two chlorine atoms are situated on the axial positions (**XXXVI**):

$$\begin{array}{c} X \\ | \\ R-As \\ | \\ X \end{array} \!\!\! \begin{array}{c} R \\ \diagup \\ \diagdown \\ R \end{array}$$

XXXVI

A preliminary communication on the X-ray analysis (*110*) of Me_3AsCl_2 and Me_3AsBr_2 is in complete agreement with the earlier studies.

Muetterties *et al.* (*140*) observed that Ph_3AsF_2 is monomeric in benzene solution, and the ^{19}F-NMR spectrum is consistent with a trigonal–bipyramidal structure, with the fluorine atoms in axial positions. Augustine *et al.* (*8*) determined the crystal structure of Ph_3AsF_2 from three-dimensional X-ray diffractometer data. Crystals of Ph_3AsF_2 are orthorhombic, space group *Pbcn*, with $a = 6.270(1)$, $b = 16.593(3)$, $c = 14.519(2)$ Å, and $X = 4$; molecular symmetry C_2. The crystals contain well-separated discrete molecules. The arsenic atom has nearly regular trigonal–bipyramidal geometry with axial fluorine atoms (**XXXVI**).

Moreland and co-workers (*139*) reported ^{1}H- and ^{19}F-NMR spectra of $(PhCH_2)_3AsF_2$ as a function of temperature. At 0°C the methylene protons occurred as a triplet, which coalesced to a broad singlet and finally to a sharp singlet as the temperature was raised to 45°C. This result suggests that an intermolecular exchange of fluorines occurs with increase in temperature. This exchange has been shown to be of first order and suggests that the lifetime is independent of concentration. Since the compound is monomeric, the mechanism of exchange is believed to involve a dissociative step. The energy of activation has been found to be 12.3 ± 2.0 kcal/mol.

Verdonck *et al.* (*182*) reported both vibrational and ^{1}H- and ^{19}F-NMR spectra for the compounds $(PhCH_2)_3MX_2$ (M = As or Sb and X = F or Cl). The vibrational spectra (in the solid state and in solution) have been interpreted in terms of a slightly distorted trigonal bipyramid. The methylene NMR signal has been found to split into a triplet by coupling with two fluorine atoms in $(PhCH_2)_3MF_2$ (M = As or Sb). However, in the case of the antimony compound, some collapse was observed and interpreted as due to intermolecular exchange of the fluorine atoms.

VI. Conclusion

From this account, some general features have emerged, which can be summarized as follows. Although many compounds of pentacoordinated arsenic(V) have been synthesized, the fundamental hydrogenated skeleton, AsH_5, and the pentaalkylarsoranes are still unknown. An exception is Me_5As. The overall stabilities and reactivities of pentaorganyl arsoranes appear to be governed mainly by stereochemical factors, namely, bulkiness of the groups and angle strain when bidentate substituents are present.

Both the geometries and dynamics of compounds of pentacoordinated arsenic(V) and other Group VA elements have been studied. The stereochemistries of spirocyclic phosphoranes reveal a continuous series of angular geometries at phosphorus between the idealized trigonal–bipyramidal (symmetry D_{3h}) and square–pyramidal (symmetry C_{4v}) forms, the latter of which is reduced by chelation of the phosphorus to a rectangular pyramid (symmetry C_{2v}). Compounds of arsenic are expected to show similar stereochemical properties, but only a few experimental results exist. The stereochemistry of arsoranes and organoarsoranes can be described in terms of distorted trigonal bipyramidal ground states, which are subject to rapid intramolecular exchange processes of the Berry type, running through tetragonal–pyramidal transition states. Because of the large size of the arsenic atom, the influence of the steric hindrance in determining the geometry of pentacoordination is reduced.

In contrast to phosphorus, tetracoordinated arsenic appears to be less stable and rearranges to yield more stable pentacoordinated arsenic compounds. Thus compounds with As=N or As=O bonds dimerize or oligomerize to give compounds with pentacoordinated arsenic.

It is well known that pentacovalent cyclic phosphorus compounds play an important role as intermediates in reactions involving nucleophilic attack on tetracoordinated phosphorus in biological systems. According to this background it appears to us that it is important to prepare the arsenic derivatives, which are more stable than the corresponding phosphorus compounds and allow the study of their conformation.

ACKNOWLEDGMENTS

This work was supported by Deutsche Forschungsgemeinschaft. R.B. thanks the Alexander von Humboldt Foundation for a fellowship.

References

1. Abalonin, B. E., Gatilov, Yu. F., and Vasilenko, G. I., *Zh. Obsch. Khim.* **46,** 2734 and 2737 (1976).
2. Abalonin, B. E., Gatilov, Yu. F., Zykova, T. V., Vasilenko, G. I., Izmailova, Z. M., and Zhikhareva, N. A., *Dokl. Akad. Nauk SSSR* **226,** 1323 (1976).
3. Abalonin, B. E., Gatilov, Yu. F., and Izmailova, Z. M., *Zh. Obshch. Khim.* **47,** 624 (1977).
4. Aksnes, D. W., Amer, F. A., and Bergesen, K., *Acta Chem. Scand. Sect. A* **30,** 109 (1976).
5. Andrä, A., and Andrä, K., *Z. Anorg. Allgem. Chem.* **434,** 127 (1977).
6. Ang, H. G., and Lien, W. S., *J. Fluorine Chem.* **3,** 235 (1973).
7. Ang, H. G., and Lien, W. S., *J. Fluorine Chem.* **15,** 453 (1980).
8. Augustine, A., Ferguson, G., and March, F. C., *Can. J. Chem.* **53,** 1647 (1975).
9. Azerbaev, I. N., Abramova, Z. A., Bosyakov, Yu. G., and Alekeeva, N. N., *Izv. Akad. Nauk Kaz. SSR, Ser. Khim.* **25,** 49 (1975).
10. Backer, H. J., and Van Oosten, R. P., *Rec. Trav. Chim. Pays-Bas* **59,** 41 (1940).
11. Baeyer, A., *Liebigs Ann. Chem.* **107,** 257 (1958).
12. Banks, R. E., Haszeldine, R. N., and Hatton, R., *Tetrahedron Lett.* **41,** 3993 (1967).
13. Baumstark, A. L., Landis, M. E., and Brooks, P. J., *J. Org. Chem.* **44,** 4251 (1979).
14. Beattie, I. R., Livingston, K. M. S., Ozin, G. A., and Sabine, R., *J. Chem. Soc. Dalton Trans.* 784 (1972).
15. Beauchamp, A. L., Bennett, M. J., and Cotton, F. A., *J. Am. Chem. Soc.* **90,** 6675 (1968).
16. Bernstein, L. S., Kim, J. J., Pitzer, K. S., Abramowitz, S., and Levin, I. W., *J. Chem. Phys.* **62,** 3671 (1975).
17. Bernstein, L. S., Abramowitz, S. A., and Levin, I. W., *J. Chem. Phys.* **64,** 3228 (1976).
18. Berry, F. J., Gunduz, N., Roshani, M., and Smith, B. C., *Commun. Fac. Sci. Univ. Ankara, Ser. B* **22,** 21 (1975).
19. Beveridge, A. D., and Harris, G. S., *J. Chem. Soc.* 6076 (1964).
20. Bhattacharya, S. N., and Singh, M., *Indian J. Chem. Sect. A* **16,** 778 (1978).
21. Bhattacharya, S. N., and Singh, M., *Indian J. Chem. Sect. A* **18,** 515 (1979).
22. Blicke, F. F., and Monroe, E., *J. Am. Chem. Soc.* **57,** 720 (1935).
23. Bohra, R., and Roesky, H. W., *Inorg. Synth.*, in press (1984).
24. Bohra, R., Roesky, H. W., Lucas, J., Noltemeyer, M., and Sheldrick, G. M., *J. Chem. Soc. Dalton Trans.* 1011 (1983).
25. Bohra, R., and Roesky, H. W., *J. Fluorine Chem.* **25,** 145 (1984).
26. Bohra, R., Roesky, H. W., Noltemeyer, M., and Sheldrick, G. M., unpublished results.
27. Brabant, C., Blanck, B., and Beauchamp, A. L., *J. Organomet. Chem.* **82,** 231 (1974).
28. Braunholtz, J. T., and Mann, F. G., *J. Chem. Soc.* 3285 (1957).
29. Breindel, A. W., *U.S. Patent* 3, 317, 575 (1967).
30. Brock, C. P., and Webster, D. F., *Acta Crystallogr. Sect. B* **32,** 2089 (1976).
31. Brock, C. P., *Acta Crystallogr., Sect. B* **33,** 193 (1977).
32. Brock, C. P., *Acta Crystallogr., Sect. A* **33,** 898 (1977).
33. Burrows, G. J., and Lench, A., *J. Proc. R. Soc., N.S. Wales* **70,** 294 and 437 (1937).
34. Cadogan, J. I. G., and Gosney, I., *J. Chem. Soc. Perkin I* 466 (1974).
35. Cahours, A., *Liebigs Ann. Chem.* **122,** 327 (1862).

36. Carraher, C. E., and Moon, W. G., *Am. Chem. Soc., Div. Org. Coat. Plast. Pap.* **34**, 468 (1974); Carraher, C. E., and Moon, W. G., *Eur. Polym. J.*, **12**, 329 (1976); Carraher, C. E., Moon, W. G., and Langworthy, T. A., *Am. Chem. Soc., Div. Polym. Chem. Polymer Preprint* **17**, 1 (1976).
37. Casey, J. P., and Mislow, K., *J. Chem. Soc. Chem. Commun.* 1410 (1970).
38. Chatt, J., and Mann, F. G., *J. Chem. Soc.* 1184 and 1192 (1940).
39. Clippard, F. B., and Bartell, L. S., *Inorg. Chem.* **9**, 805 (1970).
40. Cookson, R. C., and Mann, F. G., *J. Chem. Soc.* 618 (1947).
41. Crow, J. P., and Cullen, W. R., *Int. Rev. Sci. (MTP)* **4**, 355 (1972).
42. Dahlmann, J., and Austenat, L., *J. Prakt. Chem.* **312**, 10 (1970).
43. Dahlmann, J., and Winsel, K., *J. Prakt. Chem.* **321**, 370 (1979).
44. Dale, A. J., and Frøyen, P., *Acta Chem. Scand. Sect. B* **29**, 741 (1975).
45. Dale, A. J., and Frøyen, P., *Acta Chem. Scand. Sect. B* **29**, 362 (1975).
46. Das Gupta, H. N., *J. Indian Chem. Soc.* **14**, 400 (1937).
46a. Day, R. O., Holmes, J. M., Sau, A. C., Holmes, R. R., Deiters, J., and Devillers, J. R., *J. Am. Chem. Soc.* **104**, 2127 (1982).
47. Denney, D. B., Denney, D. Z., and Tsai, J. H., *J. Chem. Res. (S)* 458 (1978).
48. Dillon, K. B., Lynch, R. J., and Waddington, T. C., *J. Chem. Soc. Dalton Trans.* 1478 (1976).
49. Doak, G. O., and Freedman, L. D., "Organometallic Compounds of Arsenic, Antimony and Bismuth." Wiley (Interscience), New York, 1970.
50. Dub, M., "Organometallic Compounds," Vol. 3, 2nd ed., 1st Suppl. Springer Verlag, Berlin and New York, 1972.
51. Dyke, W. J. C., Davies, G., and Jones, W. J., *J. Chem. Soc.* 185 (1931).
52. Eberwein, B., Ott, R., and Weidlein, J., *Z. Anorg. Allg. Chem.* **431**, 95 (1977).
53. Eméleus, H. J., Haszeldine, R. N., and Walaschewski, E. G., *J. Chem. Soc.* 1552 (1953).
54. Englund, B., *J. Prakt. Chem.* **120**, 179 (1928).
55. Friedrich, M. E. P., and Marvel, C. S., *J. Am. Chem. Soc.* **52**, 376 (1930).
56. Frøyen, P., *Acta Chem. Scand.* **23**, 2935 (1969).
57. Frøyen, P., *Acta Chem. Scand.* **25**, 983 (1971).
58. Frøyen, P., *Acta Chem. Scand.* **27**, 141 (1973).
59. Frøyen, P., and Moeller, J., *Org. Mass Spectrom.* **9**, 132 (1974).
60. Gamayurova, V. S., Kuzmin, V. K., Chernokalskii, B. D., and Shagidullin, R. R., *Zh. Obshch. Khim.* **43**, 1937 (1973).
61. Gamayurova, V. S., Gordeev, V. K., and Chernokalskii, B. D., *Zh. Obshch. Khim.* **49**, 817 (1979).
62. Gamayurova, V. S., Gordeev, V. K., and Chernokalskii, B. D., *Zh. Obshch. Khim.* **49**, 2780 (1979).
63. Gibson, C. S., Johnson, J. D. A., and Vining, D. C., *Rec. Trav. Chim. Pays-Bas* **49**, 1006 (1930).
64. Gigauri, R. D., Chernokalskii, B. D., Indzhiya, M. A., and Gvilava, L. I., *Zh. Obshch. Khim.* **48**, 1080 (1978).
65. Gigauri, R. D., Goderdzishvili, L. I., Shatakishvili, T. N., and Chernokalskii, B. D., *Zh. Obshch. Khim.* **50**, 2517 (1980).
66. Glidewell, C., Harris, G. S., Holden, H. D., Liles, D. C., and McKechnie, J. S., *J. Fluorine Chem.* **18**, 143 (1981).
67. Goldwhite, H., *J. Chem. Soc. Chem. Commun.* 651 (1970).
68. Goldwhite, H., Grey, J., and Teller, R., *J. Organomet. Chem.* **113**, C1 (1976).
69. Goldwhite, H., and Teller, R., *J. Am. Chem. Soc.* **78**, 5357 (1978).

70. Gorski, I., Schpanski, W., and Muljar, L., *Ber.* **67**, 730 (1934).
71. Gosney, I., and Lloyd, D. M. G., *Tetrahedron* **29**, 1697 (1973).
72. Götz, J., and Wieber, M., *Z. Anorg. Allg. Chem.* **423**, 239 (1976).
73. Götz, J., and Wieber, M., *Z. Anorg. Allg. Chem.* **423**, 235 (1976).
74. Goubeau, J., and Baumgärtner, R., *Z. Electrochim.* **64**, 598 (1960).
75. Haiduc, I., "The Chemistry of Inorganic Ring Systems." Wiley (Interscience), New York, 1970.
76. Hands, A. R., and Mercer, A. J. H., *J. Chem. Soc. C* 1099 (1967).
77. Harris, G. S., *Proc. Chem. Soc.* 65 (1961).
78. Harris, G. S., and Ali, M. F., *Inorg. Nucl. Chem. Lett.* **4**, 5 (1968).
79. Harris, G. S., Mack, I. M., and McKechnie, J. S., *J. Fluorine Chem.* **11**, 481 (1978).
80. Hass, D., *Z. Anorg. Allg. Chem.* **347**, 123 (1966).
81. Hass, D., *Z. Chem.* **7**, 465 (1967).
82. Hass, D., and Cech, I., *Z. Chem.* **9**, 384 and 432 (1969).
83. Hellwinkel, D., *Chem. Ber.* **99**, 3628 (1966).
84. Hellwinkel, D., *Angew. Chem.* **78**, 749 (1966).
85. Hellwinkel, D., and Kilthau, G., *Angew. Chem.* **78**, 1018 (1966).
86. Hellwinkel, D., and Kilthau, G., *Liebigs Ann. Chem.* **705**, 66 (1967).
87. Hellwinkel, D., and Kilthau, G., *Chem. Ber.* **101**, 121 (1968).
88. Hellwinkel, D., *Chimia* **22**, 488 (1968).
89. Hellwinkel, D., *Chem. Ber.* **102**, 528 (1969).
90. Hellwinkel, D., and Wilfinger, H. J., *Tetrahedron Lett.* 3423 (1969).
91. Hellwinkel, D., and Bach, M., *Naturwissenschaften* **56**, 214 (1969).
92. Hellwinkel, D., and Wünsche, C., *J. Chem. Soc. Chem. Commun.* 1412 (1969).
93. Hellwinkel, D., Knabe, B., and Kilthau, G., *J. Organomet. Chem.* **24**, 165 (1970).
94. Hellwinkel, D., and Knabe, B., *Chem. Ber.* **104**, 1761 (1971).
95. Hellwinkel, D., and Knabe, B., *Phosphorus* **2**, 129 (1972).
96. Hellwinkel, D., Wünsche, C., and Bach, M., *Phosphorus* **2**, 167 (1973).
97. Hellwinkel, D., Lindner, W., and Wilfinger, H. J., *Chem. Ber.* **107**, 1428 (1974).
98. Hellwinkel, D., Lindner, W., and Schmidt, W., *Chem. Ber.* **112**, 281 (1979).
99. Hellwinkel, D., *Top. Curr. Chem.* **109**, 1 (1983).
100. Henry, F. T., and Thorpe, T. M., *J. Chromatogr.* **166**, 577 (1978).
101. Holmes, R. R., *Acc. Chem. Res.* **5**, 296 (1972).
102. Holmes, R. R., and Deiters, J. A., *J. Am. Chem. Soc.* **99**, 3318 (1977) and references cited therein.
103. Holmes, R. R., *J. Am. Chem. Soc.* **100**, 433 (1978).
104. Holmes, R. R., *Acc. Chem. Res.* **12**, 257 (1979).
105. Holmes, R. R., "Pentacoordinated Phosphorus. Structure and Spectroscopy," Vol. I, ACS Monogr. 175. Washington, D.C., 1980.
106. Holmes, R. R., "Pentacoordinated Phosphorus. Reaction Mechanisms," Vol. II, ACS Monogr. 176. Washington, D.C., 1980.
107. Hoskins, L. C., and Lord, R. C., *J. Chem. Phys.* **46**, 2402 (1967).
108. Houlla, D., Wolf, R., Gagnaire, D., and Robert, J. B., *J. Chem. Soc. Chem. Commun.* 443 (1969).
109. Huisgen, R., and Wulff, J., *Tetrahedron Lett.* 917 (1967).
110. Hursthouse, M. B., and Steer, I. A., *J. Organomet. Chem.* **27**, C 11 (1971).
111. Hursthouse, M. B., *Mol. Struct. Differ. Methods* **4**, 393 (1976).
112. Jain, V. K., Bohra, R., and Mehrotra, R. C., *Struct. Bond.* **52**, 147 (1982).
113. Jensen, K. A., *Z. Anorg. Allg. Chem.* **250**, 257 (1943).
114. Jones, E. D., and Uehling, E. A., *J. Chem. Phys.* **36**, 1691 (1962).

115. Kappelmeier, C. P. A., *Rec. Trav. Chim. Pay-Bas* **49,** 57 (1930).
116. Keck, G. M., and Klar, G., *Z. Naturforsch.* **B27,** 591 (1972).
117. Kemmitt, R. D. W., and Sharp, D. W. A., *Adv. Fluorine Chem.* **4,** 142 (1965).
118. Keppert, D. L., "Inorganic Stereochemistry." Springer-Verlag, Berlin and New York, 1982.
119. Klippel, J., *Rocz. Chem.* **10,** 777 (1930).
120. Kustan, E. H., Smith, B. C., Sobeir, M. E., Swami, A. N., and Woods, M., *J. Chem. Soc. Dalton Trans.* 1326 (1971).
121. Kuykendall, G. L., and Mills, J. L., *J. Organomet. Chem.* **118,** 123 (1976).
122. Lacoste, W., and Michaelis, A., *Liebigs Ann. Chem.* **201,** 184 (1880).
123. Littlefield, L. B., and Doak, G. O., *J. Am. Chem. Soc.* **98,** 7881 (1976).
124. Luckenbach, R., "Dynamic Stereochemistry of Pentacoordinated Phosphorus and Related Elements." Thieme, Stuttgart, 1973.
125. Lynch, R. J., Waddington, T. C., *in* "Advances in Nuclear Quadrupole Resonance" (J. A. S. Smith, ed.), Vol. 1, p. 37. Heyden, London, 1974.
126. MacKay, K. M., Sowerby, D. B., and Young, W. C., *Spectrochim. Acta Sect. A* **24,** 611 (1968).
127. Makanova, D., Ondrejovic, G., Valigura, D., and Gazo, J., *Chem. Zvesti* **28,** 604 (1974).
128. Mallon, T., and Wieber, M., *Z. Anorg. Allg. Chem.* **454,** 31 (1979).
129. Mann, F. G., *J. Chem. Soc.* 65 (1945).
130. Mann, F. G., "The Heterocyclic Derivatives of Phosphorus, Arsenic, Antimony and Bismuth," 2nd ed. Wiley (Interscience), New York, 1970.
131. Maslowsky, E., Jr., *J. Organomet. Chem.* **70,** 153 (1974).
132. McKenzie, A., and Wood, J. K., *J. Chem. Soc.* **117,** 406 (1920).
133. Michaelies, A., and Paetow, U., *Ber.* **18,** 41 (1885).
134. Michaelies, A., and Paetow, U., *Liebigs Ann. Chem.* **233,** 60 (1886).
135. Michaelies, A., *Liebigs Ann. Chem.* **321,** 1141 (1902).
136. Mislow, K., *Acc. Chem. Res.* **3,** 321 (1970).
137. Mitschke, K. H., and Schmidbaur, H., *Chem. Ber.* **106,** 3645 (1973).
138. Monagle, J. J., *J. Org. Chem.* **27,** 3851 (1962).
139. Moreland, C. G., Beam, R. J., Wooten, C. W., and Horner, S. M., *Inorg. Nucl. Chem. Lett.* **7,** 243 (1971).
140. Muetterties, E. L., Mahler, W., Packer, K. J., and Schmutzler, R., *Inorg. Chem.* **3,** 1298 (1964).
141. Muetterties, E. L., and Schunn, R. A., *Q. Rev. Chem. Soc.* **20,** 245 (1966).
142. Murrell, J. N., and Scollary, C. E., *J. Chem. Soc. Dalton Trans.* 818 (1976).
143. Nyle'n, P., *Z. Anorg. Allg. Chem.* **246,** 227 (1941).
144. O'Brien, M. H., Doak, G. O., and Long, G. G., *Inorg. Chim. Acta* **1,** 34 (1967).
145. O'Brien, M. H., Ph.D. thesis, North Carolina State University, 1968.
146. Otero, A., and Royo, P., *J. Organomet. Chem.* **149,** 315 (1978).
147. Ott, R., Weidlein, J., and Mitschke, K. H., *Chimia* **29,** 262 (1975).
148. Preiss, H., and Jancke, H., *Z. Anorg. Allg. Chem.* **404,** 199 (1974).
149. Preiss, H., *Z. Anorg. Allg. Chem.* **404,** 175 (1974).
150. Raizees, G. W., and Gayron, J. L., "Organic Arsenic Compounds." Chemical Catalog Co., New York, 1923.
151. Ramirez, F., *Acc. Chem. Res.* **1,** 70 (1968).
152. Reiche, A., Dahlmann, J., and List, D., *Angew. Chem.* **73,** 494 (1961); *Liebigs Ann. Chem.* **678,** 167 (1964).
153. Reutov, O. A., and Ptitsyna, O. A., *Organomet. React.* **4,** 73 (1972).

154. Revitt, D. M., and Sowerby, D. B., *Spectrochim. Acta Sect. A* **26**, 1581 (1970).
155. Rochow, E. G., Hurdt, D. T., and Lewis, R. N., "The Chemistry of Organometallic Compounds," p. 295. Wiley, New York, 1957.
156. Roesky, H. W., Djarrah, H., Amirzadeh-Asl, D. A., and Sheldrick, W. S., *Chem. Ber.* **144**, 1554 (1981).
157. Roesky, H. W., Bohra, R., and Sheldrick, W. S., *J. Fluorine Chem.* **22**, 199 (1983).
158. Ruff, O., Braida, A., Bretschneider, O., Menzel, W., and Plant, H., *Z. Anorg. Allg. Chem.* **206**, 59 (1932).
159. Ruppert, I., and Bastian, V., *Angew. Chem.* **90**, 226 (1978).
160. Ruppert, I., *Chem. Ber.* **112**, 3023 (1979).
161. Salmi, E. J., Merivuori, K., and Laaksonen, E., *Suom. Kemistil.* **19B**, 102 (1946).
162. Samitov, Yu. Yu., Tazeeva, N. K., and Chernokalskii, B. D., *Zh. Obshch. Khim.* **45**, 1498 (1975).
163. Sau, A. C., and Holmes, R. R., *J. Organomet. Chem.* **217**, 157 (1981).
164. Schmidbaur, H., and Richter, W., *Angew. Chem.* **87**, 204 (1975).
165. Schmidbaur, H., and Holl, P., *Chem. Ber.* **109**, 3151 (1976).
166. Seel, F., and Detmer, O., *Z. Anorg. Allg. Chem.* **301**, 113 (1959).
167. Seppelt, K., *Angew. Chem.* **88**, 410 (1976).
168. Seppelt, K., *Z. Anorg. Allg. Chem.* **434**, 5 (1977).
169. Seyferth, D., *J. Am. Chem. Soc.* **80**, 1336 (1958).
170. Sheldrick, W. S., *Top. Curr. Chem.* **73**, 1 (1978).
171. Smith, S. L., and Brock, C. P., unpublished results.
172. Smith, W. C., Tullock, C. W., Muetterties, E. L., Hasek, W. R., Fawcett, F. S., Engelhardt, V. A., and Coffman, D. D., *J. Am. Chem. Soc.* **81**, 3165 (1959).
173. Smith, W. C., *J. Am. Chem. Soc.* **82**, 6176 (1960).
174. Smith, W. C. (to E. I. DuPont de Nemours), *U.S. Patent* 2, 950, 306 (1960).
175. Steinkopf, W., Dudek, H., and Schmidt, S., *Ber.* **61**, 1911 (1928).
176. Svergun, V. I., Rozinov, V. G., Grechkin, E. F., Timokhin, V. G., Maksyumin, Yu. K., and Semin, G. K., *Izvest. Akad. Nauk. SSSR, Ser. Khim.* 1918 (1970).
177. Tanzella, F. L., and Bartlett, N., *Z. Naturforsch.* **B36**, 1461 (1981).
178. Tzschach, A., and Heinicke, J., "Arsenheterocyclen." Deutscher Verlag für Grundstoffindustrie, VEB, Leipzig, 1978.
179. Usacheva, G. M., and Kamai, G. Kh., *Izv. Akad. Nauk SSSR, Ser. Khim.* 1432 (1970).
180. Usacheva, G. M., and Kamai, G. Kh., *Zh. Obshch. Khim.* **41**, 2705 (1971).
181. Vasile, M. J., and Falconer, W. E., *Inorg. Chem.* **11**, 2282 (1972).
182. Verdonck, L., and Van der Kelen, G. P., *Spectrochim. Acta Sect.* **A 29**, 1675 (1973).
183. Verdonck, L., and Van der Kelen, G. P., *Spectrochim. Acta Sect.* **A 33**, 601 (1977).
184. Verdonck, L., and Van der Kelen, G. P., *Spectrochim. Acta Sect.* **A 35**, 861 (1979).
185. Wells, A. F., *Z. Kristallogr.* **99**, 367 (1938).
186. Westheimer, F. H., *Acc. Chem. Res.* **1**, 70 (1968).
187. Wheatly, P. J., and Wittig, G., *Proc. Chem. Soc.* 251 (1962).
188. Wheatly, P. J., *J. Chem. Soc.* 3718 (1964).
189. Wheatly, P. J., *J. Chem. Soc.* 2206 (1964).
190. Wiberg, E., and Mödritzer, K., *Z. Naturforsch.* **B11**, 751 (1956).
191. Wiberg, E., and Mödritzer, K., *Z. Naturforsch.* **B12**, 127 (1957).
192. Wieber, M., Eichhorn, B., and Götz, J., *Chem. Ber.* **106**, 2738 (1973).
193. Wieber, M., and Götz, J., *Z. Anorg. Allg. Chem.* **424**, 56 (1976).
194. Wittig, G., and Clauss, K., *Liebigs Ann. Chem.* **577**, 26 (1952).
195. Wittig, G., and Torsell, K., *Acta Chem. Scand.* **7**, 1293 (1953).

196. Wittig, G., and Hellwinkel, D., *Angew. Chem.* **74,** 76 (1962).
197. Wittig, G., and Hellwinkel, D., *Angew. Chem.* **74,** 782 (1962).
198. Wittig, G., and Hellwinkel, D., *Chem. Ber.* **97,** 769 (1964).
199. Wittig, G., and Hellwinkel, D., *Angew. Chem.* **76,** 382 (1964).
200. Wizemann, T., Müller, H., Seybold, D., and Dehnicke, K., *J. Organomet. Chem.* **70,** 211 (1969).
201. Wood, J. S., *Prog. Inorg. Chem.* **16,** 227 (1972).
202. Woods, C., and Long, G. G., *J. Mol. Spectrosc.* **40,** 435 (1971).
203. Wunderlich, H., *Acta Crystallogr., Sect. B* **34,** 1000 (1978).
204. Wunderlich, H., *Acta Crystallogr., Sect. B* **36,** 1492 (1980).

PERCHLORATE ION COMPLEXES

N. M. N. GOWDA, S. B. NAIKAR, and G. K. N. REDDY

Department of Chemistry, Central College, Bangalore University, Bangalore, India

I. Introduction.	255
II. Perchlorate Ion Coordination and Methods of Identification	256
A. X-Ray Crystal Structure Analysis	256
B. Infrared and Raman Spectra	257
C. Electronic Spectra.	258
D. Magnetic Susceptibility and ESR Spectra.	260
E. Molar Conductivities.	260
III. Complexes of Early Transition Metals	260
IV. Complexes of Iron Group Metals	263
V. Complexes of Cobalt Group Metals	265
VI. Complexes of Nickel Group Metals	268
VII. Complexes of Copper Group Metals	273
VIII. Complexes of Zinc Group Metals	283
IX. Complexes of Lanthanides	287
X. Complexes of Other Metals.	288
XI. Conclusion	290
XII. Abbreviations	291
References	294

I. Introduction

Coordination of perchlorate to transition and nontransition metal ions has been well established, though in most of the cases the metal–perchlorate bond is rather weak and has been termed as "semicoordination" (1). The coordination that can be recognized in the solid state may frequently break down in solution, particularly in aqueous medium (the solvent molecule replacing the coordinated perchlorate).

A concise survey of perchlorate complexes was made by Rosenthal (2) in 1973. A year later, the possibility of perchlorate coordination in aqueous medium was reviewed by Johansson (3). Since the publication of these two accounts a number of papers reporting new findings on perchlorate complexes have appeared, justifying another comprehen-

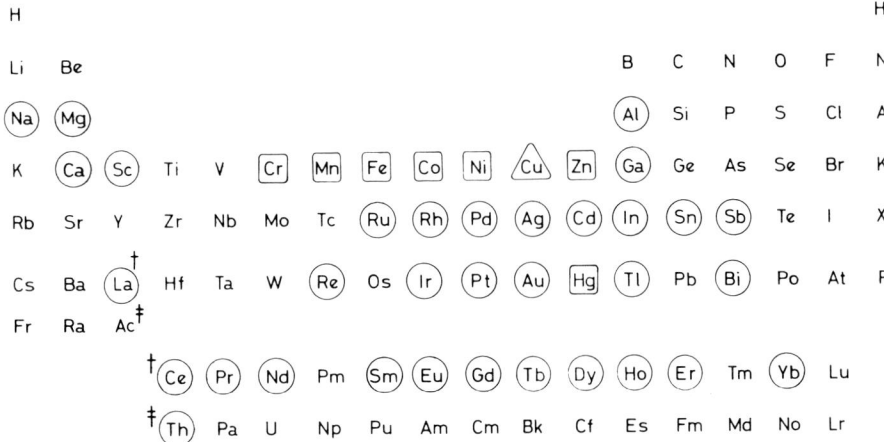

FIG. 1. Perchlorate ion complexes known: ○, a few; □, considerable; △, a large number.

sive review such as the one presented here. In this article only such perchlorate complexes where, in addition to perchlorate, at least another ligand (other than perchlorate) is also present, are considered. Figure 1 depicts the metals for which perchlorate complexes have been isolated.

II. Perchlorate Ion Coordination and Methods of Identification

The highly symmetrical perchlorate ion is a hard base and falls into the category of relatively nonpolarizable and noncoordinating anions such as BF_4^- and PF_6^- (4). Its ability to coordinate strongly to soft metal ions is therefore limited. When coordination does occur, the perchlorate group may be linked to a metal ion in a monodentate and/or bidentate manner through oxygen atoms (Fig. 2); the latter linkage can be either chelating (linked to the same metal ion) or bridging (bonded to two metal ions).

A. X-Ray Crystal Structure Analysis

X-ray crystal structure analysis has unequivocally established coordination of perchlorate in a number of perchlorate complexes. The O—Cl bond length of the coordinated part is generally greater than

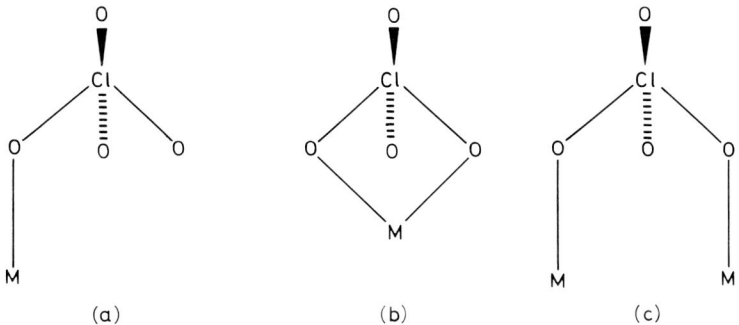

FIG. 2. Types of metal–perchlorate bonding: (a) monodentate; (b) chelating bidentate; (c) bridging bidentate.

that of the uncoordinated ones. The metal–perchlorate linkage is also manifest in the M—O—Cl bond angle, which is expected to be around 120° (5, 6), assuming that the coordinated oxygen uses sp^2 hybrid orbitals for bonding. The magnitude of the bond angle should give an idea of the metal–oxygen bond strength.

As early as 1965 Pauling et al. (7) and Cotton and Weaver (8) recognized coordinated perchlorate(s) in cobalt(II) complexes by X-ray structure analysis. In 1971, crystallographic evidence was presented by Nakai (9) to show that bridging, because of coordination by perchlorate, was responsible for the polymeric nature of $Cu(bipy)_2(ClO_4)_2$. Similarly, Hodgson and co-workers (10) identified the presence of intramolecular bidentate perchlorate in $\alpha\text{-}[Cu(dmaep)(\mu\text{-}OH)(\mu\text{-}ClO_4)]_2$. The presence of both unidentate and chelating bidentate perchlorate groups in a few lanthanide complexes has clearly been established by Ciampolini et al. (11). There have been several other reports on the X-ray crystal structure analyses of perchlorate complexes. The recorded metal–perchlorate distances vary over a wide range even for the same metal ion, depending on the other ligands present and the geometry of the complex and indicating varying degrees of metal–oxygen interaction. The M—O bond length in copper complexes, for instance, ranges from 2.2 to 2.9 Å (p. 274). There are a few instances where the M—O distances are comparable to the sum of the covalent radii (11–14).

B. INFRARED AND RAMAN SPECTRA

Hathaway and Underhill (15) were the earliest to suggest the use of IR and Raman spectroscopic methods to investigate perchlorate coordination to a metal ion. Perchlorate ion with its T_d symmetry should have its nine vibrational degrees of freedom distributed among four normal modes of vibration; the assignments of these modes are collected in Table I. While all four vibrations, ν_1 to ν_4, of ClO_4^- are Raman active, only the triply degenerate frequencies ν_3 and ν_4 are IR active. The ν_3 band appears prominently in the IR as a very strong and broad band with a poorly defined maximum.

Monodentate attachment to a metal ion lowers the symmetry of perchlorate to C_{3v} and bidentate attachment to C_{2v} (15–17). Consequently the number of vibrational modes should increase (Table I). In addition, a metal–oxygen stretching frequency would also be expected in the far-IR region and has been located in the range 360–290 cm^{-1} (18). These effects resulting on coordination, particularly the increase in the number of vibrational modes, may be used for identifying coordination of perchlorate.

It must, however, be borne in mind that minor shifts and weak splittings of the IR bands may arise on account of lowering of site symmetry because of strong lattice effects or of coupling of vibrations between perchlorate groups or from a purely isotopic effect within the group (15, 16). For example, the broad and strong band due to the ν_3 mode of ionic perchlorate is often split because of lattice effects. Despite these limitations, with a little care and caution, coordinated and noncoordinated perchlorate(s) are conveniently identified by IR spectroscopy.

C. ELECTRONIC SPECTRA

Since the nature of the ligand field around the metal ion is often reflected in the spectral features, perchlorate coordination in a metal complex may also be revealed by the electronic spectra of the complex. For instance, in chromium(III) complexes, $CrL_4(ClO_4)_3$ (L = n-Bu$_3$PO, Ph$_3$PO, mmpp), the spectral bands around 460 and 650 nm are assigned to $^4A_{2g} \rightarrow {}^4T_{1g}$ and $^4A_{2g} \rightarrow {}^4T_{2g}$ transitions, respectively, arising from an octahedral stereochemistry for the complex requiring coordination of two of the perchlorates to the metal ion (18–21). The fact that both Ni(2-i-PrIm)$_4$(ClO$_4$)$_2$ and Ni(2-i-PrIm)$_4$I(ClO$_4$) show a band around 23,700 cm^{-1} (22) suggests that the ligand field strength of perchlorate is comparable to that of iodide. However, a few other reports (12, 23) have indicated the position of perchlorate to be very close to chloride in the spectrochemical series. The spectrochemical series for the anions represented by X in the complexes [NiX(tdaa)]ClO$_4$ is found to be (X =) I < Br < Cl < NO$_3$ < ClO$_4$ < NCS < CN, based on their electronic spectra (24).

TABLE I

NORMAL MODES OF VIBRATION OF PERCHLORATE

Vibration type and infrared and Raman activity[a,b]

State of ClO_4^- (symmetry)									
Ionic (T_d)	ν_1 A_1(R) sy. str. (940–925)	ν_2 E(R) sy. bend (480–455)	ν_3 T_2(IR,R) asy. str. (1100–1070)		ν_4 T_2(IR,R) asy. bend (635–615)				
Monodentate (C_{3v})	ν_2 A_1(IR,R) ClO str. (950–870)	ν_6 E(IR,R) rocking (480–460)	ν_1 A_1(IR,R) ClO_3 sy. str. (1040–1025)	ν_4 E(IR,R) ClO—M asy. bend (1160–1020)	ν_3 A_1(IR,R) ClO_3 sy. bend (660–630)	ν_5 E(IR,R) ClO_3 asy. bend (620–600)			
Bidentate (C_{2v})	ν_2 A_1(IR,R) Cl(O—M)(O—M) sy. str. (990–900)	ν_4 A_1(IR,R) Cl(O—M)(O—M) sy. bend (480–450)	ν_5 A_2(R) torsion (440–400)	ν_1 A_1(IR,R) Cl(O)(O) sy. str. (1060–1020)	ν_6 B_1(IR,R) Cl(O)(O) asy. str. (1140–1090)	ν_8 B_2(IR,R) Cl(O—M)(O—M) asy. str. (1220–1140)	ν_3 A_1(IR,R) Cl(O)(O) sy. bend (670–635)	ν_7 B_1(IR,R) rocking (635–620)	ν_9 B_2(IR,R) rocking (620–605)

[a] sy. = symmetric; asy. = asymmetric; str. = stretching; bend. = bending; A, B = nondegenerate; E = doubly degenerate; T = triply degenerate; approximate vibrational frequency ranges (cm^{-1}) are given in parentheses.
[b] Taken by permission from B. J. Hathaway and A. E. Underhill, *J. Chem. Soc.* 3091 (1961); R. P. Scholer and A. E. Merbach, *Inorg. Chim. Acta* **15**, 15 (1975).

D. Magnetic Susceptibility and ESR Spectra

The coordination number of the metal ion, the stereochemistry of the complex, and hence the magnetic susceptibility may vary, depending on whether or not the perchlorate is coordinated to the metal ion. Thus magnetic susceptibility measurement can be of diagnostic value in determining the presence of the metal–perchlorate linkage. For example, the complex Co(β-PyBzl)$_4$(ClO$_4$)$_2$ has a magnetic moment of 4.9 μ_B, suggesting octahedral stereochemistry for the complex and hence coordination of both perchlorates to the metal ion (25). Similarly the ligand field environment around a metal ion is reflected in the ESR spectrum of a complex and hence can be used to identify metal–perchlorate bonding (26, 27).

E. Molar Conductivities

Perchlorate coordinated to a metal ion in the solid state generally is displaced in a polar solvent by a solvent (or another ligand) molecule. However, under favorable conditions, perchlorate coordination may persist in solution. In such cases, conductivity measurements are helpful in detecting the coordination of perchlorate (28). For instance, the conductance values of some lanthanide complexes Ln(hmpa)$_4$(ClO$_4$)$_3$ (Ln = La, Nd, Gd, Dy, Yb) in nitrobenzene correspond to those of uni-univalent electrolytes, implying coordination of two of the perchlorates (29). The coordinated perchlorates are, however, easily displaced by the addition of hmpa as shown by the conductance values, which gradually increase to that corresponding to a 1:3 electrolyte.

III. Complexes of Early Transition Metals

Except for the scandium group, investigations on perchlorate complexes of the early transition metals, namely, the titanium and vanadium groups, appear not to have been carried out. Scandium perchlorate reacts with Ph$_3$PO in ethanol to give white crystals of composition Sc(Ph$_3$PO)$_4$(ClO$_4$)$_3$. On the basis of IR data, a distorted octahedral geometry with two monodentately coordinated perchlorates has been postulated for the complex. However, the complex is a 1:3 electrolyte in MeCN, MeNO$_2$, and dmf, owing to solvolysis (30). Treatment of lanthanum perchlorate in EtOH–teof (or dmp) with hmpa produced La(hmpa)$_4$(ClO$_4$)$_3$. Based on IR (Table II) and conductivity data, monodentate coordination of two of the perchlorates has been postulated

(29). However, the complex shows a band at 436 cm^{-1} in the Raman (it is absent in the IR) spectrum, which has been assigned to the ν_5 mode of bidentate perchlorate. Hence the lanthanide ion has been suggested to have a coordination number between six and eight (17). The La(III) ion attains coordination number 10 in La(tmp)$_7$(H$_2$O)$_2$(ClO$_4$)$_3$, with one of the perchlorates coordinated to the metal ion as indicated by the perchlorate band splittings in the IR. The complex behaves as a 1 : 2 electrolyte in MeNO$_2$, supporting the above deductions (28). The X-ray crystal structure analysis of La(tdod)(H$_2$O)(ClO$_4$)$_3$ has indicated a decatetrahedral geometry for the complex, with the metal ion attaining a coordination number of 10 (31). Two of the perchlorates are bound to the metal ion, one as a monodentate ligand (La—O = 2.494 Å) and the other as a bidentate ligand (La—O = 2.683, 2.738 Å). Similarly, the metal ion in La(tdhc)(ClO$_4$)$_3$ is contemplated to be 10-coordinate with two perchlorates behaving as unidentate and the third one as a bidentate ligand (11).

Karayannis and co-workers (18–21, 32–36) have reported the syntheses of paramagnetic, green chromium(III) perchlorate complexes of the compositions CrL$_4$(ClO$_4$)$_3$ (L = n-Bu$_3$PO, Ph$_3$PO, mmpp, ttp) and CrL$_2$S$_2$(ClO$_4$)$_3$ (L = tetp; S = H$_2$O; L = adHNO, puHNO; S = EtOH). Chromium(III) complexes are, in general, six-coordinate, and therefore two of the perchlorates in each of the above complexes would be acting as monodentate ligands. This is supported by the IR spectra (Table II) of the complexes. Bidentate nature of perchlorate is observed in the six-coordinate complex Cr(depf)$_2$ClO$_4$ (37). Perchlorate coordination has also been recognized in a variety of six-coordinate chromium(III) complexes containing N-oxides such as miqNO, phzNO, phzNO$_2$, quxNO, and quxNO$_2$ (38–41). The existence of [Cr(H$_2$O)$_5$OClO$_3$]$^{2+}$ ion in perchloric acid medium has been detected spectrophotometrically (42). A trinuclear complex of composition [Cr$_3$(Ph$_3$PO)$_8$(OClO$_3$)$_3$](ClO$_4$)$_6$ is also known (20).

Wimmer and Snow (43, 44) have prepared a few octahedral manganese(I) carbonyl complexes of the formulas Mn(CO)$_5$OClO$_3$ and Mn(CO)$_3$L$_2$(OClO$_3$) (L = Ph$_3$P, (PhO)$_3$P; L$_2$ = bipy) by treating the corresponding bromides with AgClO$_4$ in dichloromethane. The pentacarbonyl undergoes substitution of perchlorate with Q (Ph$_3$P, m-tolyl$_3$P, p-tolyl$_3$P, dppe, dppm, MeCN, PhCN, Py, H$_2$O) to produce [Mn(CO)$_5$Q]ClO$_4$ complexes. Manganese(II) complexes of the type MnL$_4$(ClO$_4$)$_2$ (L = Me$_3$PO, Ph$_3$PO, Me$_3$AsO, Ph$_3$AsO, Ph$_2$MeAsO, MeCN, Py, 4-MePy, pyzNO, dpNO) have been reported (15, 26, 45–50). In the tertiary arsine oxide and phosphine oxide complexes, Mn has been postulated to have coordination number 5 on the basis of IR

TABLE II

PERCHLORATE FREQUENCIES OF EARLY TRANSITION METAL COMPLEXES

Complex	Perchlorate bands (cm^{-1})				Reference
	ν_3	ν_1	ν_4	ν_2	
[Sc(Ph$_3$PO)$_4$(OClO$_3$)$_2$]ClO$_4$	1190, 1082, 1068, 1022, 1015, 994	930, 895	623, 610	445	30
[La(hmpa)$_4$(OClO$_3$)$_2$]ClO$_4$	1119, 1112, 1099	916	644, 623, 614	480, 346^a	29
[Cr(n-Bu$_3$PO)$_4$(OClO$_3$)$_2$]ClO$_4$	1150, 1030	930	642, 616, 600	480, 460, 337^a	18
[Cr(mmpp)$_4$(OClO$_3$)$_2$]ClO$_4$	1125, 1090, 1039	923	652, 634, 618	327^a	21
[Cr(miqNO)$_4$(H$_2$O)OClO$_3$]ClO$_4$	1118, 1098, 1035		640, 629, 621, 616, 611	322^a	38
Cr(depf)$_2$(O$_2$ClO$_2$)	1212, 1115, 1030	932	655, 633	480, 460, 355^a	37
Mn(CO)$_5$(OClO$_3$)	1161, 1020	880	625		43
Mn(CO)$_3$bipy(OClO$_3$)	1146, 1021	884	633, 615		43
Mn(CO)$_3$(Ph$_3$P)$_2$(OClO$_3$)	1148, 1021	890	625		43
[Mn(Ph$_3$PO)$_4$(OClO$_3$)]ClO$_4$	1099, 1044	925	621, 614		26
[Mn(miqNO)$_5$(OClO$_3$)]ClO$_4$	1114, 1110, 1088, 1045		639, 628, 621, 614	284^a	38
Mn(Py)$_4$(OClO$_3$)$_2$	1130, 1030	936	625, 617		47
Mn(qun)$_2$(O$_2$ClO$_2$)$_2$	1107, 1090	937	633, 622	484	47
[Mn(bdpm)$_2$(O$_2$ClO$_2$)]ClO$_4$	1155, 1100, 1035	935	627, 619, 615		51
Re(CO)$_5$(OClO$_3$)	1190, 1162, 1150, 1021	910, 870	630		54
Re(CO)$_3$bipy(OClO$_3$)	1150, 1010	855	630		54

^{a} ν[M—O(ClO$_4$)].

spectra, which indicate coordination of one of the perchlorates to the metal ion. ESR spectral study of the above complexes (L = Ph_3PO, Ph_3AsO) also indicates five-coordination for the metal ion. Comparison with the ESR data of the analogous complexes MnL_4X_2 (X = I, NCS) suggests that the ligand field strength of perchlorate is comparable to that of iodide and isothiocyanate (26).

The electronic spectral results for the complex $Mn(qun)_2(ClO_4)_2$ are in consonance with a distorted octahedral environment around the metal ion, and as a consequence the perchlorates are bidentate (47); the IR spectrum of the complex, however, does not provide supporting evidence (Table II). On the other hand, bidentate behavior of one of the perchlorates in the $Mn(bdpm)_2(ClO_4)_2$ complex has been revealed by IR data (51). Infrared spectral measurements suggest the presence of both monodentate and ionic perchlorates in several manganese(II) complexes possessing such ligands as adHNO, miqNO, phzNO, $phzNO_2$, quxNO, $quxNO_2$, and adH (34, 38–40, 52, 53).

The six-coordinate $Re(CO)_3(bipy)OClO_3$ and $Re(CO)_5OClO_3$ are probably the only perchlorate complexes that are known for rhenium. They are prepared by treating the respective bromides with $AgClO_4$ in dichloromethane (54).

IV. Complexes of Iron Group Metals

Ferrous and ferric complexes of the formulas $FeL_4(ClO_4)_2$ (L = Me_3PO, n-Bu_3PO, Me_3AsO, miqNO) and $FeL_4(ClO_4)_3$ (L = n-Bu_3PO, Ph_3AsO, tetp, ttp, miqNO) have been reported (18, 19, 32, 36, 38, 45, 55). They are all paramagnetic and in most of them the metal ion is six-coordinate with two monodentately bound perchlorates as per the IR evidence (Table III). In a few of these complexes (L = Me_3PO, Me_3AsO, Ph_3AsO, miqNO), however, the metal ion is five-coordinate with a monodentate perchlorate. The electrical conductivity of many of these complexes, measured in nonaqueous polar solvents, suggests partial to complete replacement of coordinated perchlorates by solvent molecules. Ferrous perchlorate reacts with bdpm to produce the six-coordinate $Fe(bdpm)_2(ClO_4)_2$ complex wherein both the perchlorates are monodentate (51).

Reed *et al.* (12) and Masuda *et al.* (56) have elucidated the structures of five-coordinate iron(III) complexes $Fe(tpp)ClO_4$ (Fig. 3) and $Fe(oep)ClO_4$ by X-ray crystallography. In both complexes, the perchlorate is linked to the metal ion through one of the oxygen atoms at the apical site of the square pyramid. The Fe—O distance is slightly shorter in

TABLE III

PERCHLORATE FREQUENCIES OF IRON COMPLEXES

Complex	Perchlorate bands (cm^{-1})				Reference
	ν_3	ν_1	ν_4	ν_2	
[Fe(miqNO)$_4$OClO$_3$]ClO$_4$	1120, 1110, 1086, 1043		640, 629, 621, 614	326[a]	38
Fe(n-Bu$_3$PO)$_4$(OClO$_3$)$_2$	1150, 1030	930	640, 620, 602	480, 460, 290[a]	18
Fe(bdpm)$_2$(OClO$_3$)$_2$	1115, 1050	930			51
[Fe(mmpp)$_2$(H$_2$O)$_2$OClO$_3$]$_2$(ClO$_4$)$_2$	1145, 1100, 1034	920	639, 627, 613	295[a]	21
[Fe(pyzNO)$_3$(H$_2$O)OClO$_3$]ClO$_4$	1098, 1094	930	621	278[a]	49
[Fe(n-Bu$_3$PO)$_4$(OClO$_3$)$_2$]ClO$_4$	1140, 1030	910	635, 619, 599	480, 455, 315[a]	18
[Fe(tetp)$_4$(OClO$_3$)$_2$]ClO$_4$	1142, 1079, 1044	923	655, 637, 624		32
[Fe(ttp)$_4$(OClO$_3$)$_2$]ClO$_4$	1107, 1046		621	328[a]	36
[Fe(miqNO)$_4$(OClO$_3$)$_2$]ClO$_4$	1117, 1095, 1047, 1020		637, 629, 621, 612	340[a], 330[a], 320[a]	38
[Fe(mmpp)$_2$(H$_2$O)$_2$OClO$_3$]$_2$(ClO$_4$)$_4$	1127, 1112, 1039	915	655, 640, 617	338[a]	21
[Fe(phzNO)$_{2/3}$(H$_2$O)$_2$OClO$_3$](ClO$_4$)$_2$	1123, 1068, 1038, 1021	926	652, 622, 617, 611	486	41
[Fe(puHNO)$_3$(O$_2$ClO$_2$)](ClO$_4$)$_2$	1133, 1095, 1077, 1050		648, 628, 619		35

[a] ν[M—O(ClO$_4$)].

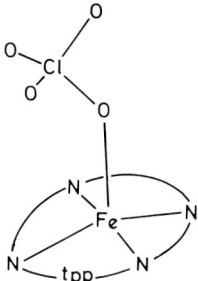

FIG. 3. Structure of Fe(tpp)OClO$_3$ complex. [Redrawn by permission from C. A. Reed et al., J. Am. Chem. Soc. **101**, 2948 (1979).]

the tpp complex (2.029 Å) than in the oep complex (2.067 Å). Likewise, the coordinated O—Cl length (1.471 Å) in the former is shorter than that in the latter (1.515 Å). Further the coordinated O—Cl distances in both the complexes are appreciably longer than that of any of the uncoordinated ones (1.375, 1.406, and 1.408 Å for the tpp complex; 1.38, 1.41, and 1.45 Å for the oep complex). It is suggested that these results are consistent with considerable electron donation by the coordinated oxygen to the metal ion as well as Fe→(OCl)π^* back bonding. The Fe—O—Cl bond angles of 131.2 (tpp) and 125.7° (oep) indicate an sp^2 character for the oxygen bonding orbitals.

Reaction of ferrous and ferric perchlorates with pyzNO (L) in EtOH–teof yields complexes of the formulas FeL$_3$(ClO$_4$)$_2$ (red) and FeL$_6$(ClO$_4$)$_3$ (straw yellow), respectively (*48, 49*). The ferrous complex, on exposure to moisture, adds on a molecule of water, turning orange. The resulting aquo complex shows IR bands characteristic of coordinated perchlorate (Table III). A number of Fe(II) and Fe(III) complexes, containing donor molecules like mmpp, puHNO, phzNO, phzNO$_2$, quxNO, quxNO$_2$, and adH, have been proposed to involve unidentately or bidentately coordinated perchlorates (*21, 35, 39–41, 53*).

Christian and Roper (*57*) have isolated white crystals of a six-coordinate ruthenium(II) complex, *trans*-RuH(CO)(*p*-tNC)(Ph$_3$P)$_2$OClO$_3$ by treating the corresponding chloro complex with AgClO$_4$ in EtOH. The coordinated perchlorate can be readily displaced by CO or Ph$_3$P.

V. Complexes of Cobalt Group Metals

Peone and Vaska (*58, 59*) have reported the synthesis of a five-coordinate cobalt(I) complex, Co(CO)$_2$(Ph$_3$P)$_2$OClO$_3$, by treating the corresponding chloro complex with AgClO$_4$ in benzene. In a tetrahe-

TABLE IV

Perchlorate Frequencies of Cobalt Group Metal Complexes

Complex	Perchlorate bands (cm^{-1})			Reference
	ν_3	ν_1	ν_4	
Co(CO)$_2$(Ph$_3$P)$_2$(OClO$_3$)	1160, 1125, 1080	928	623	58
Co(qun)$_2$(OClO$_3$)$_2$	1105, 1095	920	634, 623, 463^a	47
[Co(mmpp)$_4$OClO$_3$]ClO$_4$	1150, 1085, 1032	926	651, 637, 614, 312^b	21
[Co(tetp)$_4$OClO$_3$]ClO$_4$	1141, 1105, 1050	920	650, 629, 619	32
[Co(topo)$_4$OClO$_3$]ClO$_4$	1149, 1119, 1085, 1029, 1011	977, 926	644, 621, 616, 467^a, 330^b	60
Co(Py)$_4$(OClO$_3$)$_2$	1130, 1030	930	623, 615	47
Co(4-MePy)$_4$(OClO$_3$)$_2$	1137, 1020	925	625, 615	47
Co(ad)(EtOH)(H$_2$O)$_2$OClO$_3$	1120, 1085	920	640, 620	53
[Co(bdpm)$_2$(O$_2$ClO$_2$)]ClO$_4$	1185, 1100	985, 907	640, 623, 607	51
[Co(NH$_3$)$_5$(OClO$_3$)](ClO$_4$)$_2$	1190, 1115, 1035		658, 640, 620	66
Rh(CO)(Ph$_3$P)$_2$(OClO$_3$)	1160, 1130, 1070	920	620	58
Ir(CO)(Ph$_3$P)$_2$(OClO$_3$)	1160, 1130, 1050	920	620	58
Ir(cod)(tfb)(OClO$_3$)	1130, 1025		620	69

^{a} ν_2.
^{b} ν[M—O(ClO$_4$)].

dral cobalt(II) complex, Co(qun)$_2$(ClO$_4$)$_2$, the perchlorates are monodentately linked to the metal ion (47). A number of five-coordinate Co(II) complexes of the formulas CoL$_4$(ClO$_4$)$_2$ (L = Ph$_2$MeAsO, mmpp, tetp, topo; L$_4$ = tdaea) and Co(adHNO)$_2$(EtOH)$_2$(ClO$_4$)$_2$, in which one of the perchlorates is coordinated to the metal ion, have been characterized (Table IV) (21, 23, 32–34, 46, 60). Co(Ph$_2$MeAsO)$_4$(ClO$_4$)$_2$ is the first perchlorate complex whose crystal structure has been solved by X-ray analysis (7). It has a distorted tetragonal–pyramidal geometry (Fig. 4) wherein the apical position is occupied by one of the perchlorates with a Co—O distance of 2.10 Å and a Co—O—Cl angle of 130°.

A variety of six-coordinate cobalt(II) complexes of the compositions Co(L—L)$_2$(ClO$_4$)$_2$ (L—L = dth, bdpm, diars) (8, 51, 61), CoL$_3$(H$_2$O)$_2$-(ClO$_4$)$_2$ · xH$_2$O (L = phzNO$_2$; x = 4; L = pyzNO; x = 0) (41, 49), CoL$_4$(ClO$_4$)$_2$ (L = β-PyBzl, Py, substituted pyridines; L$_4$ = bbtb) (25, 47, 47a, 62), and CoL$_5$(ClO$_4$)$_2$ (L = miqNO, NH$_3$, PhNC) (38, 63, 64) have been reported to contain coordinated perchlorate(s) (Table IV). The X-ray crystal structure analysis of Co(dth)$_2$(ClO$_4$)$_2$ complex (Fig. 5) shows monodentate coordination of both perchlorates at the apical positions with a Co—O bond length of 2.34 Å (8). The coordinated O—Cl length (1.47 Å) is longer than the uncoordinated ones (1.41 and

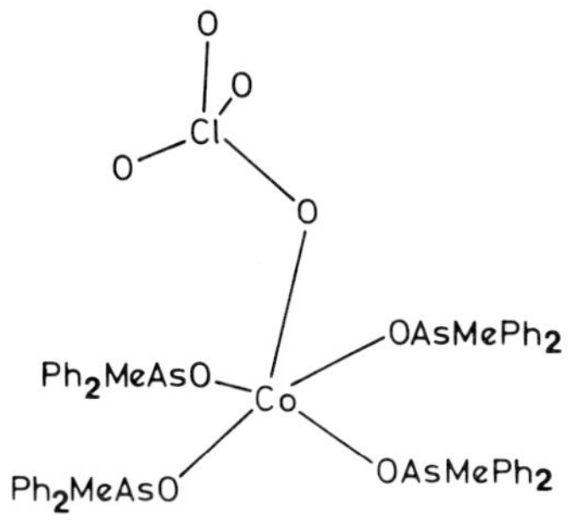

FIG. 4. Structure of [Co(Ph₂MeAsO)₄OClO₃]⁺ complex ion. [Redrawn by permission from P. Pauling et al., Nature (London) **207**, 73 (1965).]

1.43 Å). The diars complex has been isolated in two crystalline forms—monoclinic and orthorhombic. While the former has IR support for perchlorate coordination (*65*), the latter has X-ray crystal structure evidence (*61*) for very weak association of perchlorates with the metal ion in the axial positions. The PhNC complex exists in three isomeric forms, and the X-ray crystal structure of one of the forms (green), determined by Jurnak et al. (*64*), has indicated a perchlorate oxygen at the sixth coordination position around the metal ion with a Co—O distance of 2.594 Å and a Co—O—Cl angle of 135.6°. The coordinated O—Cl bond (1.396 Å) is longer than two of the uncoordinated ones (1.319, 1.327 Å) but is, however, shorter than the third one (1.413 Å). Karayannis and co-workers (*33, 34, 39, 40, 52, 53*) have characterized a series of perchlorate coordinated cobalt(II) complexes containing the ligands adHNO, phzNO, quxNO, quxNO₂, and adH (Table IV).

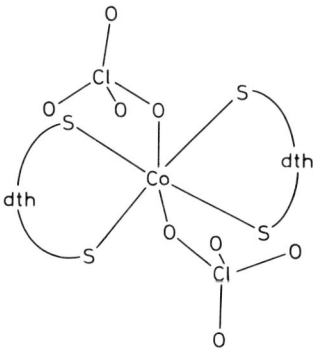

FIG. 5. Structure of Co(dth)₂(OClO₄)₂ complex. [Redrawn by permission from F. A. Cotton and D. L. Weaver, J. Am. Chem. Soc. **87**, 4189 (1965).]

Harrofield et al. (66) have reported the isolation of a cobalt(III) complex, [Co(NH$_3$)$_5$OClO$_3$](ClO$_4$)$_2$, by treating [Co(NH$_3$)$_5$N$_3$](ClO$_4$)$_2$ with HNO$_3$–NaNO$_2$ in perchloric acid. Based on the changes in the visible spectra on addition of perchloric acid to the solutions of [L$_4$Co-μ-(NH$_2$, OH)CoL$_4$]$^{4+}$ (L = NH$_3$; L$_2$ = en), formation of perchlorate complex cations of the type [L$_4$Co-μ-(NH$_2$, ClO$_4$)CoL$_4$]$^{4+}$ has been envisaged (67).

Treatment of MCl(CO)L$_2$ (M = Rh, Ir; L = Ph$_3$P) complexes in benzene with AgClO$_4$ has yielded four-coordinate complexes of the type M(CO)L$_2$(OClO$_3$) (58, 59). The perchlorate complexes undergo oxidative addition reactions with XY (H$_2$, HCl, Cl$_2$, organic halides, carboxylic acids, etc.) to produce six-coordinate complexes of the formula M(CO)L$_2$(XY)OClO$_3$ (68). In the parent complexes, the coordinated perchlorate is easily displaced by neutral molecules Q (C$_2$H$_4$, amines, Group V ligands, etc.) to produce cationic complexes of the composition [M(CO)L$_2$Q]ClO$_4$. Ir(cod)(tfb)OClO$_3$, isolated from the reaction mixture of [IrCl(cod)]$_2$, AgClO$_4$, and tfb in dichloromethane, also undergoes perchlorate substitution with Q (MeCN, PhCH$_2$CN) to yield [Ir(cod)(tfb)Q]ClO$_4$ (69). In complexes of type M(CO)L$_2$A (A = OClO$_3$, anionic ligand) a relation between the carbonyl stretching frequency and the total electronegativity of A has been discerned (70). Accordingly, the total electronegativity of the perchlorate is found to be comparable to that of NCS$^-$.

VI. Complexes of Nickel Group Metals

Four-coordinate nickel(II) complexes of types Ni(qun)$_2$(ClO$_4$)$_2$ and Ni(C$_6$Cl$_5$)L$_2$ClO$_4$ (L = PhMe$_2$P, Ph$_2$MeP) involving monodentately coordinated perchlorates have been reported (47, 71). The latter (L = PhMe$_2$P) undergoes substitution of the perchlorate ligand with Q (CO, H$_2$O) to afford [Ni(C$_6$Cl$_5$)L$_2$Q]ClO$_4$. One of the two perchlorates is unidentately bound to the metal ion in several five-coordinate complexes of the formula NiL$_4$(ClO$_4$)$_2$ (L = Me$_3$PO, Ph$_3$PO, Me$_3$AsO, Ph$_2$MeAsO, Ph$_3$AsO, topo; L$_4$ = tdaea, tdpp, tdaa, dthp) (23, 24, 26, 45, 46, 60, 72, 73). While the topo complex is a 1:1 electrolyte in MeNO$_2$ (60), those containing the quadridentate ligands tdpp and tdaa show a similar behavior in PhNO$_2$ (24), supporting coordination of one of the perchlorates. The electronic spectral features of the latter complexes are characteristic of five-coordinate nickel(II). In the case of the dthp complex, the X-ray structure analysis shows perchlorate coordination with an Ni—O distance of 2.77 Å (73).

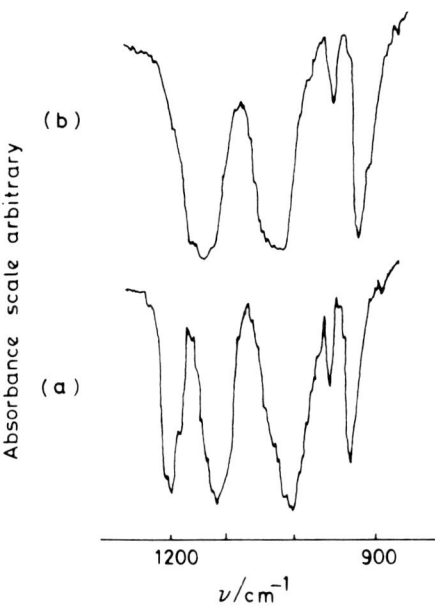

FIG. 6. IR spectra of (a) Ni(MeCN)$_2$(ClO$_4$)$_2$ and (b) Ni(MeCN)$_4$(ClO$_4$)$_2$ in the 1100 cm^{-1} region. [Reproduced by permission from A. E. Wickenden and R. A. Krause, *Inorg. Chem.* **4**, 404 (1965).]

A variety of six-coordinate complexes of formulas Ni(MeCN)$_2$(ClO$_4$)$_2$ (74), Ni(L—L)$_2$(ClO$_4$)$_2$ (L—L = en, men, dmen, tmen, bada, bipy, phen, bdpm, dap, depNO, diars) (51, 65, 75–79), NiL$_4$(ClO$_4$)$_2$ (L = MeCN, Py, substituted pyridines, 2-i-PrIm, PhNH$_2$, aniline derivatives; L$_4$ = bbtb), and Ni(2-i-PrIm)$_4$XClO$_4$ (X = Cl, Br, I, SCN) (22, 47, 62, 80–86) involve coordination of perchlorate groups. Perchlorate frequencies of some of these complexes are listed in Table V. While monodentate coordination of both perchlorates has been shown in the men and dmen complexes by IR and electronic spectra and magnetic susceptibility measurements, bidentate coordination of both perchlorates has been postulated in Ni(MeCN)$_2$(ClO$_4$)$_2$ (Fig. 6) and Ni(tmen)$_2$(ClO$_4$)$_2$ complexes (74, 75). The solid-state reflectance spectrum of the latter complex indicates an octahedral environment for the metal ion and hence the following structure (Fig. 7a) has been suggested. However, an alternative structure (Fig. 7b) for the complex has been proposed, considering the fact that the tmen groups could also function as bidentate ligands. Aubry and Brown (85) have found a distorted octahedral structure for the 3,5-Me$_2$Py complex with monodentately linked per-

TABLE V

PERCHLORATE FREQUENCIES OF NICKEL GROUP METAL COMPLEXES

Complex	Perchlorate bands (cm^{-1})			Reference
	ν_3	ν_1	ν_4	
Ni(C$_6$Cl$_5$)(Ph$_2$MeP)$_2$OClO$_3$	1170, 1153, 1019			71
Ni(qun)$_2$(OClO$_3$)$_2$	1100, 1063	920	624, 616, 467^a	47
[Ni(Ph$_3$AsO)$_4$OClO$_3$]ClO$_4$	1104, 1026		625, 614	26
[Ni(topo)$_4$OClO$_3$]ClO$_4$	1130, 1090, 1035	929	650, 621, 614, 466^a, 333^b	60
Ni(men)$_2$(OClO$_3$)$_2$	1130, 1025	933	630, 616	75
Ni(2-i-PrIm)$_4$(OClO$_3$)$_2$	1110, 1055	930	620	22
Ni(2-i-PrIm)$_4$Cl(OClO$_3$)	1115, 1063	930	620	22
Ni(en)$_2$(OClO$_3$)$_2$	1130, 1093, 1058	962	638, 621	77
Ni(depNO)$_2$(OClO$_3$)$_2$	1075	918	628, 616, 439^a, 310^b	78
Ni(ad)(H$_2$O)$_2$(EtOH)OClO$_3$	1125, 1089	921	638, 627	53
Ni(PhNH$_2$)$_4$(OClO$_3$)$_2$	1129, 1009	900	633, 613	84
Ni(3,4-xylidine)$_4$(OClO$_3$)$_2$	1139, 1034	928	638, 625	84
[Ni(bdpm)$_2$(O$_2$ClO$_2$)]ClO$_4$	1185, 1100	985, 907	640, 623, 607	51
Ni(MeCN)$_2$(O$_2$ClO$_2$)$_2$	1195, 1106, 1000	950, 920		74
Ni(tmen)$_2$(O$_2$ClO$_2$)$_2$	1170, 1125	928	635, 623, 617	75
Pd(dppe)(OClO$_3$)$_2$	1145, 1020	895	620, 610	87
Pd(C$_6$F$_5$)(Ph$_3$As)$_2$OClO$_3$	1150, 1015	870	630, 610	88
Pt(C$_6$F$_5$)(Et$_3$P)$_2$OClO$_3$	1170, 1020	870–850	650, 610	92

^{a} ν_2.
^{b} ν[M—O(ClO$_4$)].

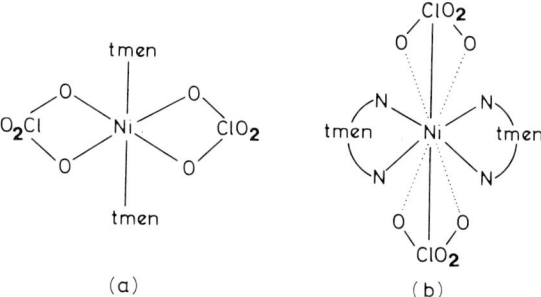

FIG. 7. Possible structures of Ni(tmen)$_2$(ClO$_4$)$_2$ complex. [Redrawn by permission from S. F. Pavkovic and D. W. Meek, *Inorg. Chem.* **4**, 1091 (1965).]

chlorates having an Ni—O length of 2.187 Å and an Ni—O—Cl angle of 158°. The bonded O—Cl length is 1.40 Å as against the nonbonded ones, 1.24, 1.34, and 1.45 Å, the last one being anomalous. By contrast, the X-ray structure analysis of the 3,4-Me$_2$Py complex indicates the ionic nature of the perchlorates (*86*). Coordination of perchlorate has been identified in a number of six-coordinate nickel(II) complexes containing donor molecules like adHNO, miqNO, phzNO, phzNO$_2$, quxNO, quxNO$_2$, and adH (*33, 34, 38–41, 53*).

Unidentately bound perchlorates satisfy coordination number 4 for the metal ion in the Pd(dppe)(ClO$_4$)$_2$ complex, which is made by treating the corresponding dichloro complex with AgClO$_4$ in dichloromethane–benzene (*87*). The complex shows a very low molar conductance in dichloromethane, but behaves as a 1:2 electrolyte in such solvents as thf, MeCN, and MeOH, obviously owing to displacement of coordinated perchlorates by solvent molecules. Synthesis of PdXL$_2$-(OClO$_3$) (X = C$_6$Cl$_5$; L = Ph$_3$P; X = C$_6$F$_5$; L = Et$_3$P, PhEt$_2$P, Ph$_2$MeP, Ph$_3$P, Ph$_3$As; L$_2$ = dppe, bipy, phen, tamen) complexes has been accomplished by several workers (*88–90*). The coordinated perchlorate in such complexes is easily displaced by neutral molecules like 2-MePy, 3-MePy, 4-MePy, 2,4-Me$_2$Py and 3,5-Me$_2$Py to produce cationic complexes (*90*).

In PtH(Ph$_3$P)$_2$ClO$_4$, the perchlorate is presumably coordinated, but supporting evidence is not available (*91*). Monodentate coordination of perchlorate has been visualized in the case of Pt(C$_6$F$_5$)(Et$_3$P)$_2$ClO$_4$ (*92*). Since the complex has a perchlorate band at 1170 cm^{-1} (Table V), it is likely that ClO$_4^-$ is bidentate. The complex is a 1:1 electrolyte in Me$_2$CO and MeNO$_2$. Substitution of the coordinated perchlorate with a variety of donor groups readily takes place to produce

TABLE VI
PERCHLORATE FREQUENCIES OF COPPER GROUP METAL COMPLEXES

Complex	Perchlorate bands (cm^{-1})			Reference
	ν_3	ν_1	ν_4	
Cu(Cy$_3$P)$_2$OClO$_3$	1120, 1020	910	618, 610	95
Cu(Cy$_3$As)$_2$OClO$_3$	1110, 1100, 1022	906	625, 620	94
Cu(triphos)OClO$_3$	1130, 1025	905	615, 610	95
Cu(Ph$_3$P)$_3$OClO$_3$	1130, 1050	930		97
[Cu(Ph$_3$P)$_2$(O$_2$ClO$_2$)]$_2$	1165, 1140, 1035	920		97
Cu(Ph$_2$EtAs)$_3$OClO$_3$	1122, 1077, 1033	911	616, 613	94
Cu(Ph$_2$-n-PrAs)$_3$OClO$_3$	1117, 1078, 1033	916	619, 614	94
[Cu(quxNO$_2$)$_4$OClO$_3$]ClO$_4$	1134, 1093, 1065, 1050	932	649, 623, 618, 454^c	40
Cu(NH$_3$)$_4$(OClO$_3$)$_2$	1105, 1050	914		110
Cu(Py)$_4$(OClO$_3$)$_2$	1107, 1040	929	625, 619	47
Cu(en)$_2$(OClO$_3$)$_2$	1130, 1050	927	624, 464^a	118b
Cu(1-ViIm)$_4$(OClO$_3$)$_2$	1130, 1045	932	625	112
Cu(dmpNO)$_2$(OClO$_3$)$_2$	1143, 1090	941	639, 631	122
[Cu(dpa)(MeCO$_2$)(O$_2$ClO$_2$)]ClO$_4$	1145, 1110, 1070	935, 930	637, 630, 625, 460^a	106
Cu(qun)$_2$(O$_2$ClO$_2$)$_2$	1100, 1050	930	633, 620	47
Cu(arg)$_2$(O$_2$ClO$_2$)$_2$	1140, 1115, 1085		635, 630, 625	127
Ag(Cy$_3$P)OClO$_3$	1150, 1000	850		131
Ag(Cy$_3$As)$_2$(OClO$_3$)	1105, 1083, 1040	910	623, 616	94
Ag(Nap$_2$EtAs)$_2$(OClO$_3$)	1117, 1112, 1035	915	626, 622	94
Ag(Ph$_2$EtAs)$_3$(OClO$_3$)	1104, 1075, 1040	920	620, 614	94
Ag(Iotz)$_2$(O$_2$ClO$_2$)	1140, 1095, 1050	930		132

^{a} ν_2.

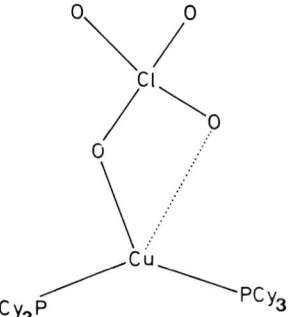

FIG. 8. The geometry about the copper atom in Cu(Cy$_3$P)$_2$ClO$_4$. [Reproduced by permission from R. J. Restivo et al., Can. J. Chem. **53**, 1949 (1975).]

[Pt(C$_6$F$_5$)(Et$_3$P)$_2$Q]ClO$_4$ (Q = H$_2$O, Pr$_2$NH, Py, CO, Et$_3$P, Ph$_3$P, Ph$_3$As, Ph$_3$PO, Ph$_3$PS) complexes. On the basis of molar conductance and UV spectral results, it is presumed that the complex Pt$_2$(pdpa)$_2$(NCS)$_2$-(ClO$_4$)$_2$ contains one monodentate perchlorate (*92a*).

VII. Complexes of Copper Group Metals

Among the perchlorate ion complexes of metals, those of copper have been widely studied, especially their X-ray crystal structure analyses. Copper(II) perchlorate reacts with bulky ligands in an ethanol medium to produce colorless copper(I) complexes of the composition CuL$_2$ClO$_4$ (L = Cy$_3$P, Cy$_3$As) (*93, 94*). The complexes have monodentately bound perchlorate as indicated by their IR spectra (Table VI), but they behave as 1:1 electrolyes in PhNO$_2$. Ferguson and co-workers (*95*) have confirmed the coordination of perchlorate to the metal ion in the Cy$_3$P complex by X-ray crystal structure analysis. The perchlorate is found to be unidentate but disordered over two sites (Fig. 8), with a bonded Cu—O length of 2.22 Å and a nonbonded intramolecular contact Cu ⋯ O of 3.203 Å. The coordinated O—Cl distance (1.46 Å) is close to the average (1.43 Å) of the uncoordinated ones (1.38, 1.41, and 1.50 Å). The coordination geometry around the metal ion is distorted trigonal. In another complex, Cu(triphos)ClO$_4$, containing a tridentate phosphine, it has been envisaged that the metal ion is either three- or four-coordinate involving monodentate perchlorate.

A few four-coordinate complexes of the formulas CuL$_3$ClO$_4$ (L = Ph$_3$P, Ph$_2$EtAs, Ph$_2$-*n*-PrAs; L$_3$ = tox) and Cu(Ph$_3$P)$_2$ClO$_4$ have been reported (*94, 96, 97*); the last one is a dimer with bridging bidentate

TABLE VII

BOND LENGTHS AND BOND ANGLES OF COPPER PERCHLORATE COMPLEXES

Complex	Coordination kernel	Bond length (Å)		Bond angle (°)	Reference
		Cu—O	O—Cl	Cu—O—Cl	
$Cu(Cy_3P)_2(OClO_3)$	CuP_2O	2.22	1.46	122.4	95
$Cu(tox)_3(OClO_3)$	CuS_3O	2.278	1.478	126.9	96
$[Cu(bpdt)OClO_3]ClO_4$	CuS_2N_2O	2.264	1.448	—	99
$\{[Cu(tmdt)OClO_3]_2Imz\}ClO_4$	CuN_4O	2.412	1.39	132.6	101
		2.379	1.41	133.6	
$[Cu(eaep)(OH)OClO_3]_2$	CuN_2O_3	2.562	1.391	139.8	102
		2.618	1.406	130.7	
$[Cu(dmaep)(OH)OClO_3]_2$ (β)	CuN_2O_3	2.721	1.422	123.0	103
$[Cu(dmaep)(OH)O_2ClO_2]_2$ (α)	CuN_2O_4	2.716	1.429	117.7	10
		2.782	1.437	121.2	
$[Cu(2-MeIm)_2(OH)(O_2ClO_2)]_2 \cdot 2H_2O$	CuN_2O_4	2.827			104
$Cu(dpa)(MeCO_2)(O_2ClO_2) \cdot H_2O$	CuN_2O_4	2.541	1.396		106
		2.638	1.387		
$Cu(pam)_2(OClO_3)_2$	CuN_2O_4	2.52	1.38	143.1	120
$[Cu(bipy)_2(O_2ClO_2)]ClO_4$	CuN_4O_2	2.73	1.48	124	9
		2.45	1.48		
		2.746	1.438		105
		2.512	1.418		

Cu(Im)₄(OClO₃)₂	CuN₄O₂	2.625			111
Cu(aep)₂(OClO₃)₂	CuN₄O₂	2.833	1.433	148.8	6
Cu(en)₂(OClO₃)₂	CuN₄O₂	2.611	1.414	122.6	108
Cu(dab)₂(OClO₃)₂ (blue-violet)	CuN₄O₂	2.579	1.451	126.2	116
(red-violet)	CuN₄O₂	2.676	1.387	125.1	
Cu(dap)₂(OClO₃)₂	CuN₄O₂	2.61	1.45	120	116a
Cu(men)₂(OClO₃)₂	CuN₄O₂	2.575	1.431	121.0	118
Cu(pyz)₂(OClO₃)₂	CuN₄O₂	2.373	1.426	170.6	119
Cu(dmpd)₂(OClO₃)₂	CuN₄O₂	2.602	1.441	123.7	121
Cu(hism)₂(OClO₃)₂	CuN₄O₂	2.616	1.422		123
Cu(bpda)(OClO₃)₂	CuN₄O₂	2.693	1.438	142.1	5
Cu(tamn)(OClO₃)₂	CuN₄O₂	2.667	1.454	118.9	113
		2.527	1.448	121.1	
Cu(Im)₂(Gly—GlyH)OClO₃	CuN₃O₃	2.97			128
Cu(mmea)₂(OClO₃)₂	CuN₂S₂O₂	2.599	1.447	130.1	124
Cu(pmmi)₂(OClO₃)₂	CuN₂S₂O₂	2.594	1.436	139.3	126
Cu(dth)₂(OClO₃)₂	CuS₄O₂	2.549	1.446		117
Cu(tctd)(OClO₃)₂	CuS₄O₂	2.652	1.441	130.3	114
Cu(ompa)₂(OClO₃)₂	CuO₆	2.546	1.407	149.5	125
[Cu(Imzn)₅(OClO₃)]ClO₄	CuO₆	2.994			109

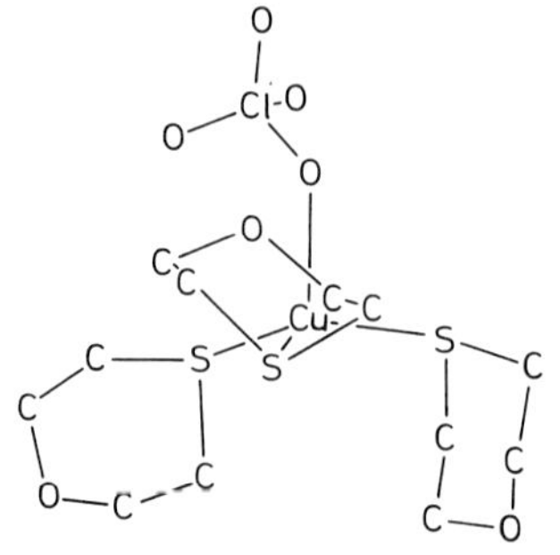

FIG. 9. A perspective drawing of Cu(tox)$_3$OClO$_3$. [Reproduced by permission from M. M. Olmstead et al., Transition Met. Chem. **7**, 140 (1982).]

perchlorates. The X-ray crystal structure analysis of the tox complex (96) shows trigonal–pyramidal geometry around copper(I) with strongly coordinated unidentate perchlorate (Fig. 9; Table VII). The bonded O—Cl length (1.478 Å) is longer than the nonbonded ones (1.40, 1.413, and 1.42 Å).

Four-coordination involving one monodentate perchlorate in each case has been visualized for the metal ion in copper(II) complexes of the compositions Cu(bpes)(ClO$_4$)$_2$ (98) and Cu(phzNO$_2$)$_3$(ClO$_4$)$_2$ · H$_2$O (41). A variety of five-coordinate complexes of the formulas Cu(bpdt)(ClO$_4$)$_2$ (99), Cu(deta)XClO$_4$ (X = Cl, Br, OAc, etc.) (100), Cu$_2$(tmdt)$_2$(μ-Imz)-(ClO$_4$)$_3$ (101), [Cu(N—N)(μ-OH)ClO$_4$]$_2$ (N—N = eaep, dmaep) (102, 103), Cu(quxNO$_2$)$_2$(H$_2$O)$_2$(ClO$_4$)$_2$ · H$_2$O, Cu(quxNO$_2$)$_3$(ClO$_4$)$_2$ · 10H$_2$O, and CuL$_4$(ClO$_4$)$_2$ (L = miqNO, quxNO$_2$; L$_4$ = tdaea) (23, 38–40), containing coordinated perchlorate, have been reported. The X-ray crystal structure of the bpdt complex, solved by Brubaker et al. (99), indicates that the metal ion assumes a square–pyramidal configuration with two nitrogen and two sulfur atoms of the tetradentate ligand occupying the basal plane and a perchlorate oxygen the apex. Similarly, pentacoordination for each of the metal ions in the binuclear tmdt complex is satisfied by four nitrogen atoms (three from the tridentate tmdt ligand and one from the bridging Imz ion) and a monodentate perchlorate (101). Hodgson and co-workers (102, 103) have described the geometry around the metal ion in each of the dimeric complexes containing eaep (violet) and dmaep (dark blue, monoclinic, β-form) as two adjoining distorted tetragonal pyramids with a common edge defined by the two bridging hydroxyl oxygen atoms. Thus the basal plane at each copper consists of two nitrogens of the bidentate ligand (Fig. 10; N—N = eaep or dmaep) and two bridging oxygens (of OH groups)

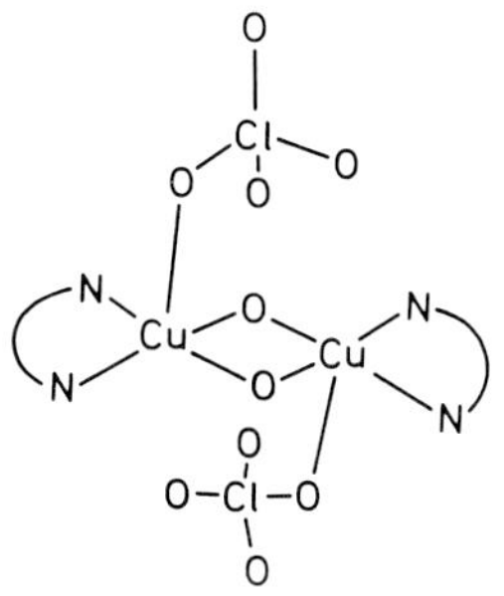

FIG. 10. Structure of [Cu(N—N)(OH)ClO$_4$]$_2$. [Redrawn by permission from D. L. Lewis et al., Inorg. Chem. **13**, 1013 (1974).]

while the fifth coordination site is occupied by a perchlorate oxygen. The metal–perchlorate interaction has been proposed to be weak.

The α isomer (deep blue, triclinic) of the dmaep complex is also dinuclear but six-coordinate. The IR spectrum of this form shows very intense peaks at 1145, 1128, and 1072 cm^{-1}, which have been assigned to bidentate perchlorate. This deduction has been confirmed by crystallographic evidence (10). The perchlorates are positioned directly above and below the planar central dimeric [(dmaep)Cu(μ-OH)$_2$Cu-(dmaep)]$^{2+}$ unit (Fig. 11), each linked to the two copper atoms via two

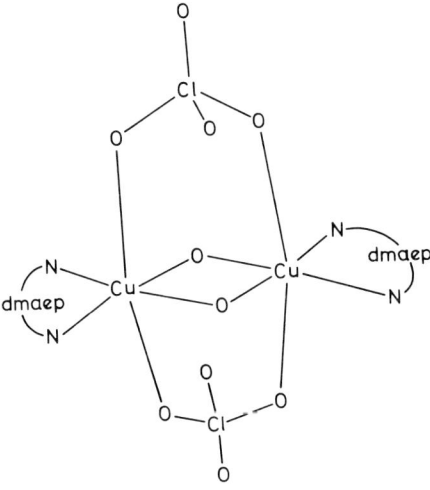

FIG. 11. Structure of α-[Cu(dmaep)(OH)ClO$_4$]$_2$. [Redrawn by permission from D. L. Lewis et al., Inorg. Chem. **13**, 147 (1974).]

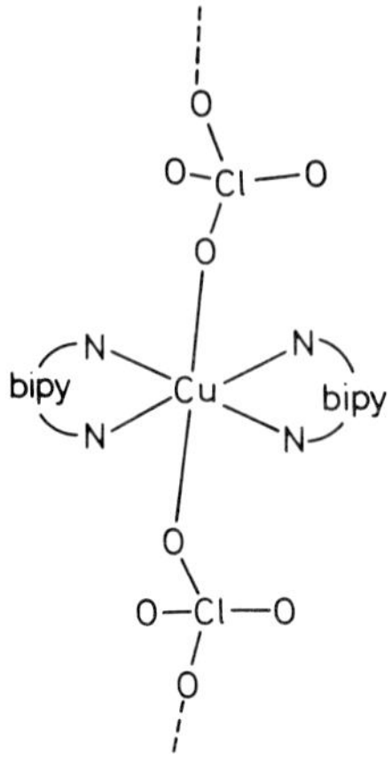

FIG. 12. Structure of a part of the infinite chain, [Cu(bipy)$_2$ClO$_4^+$]$_\infty$. [Redrawn by permission from H. Nakai, *Bull. Chem. Soc. Jpn.* **44**, 2412 (1971).]

oxygens. Further, one of the two remaining oxygen atoms of each perchlorate is hydrogen bonded to the bridging hydroxyl hydrogen with OH ⋯ O contact of 2.36 Å. The coordinated O—Cl distances (Table VII) are longer than the uncoordinated ones, 1.409 and 1.428 Å. Intramolecular bidentate coordination of perchlorates has also been confirmed in another dihydroxo-bridged complex of the formula [Cu(2-MeIm)$_2$(OH)ClO$_4$]$_2$ · 2H$_2$O by X-ray structure analysis (*104*). The distorted octahedral coordination around each metal ion is completed by two nitrogen atoms of 2-MeIm.

Nakai (*9*) has recognized an intermolecular bidentate perchlorate of a polymeric nature in Cu(bipy)$_2$(ClO$_4$)$_2$, based on X-ray photographic data collection. The coordination polyhedron around the copper atom is a tetragonally distorted octahedron (Fig. 12) wherein the four N atoms of bipy molecules are arranged in a flattened tetrahedral manner, the least-squares plane of these atoms making an equatorial plane, while the axial positions are occupied by the oxygen atoms of perchlorates. However, there is coexistence of coordinated and noncoordinated perchlorates in the crystal. Hathaway and co-workers (*105*) have redetermined the bond-length parameters more accurately in the above complex, using X-ray diffractometer data collection (Table VII). The crystal structure of Cu(dpa)(MeCO$_2$)ClO$_4$ · H$_2$O, solved by the above workers (*106*), has also indicated a polymeric nature with an elongated rhombic octahedral geometry for the complex in which the axial positions are occupied by oxygen atoms from the unsymmetrically bridging perchlorate groups. X-ray structure analysis of the

Cu(en)$_2$(ClO$_4$)$_2$ complex by Pajunen (*107, 108*) has shown unidentate coordination of both perchlorates in the long tetragonal positions above and below the planar [Cu(en)$_2$]$^{2+}$ unit.

A wide range of other six-coordinate complexes of copper(II) of the types Cu(Imzn)$_5$(ClO$_4$)$_2$ (*109*), CuL$_4$(ClO$_4$)$_2$ (L = MeCN, NH$_3$, Py, 4-MePy, Im, 1-ViIm, 1-Vi-2-MeIm, 2-EtIm, 2-*i*-PrIm; L$_4$ = bpda, bbtb, tamn, tctd) (*5, 15, 47, 47a, 62, 110–115*), Cu(L—L)$_2$(ClO$_4$)$_2$ (L—L = en, aep, dab, dap, dth, men, pyz, pam, bdpm, dmpd, dmpNO, hism, mmea, ompa, pmmi) (*1, 6, 51, 116–126*), CuL$_2$(ClO$_4$)$_2$ (L = qun, arg) (*47, 127*), Cu(Im)$_2$(Gly—GlyH)ClO$_4$ (*128*), and Cu(tbma)XClO$_4$ (X = Cl, Br) (*27*), involving coordinated perchlorates, have been reported. The X-ray studies on the blue crystals of the Imzn complex, carried out by Majeste and Trepons (*109*), have shown that one of the perchlorates completes the distorted octahedral stereochemistry for the metal ion. The Cu—O distance of 2.994 Å found for this complex appears to be the longest bond length observed among copper perchlorate complexes (Table VII). According to crystal structure analysis by Ivarsson (*111*), the perchlorates in the Im complex have been located at the apical positions of an octahedron around the metal ion. On the basis of IR data (Table VI), the monodentate nature of perchlorates has been envisaged in the complexes containing Py, 4-MePy, and substituted Im (*47, 112*). As indicated by X-ray crystal structure studies, the complexes involving the tetradentate ligands bpda (*5*), tamn (*113*), and tctd (*114*) have a tetragonal coordination for the metal ion with unidentate perchlorates at the axial sites. Likewise, the perchlorates in Cu(L—L)$_2$(ClO$_4$)$_2$ complexes are shown to occupy elongated trans positions of an octahedron. In this series, the structure of the pyrazine complex has been found to be made up of parallel sheets each having an infinite square array of the metal ions bridged by the nonchelating bidentate pyz molecules (Fig. 13). Each metal ion, with a tetragonally elongated octahedral environment, is linked strongly to four nitrogen atoms of four pyz groups and weakly to two monodentate perchlorates. The large Cu—O—Cl angle lying in the range 160–170° is thought to minimize nonbonded interactions between ClO$_4^-$ ions, which are linked to the metal ions on adjacent sheets (*119*). A weak coordination of perchlorates has been visualized in the complex having the chelating bidentate ligand pam (*120*). The coordinated O—Cl length, 1.38 Å, in the latter complex is surprisingly shorter than the average (1.41 Å) of the uncoordinated ones (1.31, 1.43, and 1.50 Å). However the metal ion bound O—Cl distance (1.441 Å) in the dmpd complex (Fig. 14) is longer than the nonbonded ones (1.401, 1.418, and 1.429 Å) (*121*). For this complex, the effect of temperature on the IR bands of the perchlorate

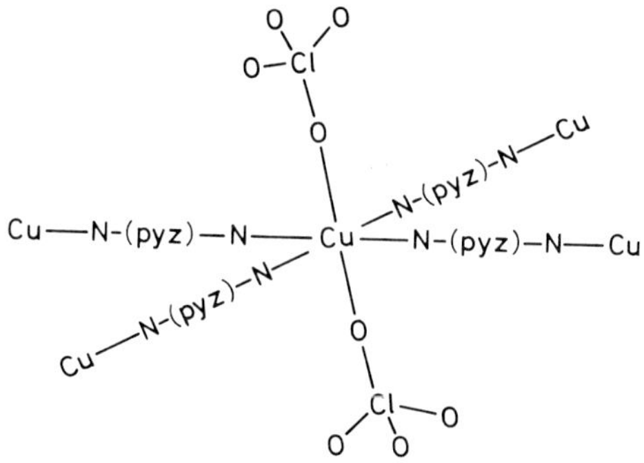

FIG. 13. A section of the sheet structure in Cu(pyz)$_2$(ClO$_4$)$_2$. [Redrawn by permission from J. Darriet et al., Inorg. Chem. **18**, 2679 (1979).]

has been studied. As seen in Fig. 15, the splitting of the ν_3 band decreases with increasing temperature.

Electronic and IR spectra and other physical techniques are in consonance with a distorted octahedral structure for CuL$_2$(ClO$_4$)$_2$ (L = qun, arg) complexes, implying that the perchlorates are bidentate *(47, 127)*. Bell et al. *(128)* have worked out the crystal structure of a distorted octahedral complex, Cu(Im)$_2$(Gly—GlyH)ClO$_4$, containing a dipeptide; the perchlorate occupies an apical position. Infrared spectra have revealed monodentate coordination of perchlorates in copper(II) complexes having ligands such as adH, adHNO, pyzNO, quxNO, and an octadentate Schiff base *(33, 34, 39, 49, 52, 53, 129)*.

The structure of a trinuclear complex, Cu$_3$(Im)$_8$(Imz)$_2$(ClO$_4$)$_4$ *(130)*, has been described to consist of three octahedrally (distorted) coordi-

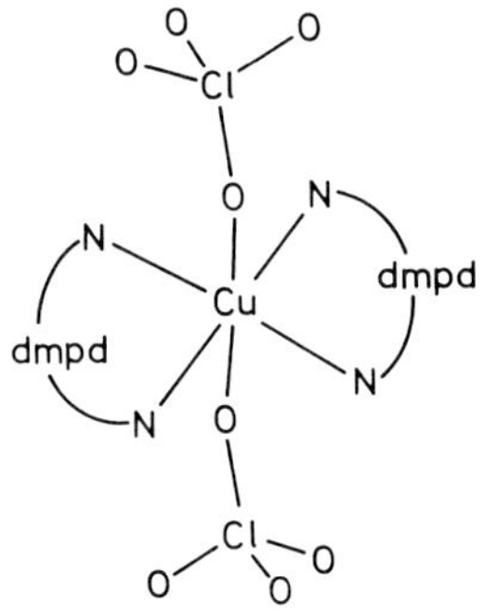

FIG. 14. Structure of Cu(dmpd)$_2$(ClO$_4$)$_2$. [Redrawn by permission from L. P. Battaglia et al., J. Chem. Soc. Dalton Trans. 8 (1981).]

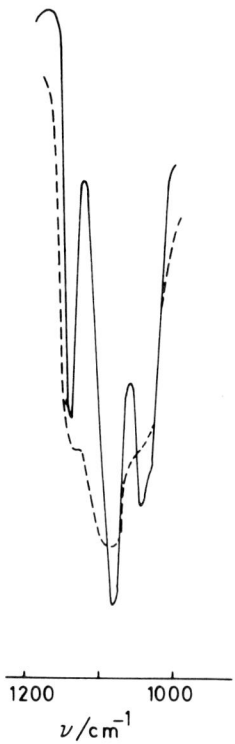

FIG. 15. IR spectra of Cu(dmpd)$_2$(ClO$_4$)$_2$ at 293 (———) and 393 K (- - -) in the 1200–950 cm^{-1} region. [Reproduced by permission from L. P. Battaglia et al., J. Chem. Soc. Dalton Trans. 8 (1981).]

nated copper atoms, one of them connected through two Imz bridges to the other two (Fig. 16). Thus the centrosymmetric metal atom (Cu-1) is attached to four N atoms (two from each of the Im and Imz groups) and two O atoms of bridging perchlorates, while each of the two symmetry-related metal atoms (Cu-2) are also linked to four N atoms (three from Im and one from Imz ligands) and two O atoms of terminal and bridging perchlorates. Consequently, there are two types of perchlorates— one is bidentate, bridging to both Cu-1 and Cu-2 with Cu—O distances 2.755 and 2.93 Å, respectively, and the other is unidentately bound to Cu-2 with Cu—O length 2.581 Å. The reason for the difference between the first two values has been ascribed to the fact that the 2.755 Å refers to the Cu-1 octahedron having both perchlorates as bidentate bridging, while 2.93 Å refers to the Cu-2 octahedron with two different perchlorates, one shared and the other unshared.

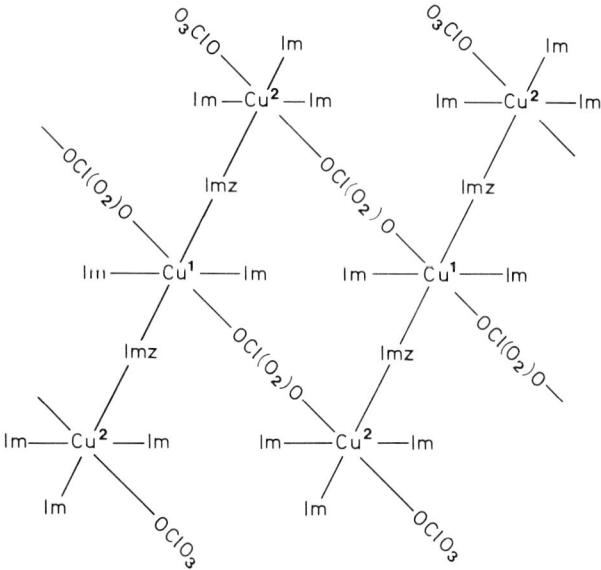

FIG. 16. Structure of $Cu_3(Im)_8(Imz)_2(ClO_4)_4$. [Redrawn by permission from G. Ivarsson et al., Acta Chem. Scand. **26**, 3005 (1972).]

A number of two-coordinate silver(I) complexes of the type $AgLOClO_3$ (L = t-Bu_3P, Cy_3P, o-$tolyl_3P$, (p-$XC_6H_4)_3P$ (X = F, Cl, OMe) have been prepared by Dikhoff and Goel (131). The complexes are stable in the solid state and are unionized in dichloromethane. Naikar et al. (94) have obtained three- and four-coordinate complexes (Table VI) of the formulas $AgL_2(OClO_3)$ (L = Nap_2EtAs, Cy_3As) and $AgL_3(OClO_3)$ (L = Ph_2EtAs, Ph_2-n-$PrAs$) by treating $AgClO_4$ with tertiary arsines in alcohol. Stoichiometric quantities of $AgClO_4$ and Iotz in aqueous medium has yielded four-coordinate $Ag(Iotz)_2(O_2ClO_2)$ complex (132). $AgClO_4$ reacts with tmb in MeOH–MeCN to give colorless crystals of $Ag_2(tmb)_3(ClO_4)_2$. According to X-ray diffraction studies, the complex has infinite chains wherein each metal ion is bonded to three CN groups of tmb ligands and weakly to an oxygen atom of perchlorate. Thus the metal ion has a trigonal–pyramidal geometry with Ag—O 2.733, O—Cl (bonded) 1.416 Å, and Ag—O—Cl angle 129.2° (133). The structure of another complex, $Ag(dmso)_2ClO_4$, is also shown by X-ray studies to be built up of infinite chains of metal ions joined by double bridging dmso oxygen atoms. For this complex, a distorted trigonal–bipyramidal geometry around each metal ion with a chelating bidentate perchlorate (Ag—O = 2.411, 2.741 Å) has been described (134).

Syntheses of perchlorate-coordinated gold(I) and gold(III) complexes of the types, AuLClO$_4$, Au$_2$(dppm)(ClO$_4$)$_2$, and Au(C$_6$F$_5$)$_2$LClO$_4$ (L = Ph$_3$P) have been described by Uson and co-workers (*135–137*). The coordinated perchlorate can easily be displaced by cod in the former Ph$_3$P complex (135) and by anionic (N$_3$$^-$, HCO$_3$$^-$) or neutral monodentate (Et$_3$P, Bu$_3$P, Ph$_2$MeP, Ph$_3$PO, Ph$_3$AsO, PyNO, etc.) ligands in the latter (*138*).

VIII. Complexes of Zinc Group Metals

Reaction of zinc(II) perchlorate with tertiary arsines in thf–teof yields complexes of compositions Zn(Cy$_3$As)$_2$(ClO$_4$)$_2$ and ZnL$_3$(ClO$_4$)$_2$ (L = Ph$_2$MeAs, Ph$_2$EtAs) (*94*). Unidentate coordination of one of the perchlorates to the metal ion in each of the above complexes has been inferred from the solid-state IR spectra (Table VIII). In the case of the Cy$_3$As complex, perchlorate coordination continues to exist in dichloromethane solution (*139*). A planar three-coordinate structure for the Cy$_3$As complex and a tetrahedral configuration for the other two complexes in the solid state have been visualized. Tetrahedral complexes of stoichiometries Zn(qun)$_2$(OClO$_3$)$_2$ (*47*) and Zn(miqNO)$_3$(OClO$_3$)-ClO$_4$ · H$_2$O (*38*) have also been characterized.

Several zinc(II) complexes of the formulas ZnL$_4$(ClO$_4$)$_2$ (L = mmpp, tetp, Ph$_2$MeAsO; L$_4$ = taea) (*21, 32, 46, 140*), Zn(adHNO)$_2$-(H$_2$O)$_2$(ClO$_4$)$_2$ (*34*), Zn$_2$(phzNO)$_4$(H$_2$O)$_2$(ClO$_4$)$_4$ · 6H$_2$O (*39*), and Zn(Tpp)-ClO$_4$ (*13*), where the metal ion is five-coordinate, have been reported. Infrared evidence has been presented for monodentate coordination of one of the perchlorates to the metal ion in each of these complexes. Perchlorate coordination to the metal ion persists in dichloromethane, as shown by the IR spectrum of the Tpp complex. Spaulding *et al.* (*13*) have elucidated the structure of the above complex by the X-ray diffraction method. The complex has unidentate perchlorate at the apex of a square pyramid (Fig. 17) with a Zn—O distance of 2.07 Å and a Zn—O—Cl angle of 130.2°. The metal-bound O—Cl length is 1.451 Å, as against the nonbonded ones of 1.328, 1.374, and 1.525 Å. In the phzNO complex, the metal ions are bridged through two molecules of the *N*-oxides (*39*). Six-coordinate zinc(II) complexes of the types Zn(MeCN)$_4$(ClO$_4$)$_2$ (*15*), Zn(pyzNO)$_4$(ClO$_4$)$_2$ · H$_2$O (*49*), Zn(phzNO$_2$)$_3$(ClO$_4$)$_2$ · 5H$_2$O (*41*), Zn(bdpm)$_2$(ClO$_4$)$_2$ (*51*), and Zn$_2$L$_6$(ClO$_4$)$_4$ · *x*H$_2$O (L = quxNO, *x* = 10; L = quxNO$_2$, *x* = 12) (*39, 40*) are considered to involve one or two monodentate perchlorates. In the dimeric complexes, the metal ions are again bridged by two of the *N*-oxide ligands.

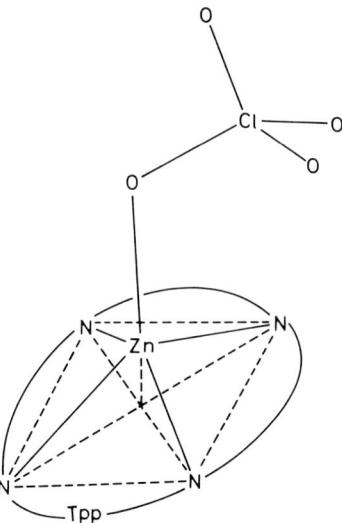

FIG. 17. Geometry about the zinc atom in Zn(Tpp)OClO$_3$. [Reproduced by permission from L. D. Spaulding et al., J. Am. Chem. Soc. **96**, 982 (1974).]

Perchlorate coordination has been identified in only a few complexes in the case of cadmium. Cadmium(II) perchlorate reacts with tertiary arsines in thf–teof to afford Cd(Cy$_3$As)$_2$(ClO$_4$)$_2$ and Cd(Ph$_2$RAs)$_3$-(ClO$_4$)$_2$ (R = Me, Et, n-Pr) complexes (94). It has been envisaged on the basis of IR data (Table VIII) that one of the perchlorates is serving as a monodentate ligand. As a consequence, a planar three-coordinate structure for the Cy$_3$As complex and a tetrahedral configuration for the rest of the complexes have been proposed. Cd(bdpm)$_2$(ClO$_4$)$_2$ is suggested to involve a bidentate perchlorate (51) and hence coordination number 6 for the metal ion. A variable coordination number (6 or 7) for the metal ion is postulated in the CdL(ClO$_4$)$_2$ (L = pentagonal Schiff base) complex (129).

A series of mercury(II) complexes of stoichiometries HgL$_2$(ClO$_4$)$_2$ (L = Cy$_3$P, t-Bu$_3$P, o-tolyl$_3$P, mesityl$_3$P, Cy$_3$As, Nap$_2$MeAs, Nap$_2$EtAs), Hg(Ph$_2$EtAs)$_2$(H$_2$O)(ClO$_4$)$_2$, and HgL$_3$(ClO$_4$)$_2$ (L = Ph$_3$P, p-tolyl$_3$P, Ph$_2$MeAs, Ph$_2$EtAs, Ph$_2$-n-PrAs) have been prepared by treating mercuric perchlorate with tertiary phosphines or arsines in alcohol (141, 142). Coordination of one or two perchlorates in each of these complexes is indicated by the IR spectra (Table VIII). Thus the metal ion is thought to have planar three-coordination in HgL$_2$(ClO$_4$)$_2$ (L = tertiary arsine) complexes, and a tetrahedral geometry in the other series. However, HgL$_2$(ClO$_4$)$_2$ complexes containing tertiary phosphines

TABLE VIII

Perchlorate Frequencies of Zinc Group Metal Complexes

Complex	Perchlorate bands (cm^{-1})			Reference
	ν_3	ν_1	ν_4	
[Zn(Cy$_3$As)$_2$OClO$_3$]ClO$_4$	1127, 1100, 1044	927	625, 620	94
[Zn(Ph$_2$MeAs)$_3$OClO$_3$]ClO$_4$	1128, 1110, 1084, 1047	920	622, 612	94
[Zn(miqNO)$_3$OClO$_3$]ClO$_4$ · 3H$_2$O	1110, 1090, 1053, 1015		641, 626, 619, 612, 328[a]	38
Zn(qun)$_2$(OClO$_3$)$_2$	1107, 1085	925	635, 624, 483[b]	47
[Zn(taea)OClO$_3$]ClO$_4$	1147, 1094, 1001	930	654, 621	140
[Zn(mmpp)$_4$OClO$_3$]ClO$_4$	1141, 1085, 1040	927	652, 647, 616, 310[a]	21
[Zn(tetp)$_4$OClO$_3$]ClO$_4$	1140, 1095, 1048	924	647, 631, 622	32
[Zn(pyzNO)$_4$(H$_2$O)OClO$_3$]ClO$_4$	1115, 1097, 1081		621, 293[a]	49
[Cd(Cy$_3$As)$_2$OClO$_3$]ClO$_4$	1122, 1105, 1082, 1050	930	624, 619	94
[Cd(Ph$_2$MeAs)$_3$OClO$_3$]ClO$_4$	1127, 1089, 1079, 1033	911	623, 620	94
[Cd(Ph$_2$EtAs)$_3$OClO$_3$]ClO$_4$	1130, 1085, 1038	928	624, 622	94
[Cd(Ph$_2$-n-PrAs)$_3$OClO$_3$]ClO$_4$	1135, 1105, 1082, 1050	910	623, 613	94
[Hg(Cy$_3$As)$_2$OClO$_3$]ClO$_4$	1130, 1115, 1080, 1040	915	625, 621, 618	142
[Hg(Nap$_2$MeAs)$_2$OClO$_3$]ClO$_4$	1135, 1115, 1090, 1025	910	623, 618, 615	142
Hg(Ph$_3$F)$_2$(OClO$_3$)$_2$	1190, 1022	912	618, 612	144
[Hg(Ph$_3$P)$_3$OClO$_3$]ClO$_4$	1130, 1085, 1025	910	623, 614	142
[Hg(p-tolyl$_3$P)$_3$OClO$_3$]ClO$_4$	1125, 1090, 1030	920	620, 615	142
[Hg(Ph$_2$EtAs)$_2$(H$_2$O)OClO$_3$]ClO$_4$	1120, 1115, 1085, 1030	915	628, 625, 620	142
[HgCl(o-tolyl$_3$P)OClO$_3$]$_2$	1110, 1070, 1045	925, 917	686, 666, 622	145
[Hg(Ph$_3$P)OClO$_3$(O$_2$ClO$_2$)]$_2$	1120, 1080, 1020	906, 896	620, 616, 580	144

[a] ν[M—O(ClO$_4$)].
[b] ν_2.

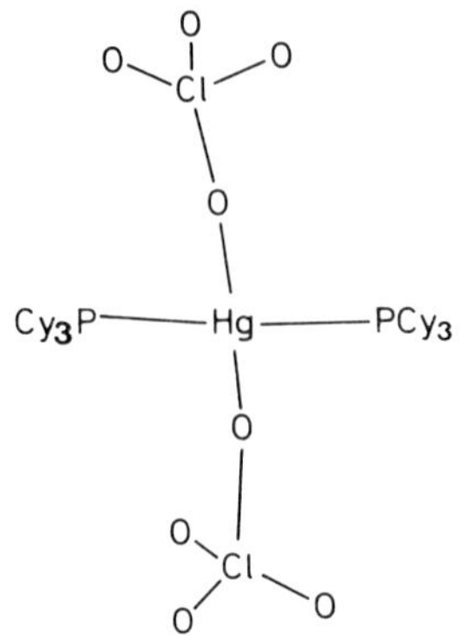

FIG. 18. Structure of Hg(Cy$_3$P)$_2$(OClO$_3$)$_2$ molecule. [Redrawn by permission from E. C. Aleya et al., J. Chem. Res. Synop. 360 (1979).]

are considered to be four-coordinate (*141*). Although the Cy$_3$P complex had earlier been characterized as a two-coordinate species on the basis of IR spectra and ^{31}P and ^{199}Hg chemical shift results (*143*), X-ray diffraction analysis has indicated weak coordination of perchlorates to the metal ion (Fig. 18) with Hg—O distances of 2.928 and 3.234 Å (*141*); however, the latter value appears to be too large for a covalent bond length. Tetrahedral complexes of the types HgL$_2$(OClO$_3$)$_2$ and [HgL(OClO$_3$)(O$_2$ClO$_2$)]$_2$ (L = Ph$_3$P, Ph$_3$As) have been reported by Davis et al. (*144*).

In an attempt to isolate a 1:2 complex from the reaction mixture of mercuric perchlorate and o-tolyl$_3$P, Aleya et al. (*145*) have obtained a dichloro-bridged dinuclear complex [HgCl(o-tolyl$_3$P)ClO$_4$]$_2$. The complex is a nonelectrolyte in dichloromethane, and the monodentate nature of the perchlorate is shown by the IR spectrum as well as by X-ray crystallographic analysis. The coordination geometry around each of the metal ions is described as trigonal–bipyramidal with an equatorial site unoccupied (Fig. 19). The weakly bound perchlorate is situated at one of the equatorial positions with Hg—O 2.73, O—Cl (coordinated) 1.42, and O—Cl (terminal) 1.36, 1.41, and 1.44 Å.

Sandström (*146*) has established by X-ray analysis unidentate coordination of two of the perchlorates to one of the metal ions (Hg—O length = 2.78, 3.04 Å; Hg—O—Cl angle = 161, 123°) in the binuclear complex Hg$_2$(dmso)$_8$(ClO$_4$)$_4$. The metal ions bridged through two of the eight dmso molecules assume distorted octahedral geometries. X-ray evidence has been presented by Epstein et al. (*147*) for the presence of a weakly bonded perchlorate in a seven-coordinate mercury(II) complex

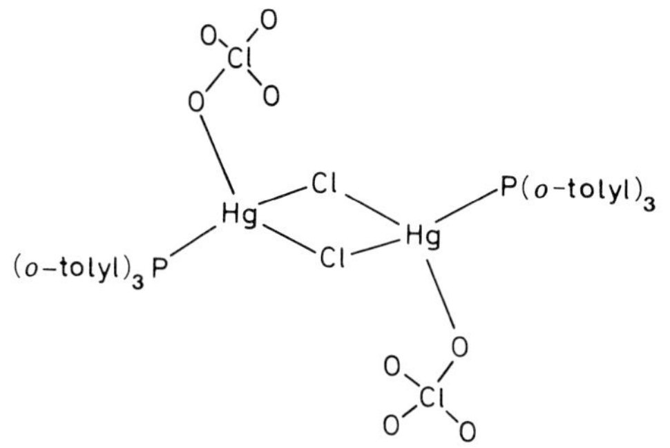

FIG. 19. Structure of [HgCl(o-tolyl$_3$P)OClO$_3$]$_2$. [Redrawn by permission from E. C. Aleya et al., Can. J. Chem. **57**, 2217 (1979).]

Hg(ntd)$_3$(ClO$_4$)$_2$ with Hg—O length = 2.93; O—Cl length = 1.34 (bonded), 1.28, 1.34, and 1.35 Å (nonbonded), and Hg—O—Cl angle = 114°. In another Hg(II) complex, Hg(tchd)(ClO$_4$)$_2$, having a tetradentate macrocyclic tetrathioether ligand, the metal ion is visualized as having heptacoordination, though its morphology is that of an irregular, tetragonally elongated octahedron. The metal ion in the complex has been located by X-ray diffraction studies near the center of the four S atoms of the macrocycle, and the inner coordination sphere is completed by both perchlorates, which are unequally bonded along the z axis, one being monodentate (Hg—O, 2.76 Å) and the other being weakly bidentate (Hg—O, 3.08 and 3.26 Å) (*148*).

IX. Complexes of Lanthanides

On treatment of lanthanide perchlorates with Q in MeOH–teof, six-coordinate complexes of the type LnQ$_4$(ClO$_4$)$_3$ (Ln = Ce; Q = *n*-Bu$_3$PO, Ph$_3$PO; Ln = Nd; Q = ttp), in which two of the perchlorates are covalently linked to the metal ion, are obtained (*18, 20, 36*). The cerium complex is a 1:1 electrolyte in MeNO$_2$, implying that the two ClO$_4$$^-$ groups continue to be coordinated in solution. On the basis of analytical results and IR and conductance measurements, several lanthanide complexes containing hmpa have been formulated as [Ln(hmpa)$_4$-(OClO$_3$)$_2$]ClO$_4$ (Ln = La, Nd, Gd, Ho, Yb), indicating hexacoordination for the metal ion. However, detailed IR and Raman spectral studies indicate bidentate coordination of one or more of the perchlorates, suggesting that the coordination number of the lanthanide ion may vary from six to eight in these systems (*17, 29*). CeQ$_6$(ClO$_4$)$_3$ (Q = tmp, dmmp) and Nd(phen)$_3$(ClO$_4$)$_3$ complexes are postulated to involve a monodentate perchlorate (28, 149–151). Seven-coordination for the metal ion in these complexes is assigned, based on the facts that the former is a 1:2 electrolyte in MeNO$_2$, while the latter shows strong IR peaks at 1100 (split) and 927 cm^{-1}. Unidentate coordination of two of the perchlorates in Ln(CypNO)$_n$(ClO$_4$)$_3$ · 2H$_2$O (n = 5; Ln = Gd, Tb; n = 6; Ln = La, Sm, Dy, Ho; n = 7; Ln = Pr, Nd, Er, Yb) complexes is revealed by IR and molar conductivity measurements (151a).

The IR spectra of Ln(III) complexes Ln(tdod)(ClO$_4$)$_3$ · xH$_2$O (Ln = Ce, Pr, x = 1; Ln = Nd, x = 0) are consistent with the presence of exclusively unidentate perchlorates (31). The complexes are 1:2 electrolytes in MeCN, and thus one of the perchlorates continues to be attached to the metal ion in solution. Lanthanide ions also form a number of other 10-coordinate complexes having coordinated perchlorates, namely, LnQ(ClO$_4$)$_3$ · S (Ln = Sm, Eu, Yb; Q = htod; S = H$_2$O; Ln = Eu; Q = dtod; S = H$_2$O; Ln = Pr, Nd, Eu; Q = hdhc; S = MeCN; Ln = Ce, Pr, Nd, Sm, Eu, Dy, Ho; Q = tdhc) (11, 31, 152). Ciampolini et al. (11, 152) have elucidated the structures of Sm(tdhc)(ClO$_4$)$_3$ and Eu(hdhc)(ClO$_4$)$_3$ by X-ray diffraction. The coordination polyhedron around the metal ion is a hexagonal bipyramid (having six oxygens of the tdhc ligand in the equatorial plane and four oxygens of the perchlorates at the apices) for the former, and a square bicapped antiprism (with two nitrogen atoms in axial positions and eight oxygen atoms at the vertices of the two squares) for the latter. Thus all three perchlorates are linked to the samarium ion, one of them being bidentate (Sm—O length = 2.64, O—Cl length = 1.46 Å, Sm—O—Cl angle = 101.2°) while the other two are monodentate (Sm—O length = 2.36, O—Cl length = 1.48 Å, Sm—O—Cl angle = 152.5°). On the other hand, only one of the perchlorates is bidentately bound to the europium ion (Eu—O lengths = 2.67, 2.71; O—Cl lengths = 1.47, 1.62 Å). A coordination geometry similar to that around the samarium ion is ascribed to the other Ln(tdhc)(ClO$_4$)$_3$ complexes. These are 1:2 electrolytes in MeCN at low concentrations (10^{-3} M), but behave as 1:1 electrolytes in slightly more concentrated solutions (10^{-2} M).

Bidentate coordination of one or two perchlorates has been proposed in a thorium(IV) complex of composition Th(Ph$_3$PO)$_4$(ClO$_4$)$_4$ (20).

X. Complexes of Other Metals

The crystal structure studies on Na(ebsmCu)$_2$ClO$_4$ by Milburn et al. (153) have indicated an octahedral environment for the sodium ion, with the perchlorate serving as a chelating bidentate ligand (Na—O =

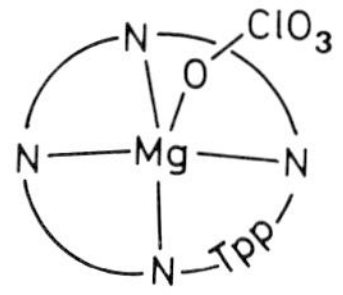

FIG. 20. Structure of Mg(Tpp)OClO$_3$. [Redrawn by permission from K. M. Barkigia et al., Inorg. Chem. **22**, 349 (1983).]

2.60 Å). Barkigia et al. (154) have deduced the crystal structure of Mg(Tpp)ClO$_4$, wherein the metal ion has a square–pyramidal geometry, with one of the oxygens of the perchlorate at the apex (Fig. 20). The Mg—O length (2.012 Å) appears to be the shortest among the M—O distances known for metal perchlorato complexes; the Mg—O—Cl angle = 134.5°. The preparation of a calcium(II) complex [Ca(Ph$_3$PO)$_4$(O$_2$ClO$_2$)]ClO$_4$ {ν(ClO$_4$) 1190, 1144, 1025, 935, 656, 625, 611, 490, and 460; ν[Ca—O(ClO$_4$)] 318 cm^{-1}} has been reported (20). In Al(tetp)$_4$(ClO$_4$)$_3$, unidentate coordination of two of the perchlorates has been suggested, based on its molar conductance in MeNO$_2$ and the occurrence of perchlorate bands at 1120, 1090, 1042, 920, 652, 638, and 621 cm^{-1} in the IR spectrum (32). Matyashin et al. (155) have briefly mentioned the preparation of compounds of type R$_4$N(MX$_3$ClO$_4$) (R = Me, Et, Bu; M = B, Al; X = F, Cl, Br, BH$_4$) having coordinated perchlorate. An indium(III) complex of the formula [In(Ph$_3$PO)$_4$(OClO$_3$)$_2$]ClO$_4$ [ν(ClO$_4$) 1140, 1030, 920, 650, 627, 609, 480, and 455; ν(In—OClO$_3$) 311 cm^{-1}] is known (20). Blundell and Powell (156) have elucidated the crystal structure of a thallium(III) complex TlMe$_2$(phen)ClO$_4$ wherein the six-coordinate metal ion is surrounded by a pentagonal–bipyramidal environment with an equatorial position vacant. The weakly bonded perchlorate behaves as a bidentate-bridging ligand between two metal ions with a Tl—O length of 2.88 Å.

X-Ray crystal structure analysis of the dianion (SnCl$_3$ClO$_4$)$^{2-}$ reveals a square–pyramidal geometry around the metal ion (157). The equatorial plane is defined by two Cl atoms, one perchlorate oxygen, and presumably the unshared electron pair of tin. The perchlorate is monodentate (Sn—O length = 2.907, coordinated O—Cl length = 1.426 Å, Sn—O—Cl angle = 135.9°), and the coordinated O—Cl distance is longer than the uncoordinated ones (1.392, 1.399, and 1.410 Å). A trigonal–bipyramidal configuration and hence pentacoordination for the metal ion in SnMe$_3$ClO$_4$ is satisfied by bidentate-bridging perchlorate (158). In another tin complex, [Sn$_3$O$_2$Cl$_4$(ClO$_4$)$_4$]$_2$, the X-ray diffraction, IR, and Raman spectral investigations are consistent with the occurrence of strongly bonded two unidentate and six biden-

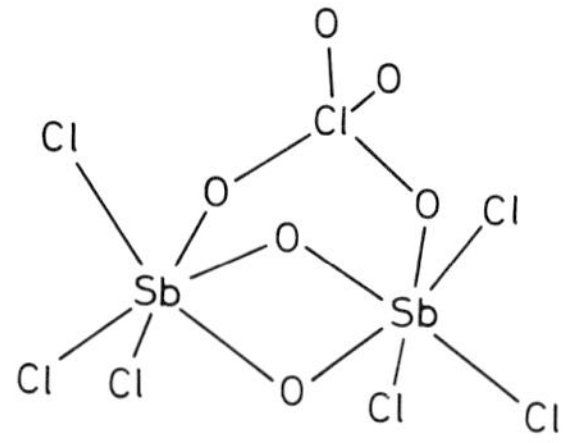

FIG. 21. Structure of $Sb_2Cl_6O(OH)(O_2ClO_2)$ molecule. [Redrawn by permission from C. H. Belin et al., Inorg. Chem. 21, 3557 (1982).]

tate bridging perchlorates with an average Sn—O distance of 2.16 Å (159).

The covalent nature of perchlorate is envisaged in several antimony complexes of formulas $(SbMe_3ClO_4)_2O$, $(SbPh_3ClO_4)_2O$, $(SbPh_3ClO_4)_2O \cdot 2H_2O$, $Sb_2Cl_6O(OH)ClO_4$, and $Sb_8Cl_{24}O_5(ClO_4)_6$ (14, 160, 161). The perchlorates in $(SbMe_3ClO_4)_2O$ are shown to be unidentate (Sb—O length = 2.60, O—Cl length = 1.443 Å, Sb—O—Cl angle = 134.7°), and the metal ions are five-coordinate with trigonal–bipyramidal geometry (161). X-Ray diffraction and IR and Raman spectra of $Sb_2Cl_6O(OH)ClO_4$ provide evidence for bidentate-bridging perchlorate (Fig. 21) with Sb—O distances of 2.207 and 2.247 Å; each antimony atom has a distorted octahedral environment. The bridging O—Cl lengths of 1.477 and 1.479 Å are appreciably longer than the terminal ones (1.390 and 1.398 Å) (14). Perchlorate coordination is also observed in a few bismuth complexes of the types $(BiPh_3ClO_4)_2O$ and $(BiPh_3ClO_4)_2O \cdot 2H_2O$ (160, 161). March and Ferguson (162) have carried out X-ray diffraction of the former, and a distorted trigonal–bipyramidal environment around each of the bismuth atoms (bridged through an oxygen atom) with equatorial phenyls was found (Fig. 22). In contrast to the case of $(SbMe_3ClO_4)_2O$, while only one of the perchlorates is monodentate, having Bi—O length = 2.655, O—Cl length = 1.48 Å, Bi—O—Cl angle = 132.5°, the other is disordered between two sites with Bi—O lengths = 2.53, 2.76; O—Cl lengths = 1.46, 1.36 Å; Bi—O—Cl angles = 137.5, 126.7°.

XI. Conclusion

It is evident that a majority of the metals form perchlorato complexes provided care is taken to eliminate from the system water and other anions, especially halides. The O—Cl bond of the M—O—Cl unit

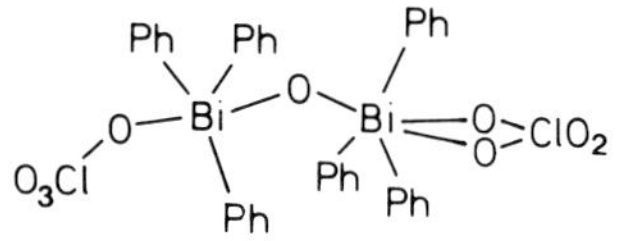

FIG. 22. Structure of (BiPh₃ClO₄)₂O. [Redrawn by permission from F. C. March and G. Ferguson, *J. Chem. Soc. Dalton Trans.* 1291 (1975).]

is generally longer than the uncoordinated ones of ClO_4^-, though there are a few exceptions. Coordination of perchlorate and the resulting change in the stereochemical environment around the metal ion is manifest where possible in the magnetic and electronic spectral properties of the metal complex. Similarly, the lowering of the symmetry of ClO_4^- owing to coordination is amply reflected in the IR and Raman spectral features. For several complexes, $\nu(M-O)$ is found around 330 cm^{-1}, implying considerable covalent character in the metal–perchlorate bond. It is likely that appreciable M→O backbonding exists in some of the complexes. Nevertheless, the M—O bond is usually weak, and the term "semicoordination" aptly describes the situation.

The coordinated oxygen atom may be visualized as having sp^2 hybridization, and the observed M—O—Cl bond angle is expected to reflect the M—O bond strength. In practice this may not be true since steric factors involving other ligands play a dominant role in perchlorate coordination. Attempts to evaluate the M—O bond strength in perchlorato complexes would be quite interesting.

Stable as well as transitory perchlorate-coordinated species are convenient intermediates for the isolation of novel complexes. With four potential donor sites, the role of ClO_4^- acting as a bridging ligand perhaps has interesting possibilities yet to be revealed.

XII. Abbreviations

aep, 2-[-2-(Aminoethyl)]pyridine
adH, Adenine
adHNO, Adenine $N(1)$-oxide
arg, Arginine

bada, Biacetyldianil
bbtb, p,p' Bis(benzoylthiourea)-biphenyl
bdpm, Bis(3,5-dimethylpyrazolyl)methane
bipy, 2,2'-Bipyridine
bpda, 1,8-Bis-(2'-pyridyl)-3,6-diazaoctane
bpdt, 1,8-Bis-(2'-pyridyl)-3,6-dithiaoctane
bpes, Bis-[β-pyridyl-(2)-ethyl] sulfide

Bu, Butyl
Bzl, Benzimidazole

cod, Cycloocta-1,5-diene
Cy, Cyclohexyl
CypNO, 4-Cyanopyridine N-oxide

dab, 1,3-Diaminobutane
dap, 1,2- or 1,3-Diaminopropane
depf, Diethylphosphonoformate anion
depNO, 2-Diethylaminopyridine N-oxide
deta, Diethylenetriamine
diars, o-Phenylenebisdimethylarsine
dmaep, 2-(2-Dimethylaminoethyl)pyridine
dmen, N,N'-Dimethylethylenediamine
dmf, N,N'-Dimethylformamide
dmmp, Dimethyl methylphosphonate
dmp, 2,2'-Dimethoxypropane
dmpd, 2,2-Dimethylpropane-1,3-diamine
dmpNO, 2-Dimethylaminopyridine N-oxide
dmso, Dimethyl sulfoxide
dpa, Di-2-pyridylamine
dpNO, 2,6-Dimethylpyridine N-oxide
dppe, 1,2-Bis(diphenylphosphino)ethane
dppm, 1,1-Bis(diphenylphosphino)methane
dth, 2,5-Dithiahexane
dthp, 2,12-Dimethyl-3,7,11,17-tetraazabicyclo[11.3.1]heptadeca-1(17),2,11,13,15-pentaene
dtod, 1,10-Dioxa-4,7,13,16-tetrathiacyclooctadecane

eaep, 2-(2-Ethylaminoethyl)pyridine
ebsm, N,N'-Ethylenebis(salicyleneiminato)
en, Ethylenediamine
ESR, Electron spin resonance
Et, Ethyl

Gly–GlyH, Glycylglycine anion

hdhc, 4,7,13,16,21,24-Hexaoxa-1,10-diazabicyclo[8.8.8]hexacosane
hism, Histamine
hmpa, Hexamethylphosphoramide
htod, 1,4,7,10,13,16-Hexathiacyclooctadecane

Im, Imidazole
Imz, Imidazolate anion
Imzn, 2-Imidazoline
Iotz, 2-Imino-4-oxa-1,3-thiazolidine
IR, Infrared

M, Metal ion
Me, Methyl
men, N-Methylethylenediamine
miqNO, 3-Methylisoquinoline N-oxide
mmea, β-Methylmercaptoethylamine
mmpp, Methyl methylphenylphosphinate

Nap, Naphthyl
ntd, 1,8-Naphthyridine

oep, Octaethylporphinato dianion
ompa, Octamethylphosphoramide

pam, Pyridine-2-acetamide
Ph, Phenyl
phen, 1,10-Phenanthroline
phzNO, Phenazine N-oxide
phzNO$_2$, Phenazine N,N-dioxide
pmmi, 4-(n-Propylmercaptomethyl)imidazole
Pr, Propyl
puHNO, Purine $N(1)$-oxide
Py, Pyridine
β-PyBzl, β-Pyridylbenzimidazole
pyz, Pyrazine
pyzNO, Pyrazine N-oxide

qun, Quinoline
quxNO, Quinoxaline N-oxide
quxNO$_2$, Quinoxaline 1,4-dioxide

taea, Tris(2-aminoethyl)amine
tamen, N,N,N',N'-Tetramethylethylenediamine
tamn, 2,3,2-Tetraamine
tbma, Tris(2-benzimidazolylmethyl)amine
tchd, 1,5,9,13-Tetrathiacyclohexadecane
tctd, 1,4,8,11-Tetrathiacyclotetradecane
tdaa, Tris(o-diphenylarsinophenyl)arsine
tdaea, Tris(2-dimethylaminoethyl)amine
tdhc, 5,6,14,15-Tetrahydrodibenzo[b,k][1,4,7,10,13,16]hexaoxacyclooctadecin
tdod, 1,4,10,13-Tetraoxa-7,16-dithiacyclooctadecane
tdpp, Tris(o-diphenylphosphinophenyl)phosphine
teof, Triethyl orthoformate
tetp, Triethyl thiophosphate
tfb, Tetrafluorobenzobarrelene
thf, Tetrahydrofuran
tmb, 2,5-Dimethyl-2,5-diisocyanohexane
tmdt, 1,1,7,7-Tetramethyldiethylenetriamine
tmen, N,N,N'-Trimethylethylenediamine
tmp, Trimethyl phosphate

p-tNC, *p*-Tolyl isocyanide
topo, Tri-*n*-octylphosphine oxide
tox, 1,4-Thioxane
tpp, *meso*-Tetraphenylporphinate dianion
Tpp, Tetraphenylporphinate anion
triphos, 1,1,1-Tris(diphenylphosphinomethyl)ethane
ttp, Tri-*p*-tolyl phosphate

UV, Ultraviolet
Vi, Vinyl

References

1. Brown, D. S., Lee, J. D., Melsom, B. G. A., Hathaway, B. J., Procter, I. M., and Tomlinson, A. A. G., *J. Chem. Soc. Chem. Commun.* 369 (1967).
2. Rosenthal, M. R., *J. Chem. Educ.* **50**, 331 (1973).
3. Johansson, L., *Coord. Chem. Rev.* **12**, 241 (1974).
4. Davies, J. A., and Hartley, F. R., *Chem. Rev.* **81**, 79 (1981).
5. Wright, D. A., and Quinn, J. D., *Acta Crystallogr. Sect. B* **30**, 2132 (1974).
6. Lewis, D. L., and Hodgson, D. J., *Inorg. Chem.* **13**, 143 (1974).
7. Pauling, P., Robertson, G. B., and Rodley, G. A., *Nature (London)* **207**, 73 (1965).
8. Cotton, F. A., and Weaver, D. L., *J. Am. Chem. Soc.* **87**, 4189 (1965).
9. Nakai, H., *Bull. Chem. Soc. Jpn.* **44**, 2412 (1971).
10. Lewis, D. L., Hatfield, W. E., and Hodgson, D. J., *Inorg. Chem.* **13**, 147 (1974).
11. Ciampolini, M., Nardi, N., Cini, R., Mangani, S., and Orioli, P., *J. Chem. Soc. Dalton Trans.* 1983 (1979).
12. Reed, C. A., Mashiko, T., Bentley, S. P., Castner, M. E., Scheidt, W. R., Spartalian, K., and Long, G., *J. Am. Chem. Soc.* **101**, 2948 (1979).
13. Spaulding, L. D., Eller, P. G., Bertrand, J. A., and Felton, R. H., *J. Am. Chem. Soc.* **96**, 982 (1974).
14. Belin, C. H., Chaabouni, M., Pascal, J. L., and Potier, J., *Inorg. Chem.* **21**, 3557 (1982).
15. Hathaway, B. J., and Underhill, A. E., *J. Chem. Soc.* 3091 (1961).
15a. Hathaway, B. J., *Curr. Contents* (45), 20 (1982).
16. Ross, S. D., *Spectrochim. Acta* **18**, 225 (1962).
16a. Pearson, G. S., *Adv. Inorg. Radiochem.* **8**, 190 (1966).
17. Scholer, R. P., and Merbach, A. E., *Inorg. Chim. Acta* **15**, 15 (1975).
18. Karayannis, N. M., Mikulski, C. M., Pytlewski, L. L., and Labes, M. M., *Inorg. Chem.* **9**, 582 (1970).
19. Karayannis, N. M., Mikulski, C. M., Pytlewski, L. L., and Labes, M. M., *Inorg. Chim. Acta* **10**, 97 (1974).
20. Karayannis, N. M., Mikulski, C. M., Strocko, M. J., Pytlewski, L. L., and Labes, M. M., *J. Inorg. Nucl. Chem.* **32**, 2629 (1970).
21. Mikulski, C. M., Unruh, J., Pytlewski, L. L., and Karayannis, N. M., *Transition Met. Chem.* **4**, 98 (1979).
22. Pujari, P., and Dash, K. C., *J. Inorg. Nucl. Chem.* **38**, 1891 (1976).
23. Ciampolini, M., and Nardi, N., *Inorg. Chem.* **5**, 41 (1966).
24. Dyer, G., Hartley, J. G., and Venanzi, L. M., *J. Chem. Soc.* 1293 (1965).

25. Shashikala, N., Leelamani, E. G., and Reddy, G. K. N., *Indian J. Chem.* **21A,** 743 (1982).
26. Goodgame, D. M. L., Goodgame, M., and Hayward, P. J., *J. Chem. Soc. A* 1352 (1970).
27. Addison, A. W., Hendriks, H. M. J., Reedijk, J., and Thompson, L. K., *Inorg. Chem.* **20,** 103 (1981).
28. Graham, P., and Joesten, M., *J. Inorg. Nucl. Chem.* **32,** 531 (1970).
29. Durney, M. T., and Marianelli, R. S., *Inorg. Nucl. Chem. Lett.* **6,** 895 (1970).
30. Crawford, N. P., and Melson, G. A., *J. Chem. Soc.* 141 (1970).
31. Ciampolini, M., Mealli, C., and Nardi, N., *J. Chem. Soc. Dalton Trans.* 376 (1980).
32. Mikulski, C. M., Chauhan, S., Rabin, R., and Karayannis, N. M., *J. Inorg. Nucl. Chem.* **43,** 2017 (1981).
33. Mikulski, C. M., Iaconianni, F. J., Pytlewski, L. L., Speca, A. N., and Karayannis, N. M., *Inorg. Chim. Acta* **46,** L47 (1980).
34. Mikulski, C. M., de Prince, R., Tran, T. B., Iaconianni, F. J., Pytlewski, L. L., Speca, A. N., and Karayannis, N. M., *Inorg. Chim. Acta* **56,** 163 (1981).
35. Mikulski, C. M., de Prince, R., Tran, T. B., and Karayannis, N. M., *Inorg. Chim. Acta* **56,** 27 (1981).
36. Mikulski, C. M., Pytlewski, L. L., and Karayannis, N. M., *Inorg. Chim. Acta* **32,** 263 (1979).
37. Mikulski, C. M., Marks, B., Tuttle, D., and Karayannis, N. M., *Inorg. Chim. Acta* **68,** 119 (1983).
38. Speca, A. N., Gelfand, L. S., Iaconianni, F. J., Pytlewski, L. L., Mikulski, C. M., and Karayannis, N. M., *Inorg. Chim. Acta* **33,** 195 (1979).
39. Karayannis, N. M., Speca, A. N., Chasan, D. E., and Pytlewski, L. L., *Coord. Chem. Rev.* **20,** 37 (1976).
40. Chasan, D. E., Pytlewski, L. L., Owens, C., and Karayannis, N. M., *J. Inorg. Nucl. Chem.* **40,** 1019 (1978).
41. Chasan, D. E., Pytlewski, L. L., Owens, C., and Karayannis, N. M., *J. Inorg. Nucl. Chem.* **41,** 13 (1979).
42. Jones, K. M., and Bjerrum, J., *Acta Chem. Scand.* **19,** 974 (1965).
43. Wimmer, F. L., and Snow, M. R., *Aust. J. Chem.* **31,** 267 (1978).
44. Snow, M. R., and Wimmer, F. L., *Inorg. Chim. Acta* **44,** L189 (1980).
45. Brodie, A. M., Hunter, S. H., Rodley, G. A., and Wilkins, C. J., *Inorg. Chim. Acta* **2,** 195 (1968).
46. Lewis, J., Nyholm, R. S., and Rodley, G. A., *Nature (London)* **207,** 72 (1965).
47. Brown, D. H., Nuttal, R. H., McAvoy, J., and Sharp, D. W. A., *J. Chem. Soc.* 892 (1966).
47a. Jones, W. C., and Bull, W. E., *J. Chem. Soc. A* 1849 (1968).
48. Speca, A. N., Pytlewski, L. L., and Karayannis, N. M., *Inorg. Nucl. Chem. Lett.* **9,** 365 (1973).
49. Speca, A. N., Karayannis, N. M., and Pytlewski, L. L., *J. Inorg. Nucl. Chem.* **35,** 3113 (1973).
50. Karayannis, N. M., Mikulski, C. M., Strocko, M. J., Pytlewski, L. L., and Labes, M. M., *J. Inorg. Nucl. Chem.* **33,** 3185 (1971).
51. Reedijk, J., and Verbiest, J., *Transition Met. Chem.* **4,** 239 (1979).
52. Speca, A. N., Gelfand, L. S., Iaconianni, F. J., Pytlewski, L. L., Mikulski, C. M., and Karayannis, N. M., *Inorg. Chim. Acta* **37,** L 551 (1979).
53. Speca, A. N., Gelfand, L. S., Iaconianni, F. J., Pytlewski, L. L., Mikulski, C. M., and Karayannis, N. M., *J. Inorg. Nucl. Chem.* **43,** 2771 (1981).

54. Horn, E., and Snow, M. R., *Aust. J. Chem.* **33**, 2369 (1980).
55. Phillips, D. J., and Tyree, S. Y., *J. Am. Chem. Soc.* **83**, 1806 (1961).
56. Masuda, H., Taga, T., Osaki, K., Sugimoto, H., Yosida, Z. I., and Ogoshi, H., *Inorg. Chem.* **19**, 950 (1980).
57. Christian, D. F., and Roper, W. R., *J. Chem. Soc. Chem. Commun.* 1271 (1971).
58. Peone, J., Jr., and Vaska, L., *Angew. Chem. Int. Ed.* **10**, 511 (1971).
59. Peone, J., Jr., Flynn, B. R., and Vaska, L., *Inorg. Synth.* **15**, 68 (1974).
60. Mikulski, C. M., Pytlewski, L. L., and Karayannis, N. M., *Synth. Inorg. Met. Org. Chem.* **9**, 401 (1979).
61. Einstein, F. W. B., and Rodley, G. A., *J. Inorg. Nucl. Chem.* **29**, 347 (1967).
62. Satpathy, K. C., Mishra, H. P., and Patel, B. N., *Indian J. Chem.* **22A**, 338 (1983).
63. Dual, R., *Ann. Chim.* **18**, 259 (1932).
64. Jurnak, F. A., Greig, D. R., and Raymond, K. N., *Inorg. Chem.* **14**, 2585 (1975).
65. Rodley, G. A., and Smith, P. W., *J. Chem. Soc. A* 1580 (1967).
66. Harrowfield, J. M., Sargeson, A. M., Singh, B., and Sullivan, J. C., *Inorg. Chem.* **14**, 2864 (1975).
67. Tayer, R. S., and Sykes, A. G., *J. Chem. Soc. Chem. Commun.* 1137 (1969).
68. Vaska, L., and Peone, J., Jr., *Suom. Kemistil. B* **44**, 317 (1971).
69. Uson, R., Oro, L. A., Carmona, D., and Esteruelas, M. A., *Inorg. Chim. Acta* **73**, 275 (1983).
70. Vaska, L., and Peone, J., Jr., *J. Chem. Soc. Chem. Commun.* 418 (1971).
71. Wada, M., and Oguro, M., *Inorg. Chem.* **15**, 2346 (1976).
72. Cotton, F. A., and Bannister, E., *J. Chem. Soc.* 1873 (1960).
73. Dewar, R., and Fleischer, E., *Nature (London)* **222**, 372 (1969).
74. Wickenden, A. E., and Krause, R. A., *Inorg. Chem.* **4**, 404 (1965).
75. Pavkovic, S. F., and Meek, D. W., *Inorg. Chem.* **4**, 1091 (1965).
76. Harris, C. M., and McKenzie, E. D., *J. Inorg. Nucl. Chem.* **29**, 1047 (1967); *J. Inorg. Nucl. Chem.* **19**, 372 (1961).
76a. Barker, N. T., Harris, C. M., and McKenzie, E. D., *Proc. Chem. Soc.* 335 (1961).
77. Farago, M. E., James, J. M., and Trew, V. C. G., *J. Chem. Soc. A* 820 (1967).
78. West, D. X., and O'Grady, T. J., *J. Inorg. Nucl. Chem.* **43**, 2725 (1981).
79. Pajunen, A., *Suom. Kemistil. B* **42**, 397 (1969); *Chem. Abstr.* **71**, 129682u (1969).
80. Moore, L. E., Gayhart, R. B., and Bull, W. E., *J. Inorg. Nucl. Chem.* **26**, 896 (1964).
81. Bull, W. E., and Moore, L. E., *J. Inorg. Nucl. Chem.* **27**, 1341 (1965).
82. Buffagni, S., Vallarino, L. M., and Quagliano, J. V., *Inorg. Chem.* **3**, 671 (1964).
83. Vallarino, L. M., Hill, W. E., and Quagliano, J. V., *Inorg. Chem.* **4**, 1598 (1965).
84. Butcher, A. V., Phillips, D. J., and Redfern, J. P., *J. Chem. Soc. A* 1064 (1968).
85. Aubry, F. M., and Brown, G. M., *Acta Crystallogr. Sect. B* **24**, 745 (1968).
86. Aubry, F. M., Busing, W. R., and Brown, G. M., *Acta Crystallogr. Sect. B* **24**, 754 (1968).
87. Davies, J. A., Hartley, F. R., and Murray, S. G., *J. Chem. Soc. Dalton Trans.* 2246 (1980).
88. Uson, R., Fornies, J., and Gonzalo, S., *J. Organomet. Chem.* **104**, 253 (1976).
89. Uson, R., Royo, P., and Fornies, J., *Synth. Inorg. Met. Org. Chem.* **4**, 157 (1974).
89a. Uson, R., Fornies, J., and Martine, F., *J. Organomet. Chem.* **112**, 105 (1976).
90. Coronas, J. M., Polo, C., and Sales, J., *Synth. Inorg. Met. Org. Chem.* **10**, 53 (1980).
91. Gavrilova, I. V., Gel'fnan, M. I., Ivannikova, N. V., and Razumovskii, V. V., *Russ. J. Inorg. Chem.* **16**, 596 (1971).
92. Uson, R., Royo, P., and Gimeno, J., *J. Organomet. Chem.* **72**, 299 (1974).
92a. Sandhu, S. S., Chavla, S., and Parmar, S. S., *Transition Met. Chem.* **6**, 246 (1981).

93. Moers, F. G., and Op Het Veld, P. H., *J. Inorg. Nucl. Chem.* **22,** 3225 (1970).
94. Naikar, S. B., Nanje Gowda, N. M., and Reddy, G. K. N., *Indian J. Chem.* **23A,** 133 (1984).
95. Restivo, R. J., Costin, A., Ferguson, G., and Carty, A. J., *Can. J. Chem.* **53,** 1949 (1975).
96. Olmstead, M. M., Musker, W. K., and Kessler, R. M., *Transition Met. Chem.* **7,** 140 (1982).
97. Cariati, F., and Naldini, L., *J. Inorg. Nucl. Chem.* **28,** 2243 (1966); *Gazz. Chim. Ital.* **95,** 3 (1965).
98. Uhlig, E., and Heinrich, G., *Z. Anorg. Allg. Chem.* **330,** 40 (1964).
99. Brubaker, G. R., Brown, J. N., Yoo, M. K., Kinsey, R. A., Kutchan, T. N., and Mottel, E. A., *Inorg. Chem.* **18,** 299 (1979).
100. Bew, M. J., Hathaway, B. J., and Feredy, R. J., *J. Chem. Soc. Dalton Trans.* 1229 (1972).
101. O'Young, C. L., Dewan, J. C., Lilienthal, H. R., and Lippard, S. J., *J. Am. Chem. Soc.* **100,** 7291 (1978).
102. Lewis, D. L., Hatfield, W. E., and Hodgson, D. J., *Inorg. Chem.* **11,** 2216 (1972).
103. Lewis, D. L., McGregor, K. T., Hatfield, W. E., and Hodgson, D. J., *Inorg. Chem.* **13,** 1013 (1974).
104. Ivarsson, G., *Acta Chem. Scand.* **A33,** 323 (1979).
105. Foley, J., Kennefick, D., Phelan, D., Tyagi, S., and Hathaway, B., *J. Chem. Soc. Dalton Trans.* 2333 (1983).
106. Ray, N., Tyagi, S., and Hathaway, B., *Acta Crystallogr. Sect. B* **38,** 1574 (1982).
107. Pajunen, A., *Suom. Kemistil. B* **40,** 32 (1967); *Chem. Abstr.* **67,** 15818e (1967).
108. Pajunen, A., *Ann. Acad. Sci. Fenn. Ser. A* **II,** 138 (1967).
109. Majeste, R. J., and Trepons, L. M., *Inorg. Chem.* **13,** 1062 (1974).
110. Hathaway, B. J., and Tomlinson, A. A. G., *Coord. Chem. Rev.* **5,** 1 (1970).
111. Ivarsson, G., *Acta Chem. Scand.* **27,** 3523 (1973).
112. Mohapatra, C. K. C., and Dash, K. C., *J. Inorg. Nucl. Chem.* **39,** 1253 (1977).
113. Fawcett, T. G., Rudich, S. M., Toby, B. H., Lalancette, R. A., Potenza, J. A., and Schugar, H. J., *Inorg. Chem.* **19,** 940 (1980).
114. Glick, M. D., Gavel, D. P., Diaddario, L. L., and Rorabacher, D. B., *Inorg. Chem.* **15,** 1190 (1976).
115. Sakaguchi, U., and Addison, A. W., *J. Chem. Soc. Dalton Trans.* 600 (1979).
116. Pajunen, A., Smolander, K., and Belinskij, I., *Suom. Kemistil. B* **45,** 317 (1972).
116a. Pajunen, A., and Lehtonen, M., *Suom. Kemistil. B* **45,** 43 (1972).
117. Olmstead, M. M., Musker, W. K., and Kessler, R. M., *Inorg. Chem.* **20,** 151 (1981).
118. Luukkonen, E., Pajunen, A., and Lehtonen, M., *Suom. Kemistil. B* **43,** 160 (1970).
118a. McWhinnie, W. R., *J. Inorg. Nucl. Chem.* **26,** 21 (1964).
118b. Procter, I. M., Hathaway, B. J., and Nicholls, P., *J. Chem. Soc. A* 1678 (1968).
119. Darriet, J., Haddad, M. S., Duesler, E. N., and Hendrickson, D. N., *Inorg. Chem.* **18,** 2679 (1979).
120. Sekizaki, M., Marumo, F., Yamasaki, K., and Saito, Y., *Bull. Chem. Soc. Jpn.* **44,** 1731 (1971).
121. Battaglia, L. P., Corradi, A. B., Marcotrigiano, G., Menabue, L., and Pellacani, G. C., *J. Chem. Soc. Dalton Trans.* 8 (1981).
122. West, D. X., *Inorg. Nucl. Chem. Lett.* **14,** 155 (1978).
123. Bonnett, P. J. J., and Jeannin, Y., *Acta Crystallogr. Sect. B* **26,** 318 (1970).
124. Ou, C. C., Miskowski, V. M., Lalancette, R. A., Potenza, J. A., and Schugar, H. J., *Inorg. Chem.* **15,** 3157 (1976).

125. Hussain, M. S., Joesten, M. D., and Lenhert, P. G., *Inorg. Chem.* **9,** 162 (1970).
126. Aoi, N., Matsubayashi, G., and Tanaka, T., *J. Chem. Soc. Dalton Trans.* 1059 (1983).
127. Chow, S. T., and McAuliffe, C. A., *Inorg. Nucl. Chem. Lett.* **8,** 913 (1972).
128. Bell, J. D., Freeman, H. C., Wood, A. M., Driver, R., and Walker, W. R., *J. Chem. Soc. Chem. Commun.* 1441 (1969).
129. Nelson, S. M., Esho, F. S., and Drew, M. G. B., *J. Chem. Soc. Dalton Trans.* 407 (1982).
130. Ivarsson, G., Lundberg, B. K. S., and Ingri, N., *Acta Chem. Scand.* **26,** 3005 (1972).
131. Dikhoff, T. G. M. H., and Goel, R. G., *Inorg. Chim. Acta* **44,** L72 (1980).
132. Udupa, M. R., and Padmanabhan, M., *Inorg. Chim. Acta* **13,** 289 (1975).
133. Guitard, A., Mari, A., Beauchamp, A. L., Dartiguenave, Y., and Dartiguenave, M., *Inorg. Chem.* **22,** 1603 (1983).
134. Bjork, N., and Cassel, A., *Acta Chem. Scand.* **A30,** 235 (1976).
135. Uson, R., Laguna, A., Laguna, M., and Perez, V., *Synth. Inorg. Met. Org. Chem.* **11,** 361 (1981).
136. Uson, R., Gimeno, J., Fornies, J., Martinez, F., and Fernandez, C., *Inorg. Chim. Acta* **63,** 91 (1982).
137. Uson, R., Royo, P., and Laguna, A., *Synth. Inorg. Met. Org. Chem.* **3,** 237 (1973).
137a. Uson, R., Laguna, A., and Vicente, J., *Synth. Inorg. Met. Org. Chem.* **7,** 463 (1977).
138. Uson, R., Laguna, A., and Sanjoaquin, J. L., *J. Organomet. Chem.* **80,** 147 (1974).
139. Naikar, S. B., Nanje Gowda, N. M., and Reddy, G. K. N., unpublished results.
140. Interrante, L. V., *Inorg. Chem.* **7,** 943 (1968).
141. Aleya, E. C., Dias, S. A., Ferguson, G., and Khan, M. A., *J. Chem. Res. Synop.* 360 (1979).
142. Naikar, S. B., Nanje Gowda, N. M., and Reddy, G. K. N., *Indian J. Chem.* **20A,** 436 (1981).
143. Aleya, E. C., Dias, S. A., Goel, R. G., Ogini, W. O., Pilon, P., and Meek, D. W., *Inorg. Chem.* **17,** 1697 (1978).
144. Davis, A. R., Murphy, C. J., and Plane, R. A., *Inorg. Chem.* **9,** 423 (1970).
145. Aleya, E. C., Dias, S. A., Ferguson, G., and Khan, M. A., *Can. J. Chem.* **57,** 2217 (1979).
146. Sandström, M., *Acta Chem. Scand.* **A32,** 527 (1978).
147. Epstein, J. M., Dewan, J. C., Kepert, D. L., and White, A. H., *J. Chem. Soc. Dalton Trans.* 1949 (1974).
148. Jones, T. E., Sokol, L. S. W. L., Rorabacher, D. B., and Glick, M. D., *J. Chem. Soc. Chem. Commun.* 140 (1979).
149. Karayannis, N. M., Bradshaw, E. E., Pytlewski, L. L., and Labes, M. M., *J. Inorg. Nucl. Chem.* **32,** 1079 (1970).
150. Karayannis, N. M., Owens, C., Pytlewski, L. L., and Labes, M. M., *J. Inorg. Nucl. Chem.* **31,** 2767 (1969).
151. Grandey, R. C., and Moeller, T., *J. Inorg. Nucl. Chem.* **32,** 333 (1970).
151a. Navaneetham, N. S., and Soundararajan, S., *Proc. Indian Acad. Sci. Sect. A* **90,** 439 (1981).
152. Ciampolini, M., Dapporto, P., and Nardi, N., *J. Chem. Soc. Dalton Trans.* 974 (1979).
153. Milburn, G. W. H., Truter, M. R., and Vickery, B. L., *J. Chem. Soc. Chem. Commun.* 1188 (1968).
154. Barkigia, K. M., Spaulding, L. D., and Fajer, J., *Inorg. Chem.* **22,** 349 (1983).

155. Matyashin, Yu. N., Kon'kova, T. S., Titova, K. V., Rasabvskii, V. Ya., and Lebedev, Yu. A., *Russ. J. Inorg. Chem.* **26,** 183 (1981).
156. Blundell, T. L., and Powell, H. M., *J. Chem. Soc. Chem. Commun.* 54 (1967).
157. Elder, R. C., Heeg, M. J., and Deutsch, E., *Inorg. Chem.* **17,** 427 (1978).
158. Clark, H. C., and O'Brien, R. J., *Inorg. Chem.* **2,** 740 (1963).
159. Belin, C., Chaabouni, M., Pascal, J. L., Potier, J., and Roziere, J., *J. Chem. Soc. Chem. Commun.* 105 (1980).
160. Ferguson, G., Goel, R. G., March, F. C., Ridley, G. A., and Prasad, H. S., *J. Chem. Soc. Chem. Commun.* 1547 (1971).
161. Ferguson, G., March, F. C., and Ridley, D. R., *Acta Crystallogr. Sect. B* **31,** 1260 (1975).
162. March, F. C., and Ferguson, G., *J. Chem. Soc. Dalton Trans.* 1291 (1975).

INDEX

A

Actinide oxide fluorides, 86–91
 hexavalent, 88–91
 pentavalent, 86–88
 tetravalent, 86
 trivalent, 86
Adamantane structure, exchangeability within, 195–197
Addition reaction, 105–107, 109–110
Alkene, muonic radicals and, 123–125
Aluminum, in paraelements, 170
Americium oxide fluorides, 87
Antimony, complexes with perchlorate ion, 290
Arene, muonic radicals and, 126–128
Argon, 3–4
Arsenic(V), C-containing compounds, 204–216
Arsorane, see Arsenic(V)
Atom gun, 3–4

B

Berkelium
 atomic properties, 34–41
 availability, 30–31
 chalcogenides of, 48, 53–54
 compounds of, 47–55
 electronic energies, 34–35
 emission spectra, 35–36
 halides and oxyhalides of, 48–49, 51–53
 ionic species, 55–56
 ions, magnetic behavior of, 54–55
 metallic state, 41–46
  chemical properties, 45
  physical properties, 42–45
  preparation, 41
 nuclear properties of isotopes of, 30–31
 oxidation–reduction behavior, 59–63
 oxides of, 48, 50–51
 pnictides of, 49, 53–54
 Raman spectra, 39–41
 separation and purification, 32–34
 solid state absorption spectra, 38–41
 solution absorption spectra, 36–38
 solution chemistry, 55–63
 stability constants of complexes, 56–59
 thermodynamic quantities, 56
Bismuth, complexes with perchlorate ion, 290
Bis(trifluoromethylthio)amino radical, similarity to fluorine, 179–180
Bk, see Berkelium
Borazole, transition to, from hexazabenzene, 194–195
Boron, in paraelements, 170

C

CAD spectra, see Collisionally activated dissociation spectra
Calcium, complexes with perchlorate ion, 289
Californium oxide fluoride, 86, 87
Carbon
 in compounds with arsenic(V), 204–216
 exchangeability with sulfur, 172–176
 in paraelements, 169–170
Chalcogenides, of berkelium, 49, 53–54
Chlorine
 in compounds with arsenic(V), 238–247
 in paraelements, 169–170
 similarity to tetrakis(trefluoromethylthio)pyrrolyl radical, 187–194
 to trifluoromethylthio radical, 180–186
 substitution reactions, 114–117
Chlorine-38 atom, thermal reactions of, 108–112
Chromium complexes
 oxide fluorides, 77
 perchlorates, 261, 262
Cobalamins, FAB analysis, 7–10
Cobalt group metals, complexes with perchlorate ion, 265–268

Collisionally activated dissociation spectra, 7–8
Coordination compounds, fast-atom bombardment mass spectroscopy and, 13–17
Copper group metals, complexes with perchlorate ion, 273–283
Crown ethers, 5
Cycloarsa(V)azanes, 216–223
Cycloheptaselenium, 142–144
Cyclohexaselenium, 137–142
Cyclooctaselenium, 144–150
Cyclopentaselenium, 136–137

D

Desorption ionization techniques, 6–7, 21
Dimethylsulfoxide, 5
Di-*tert*-amylphenol, 5
DMSO, *see* Dimethylsulfoxide

E

Electronegativity, of paraelements, 197–198
Electronic spectra, of perchlorate ion complexes, 258–259
Electron spin resonance spectra, of perchlorate ion complexes, 260
Element displacement principle, 167–198
  applications, 172–176
  theory, 169–171

F

FAB, *see* Fast-atom bombardment mass spectroscopy
Fast-atom bombardment spectroscopy, 1–22
  applications to coordination compounds, 13–17
  to inorganic systems, 13–19
  to organometallic compounds, 7–13
  to transition metal compounds, 11–13
  basic methods, 2–3
  instrumentation, 3–5
  matrix liquids for, 5–6
  relation to other soft ionization techniques, 6–7

FD, *see* Field desorption mass spectroscopy
Field desorption mass spectroscopy, 6, 21
Fluorine
  in compounds with arsenic(V), 238–247
  displacement by, 169–171
  similarity to $(CF_3S)_2N$ radical, 179–180
  to parafluorine, 176–179
  substitution reactions, 117–118
Fluorine 18 atom
  addition reactions, 105–107
  hydrogen abstraction, 102–103
  reactions with organometallic compounds, 107–108
  substitution reactions, 103–104
  thermal reactions of, 102–107

G

Gallium, in paraelements, 170
Glycerol, as matrix liquid, 5

H

Halides, of berkelium, 48, 51–53
Halogen, *see also* Pseudohalogen
  in compounds with arsenic(V), 238–247
  displacement principle and, 169–171
  substitution reactions, 117–118
p-Halogen, *see* Parahalogen
Hexamethylphosphoramide, 5
Hexazabenzene, transition to borazole, 194–195
HMPA, *see* Hexamethylphosphoramide
Hydrogen abstraction, 102–103, 108–109

I

Indium, in paraelements, 170
Infrared spectra, of perchlorate ion complexes, 257–258
Iodine, substitution reactions, 117–118
Iridium oxide fluoride, 85
Iron group metals, complexes with perchlorate ion, 263–265
Iron oxide fluoride, 84

L

LAMMA, see Laser microprobe mass spectrometry
Lanthanide elements
  complexes with perchlorate ion, 260–262, 287–288
  oxide fluorides of, 85–86
Laser desorption mass spectroscopy, 6
Laser microprobe mass spectrometry, 21
LD, see Laser desorption mass spectroscopy

M

Magnesium complexes, with perchlorate ion, 289
Manganese complexes
  oxide fluorides, 82
  perchlorates, 261, 262
Mass spectroscopy, see also Fast-atom bombardment mass spectroscopy; Secondary ion mass spectroscopy
Matrix liquids, for fast-atom bombardment mass spectroscopy, 5–6
Metal–perchlorate bonding, 256, 257
Molybdenum oxide fluorides, 78–82
  dioxide difluorides, 81–82
  oxide tetrafluorides, 78–80
Muonic radical, formation and reactions of, 122–130
Muonium chemistry
  gaseous phase, 119–120
  liquid mixtures, 120–122
  radical formation and reactions, 122–130

N

Negative ion spectra, 7
Neptunium oxide fluorides, 87, 88, 89, 91
Nickel group metals, complexes with perchlorate ion, 268–273
Niobium oxide fluorides, 76–77
Nitrogen
  in compounds with arsenic(V), 216–223
  in paraelements, 169–170
Nuclear transformations, chemical effects of, 101–130

O

Organoarsorane, see Arsenic(V)
Organometallic compounds
  fast-atom bombardment mass spectroscopy and, 7–13
  reactions with ^{18}F atoms, 107
Osmium oxide fluorides, 84–85
Oxide, of berkelium, 48, 50–51
Oxide fluorides
  of actinide elements, 86–91
  of lanthanide elements, 85–86
  of transition metals, 74–85
Oxygen, in compounds with arsenic(V), 223–238
Oxyhalides, of berkelium, 49, 51–53

P

Palladium oxide fluorides, 87, 88
Paraelement
  defined, 169
  first-order derivative, 169–171
  group electronegativities, 197–198
Parafluorine, similarity to fluorine, 176–179
Parahalogen, properties, 176–194
PD, see Plasma desorption mass spectroscopy
Perchlorate ion complexes, 255–299
  with cobalt group metals, 265–268
  coordination types, 256–260
  with copper group metals, 273–283
  with early transition metals, 260–263
  electronic spectra, 258–259
  ESR spectra, 260
  infrared and Raman spectra, 257–258
  with iron group metals, 263–265
  with lanthanides, 260–262, 287–288
  magnetic susceptibility, 260
  molar conductivities, 260
  with nickel group metals, 268–273
  X-ray crystal structure analysis, 256–257
  with zinc group metals, 283–287
Phosphorus, in paraelements, 169–170
Plasma desorption mass spectroscopy, 6
Platinum oxide fluoride, 85
Plutonium oxide fluorides, 86, 87, 88, 89, 91

Pnictides, of berkelium, 49, 53–54
Pseudohalogen, 168–171

R

Raman spectra, of perchlorate ion complexes, 257–258
Recoil atoms, thermalized, reactions of, 102–112
Rhenium complexes
  oxide fluorides, 82–84
  perchlorates, 262, 263
Ruthenium oxide fluoride, 84

S

Scandium group metals, complexes with perchlorate ion, 260–263
Scandium oxide fluorides, 74
Secondary ion mass spectrometry, applications, 19–21
Selenium, see also cyclo compounds
  allotropes, thermodynamic properties, 151–152
  dications, 155–160
    Se_4^{2+} ion, 155–157
    Se_8^{2+} ion, 157–159
    Se_{10}^{2+} ion, 160
  homocyclic cations, 152–160
  iodide cations, 161–162
  mass spectra of vapors, 154
  neutral ring molecules, 136–152
Silicon, in paraelements, 169–170
SIMS, see Secondary ion mass spectrometry
Sodium, complexes with perchlorate ion, 288–289
Substitution reactions
  of ^{38}Cl atoms, 110–111
  of ^{18}F atoms, 103–104
  stereochemistry in, 112–119
Sulfolane, 5
Sulfur
  in compounds with arsenic(V), 226–231
  in paraelements, 169–176
Sulfur(IV), exchangeability with carbon, 172–176

T

Tantalum oxide fluorides, 77
Technetium oxide fluorides, 82
Tetrakis(trifluoromethylthio)pyrrolyl radical, similarity to chlorine, 187–194
Thallium
  complexes with perchlorate ion, 289
  in paraelements, 170
Thioglycerol, 5
Thorium oxide fluorides, 86–91
Tin, complexes with perchlorate ion, 289–290
Titanium oxide fluorides, 74–75
Transition metal complexes
  oxide fluorides, 74–85
  perchlorates, 260–263
Trifluoromethylthio radical, similarity to chlorine, 180–186
Tritium, substitution reactions, 112–113
Triton X100, 5
Tungsten oxide fluorides, 78–82
  dioxide difluorides, 81–82
  oxide tetrafluorides, 78–80

U

Uranium oxide fluorides, 86–91

V

Vanadium oxide fluorides, 75–76

X

Xenon, 3–4
X-ray crystal structure analysis of perchlorate ion complexes, 256–257

Y

Yttrium oxide fluorides, 74

Z

Zinc group metals, complexes with perchlorate ion, 283–287
Zirconium oxide fluorides, 75

CONTENTS OF PREVIOUS VOLUMES

VOLUME 1

Mechanisms of Redox Reactions of Simple Chemistry
H. Taube

Compounds of Aromatic Ring Systems and Metals
E. O. Fischer and H. P. Fritz

Recent Studies of the Boron Hydrides
William N. Lipscomb

Lattice Energies and Their Significance in Inorganic Chemistry
T. C. Waddington

Graphite Intercalation Compounds
W. Rüdorff

The Szilard-Chambers Reactions in Solids
Garman Harbottle and Norman Sutin

Activation Analysis
D. N. F. Atkins and A. A. Smales

The Phosphonitrilic Halides and Their Derivatives
N. L. Paddock and H. T. Searle

The Sulfuric Acid Solvent System
R. J. Gillespie and E. A. Robinson

AUTHOR INDEX—SUBJECT INDEX

VOLUME 2

Stereochemistry of Ionic Solids
J. D. Dunitz and L. E. Orgel

Organometallic Compounds
John Eisch and Henry Gilman

Fluorine-Containing Compounds of Sulfur
George H. Cady

Amides and Imides of the Oxyacids of Sulfur
Margot Becke-Goehring

Halides of the Actinide Elements
Joseph J. Katz and Irving Sheft

Structure of Compounds Containing Chains of Sulfur Atoms
Olav Foss

Chemical Reactivity of the Boron Hydrides and Related Compounds
F. G. A. Stone

Mass Spectrometry in Nuclear Chemistry
H. G. Thode, C. C. McMullen, and K. Fritze

AUTHOR INDEX—SUBJECT INDEX

VOLUME 3

Mechanisms of Substitution Reactions of Metal Complexes
Fred Basolo and Ralph G. Pearson

Molecular Complexes of Halogens
L. J. Andrews and R. M. Keefer

Structure of Interhalogen Compounds and Polyhalides
E. H. Wiebenga, E. E. Havinga, and K. H. Boswijk

Kinetic Behavior of the Radiolysis Products of Water
Christiane Ferradini

The General, Selective, and Specific Formation of Complexes by Metallic Cations
G. Schwarzenbach

Atmosphere Activities and Dating Procedures
A. G. Maddock and E. H. Willis

Polyfluoroalkyl Derivatives of Metalloids and Nonmetals
R. E. Banks and R. N. Haszeldine

AUTHOR INDEX—SUBJECT INDEX

VOLUME 4

Condensed Phosphates and Arsenates
Erich Thilo

Olefin, Acetylene, and π-Allylic Complexes of Transition Metals
R. G. Guy and B. L. Shaw

Recent Advances in the Stereochemistry of Nickel, Palladium, and Platinum
J. R. Miller

The Chemistry of Polonium
  K. W. Bagnall
The Use of Nuclear Magnetic Resonance in Inorganic Chemistry
  E. L. Mutterties and W. D. Phillips
Oxide Melts
  J. D. Mackenzie
AUTHOR INDEX—SUBJECT INDEX

VOLUME 5

The Stabilization of Oxidation States of the Transition Metals
  R. S. Nyholm and M. L. Tobe
Oxides and Oxyfluorides of the Halogens
  M. Schmeisser and K. Brandle
The Chemistry of Gallium
  N. N. Greenwood
Chemical Effects of Nuclear Activation in Gases and Liquids
  I. G. Campbell
Gaseous Hydroxides
  O. Glenser and H. G. Wendlandt
The Borazines
  E. K. Mellon, Jr., and J. J. Lagowski
Decaborane-14 and Its Derivatives
  M. Frederick Hawthorne
The Structure and Reactivity of Organophosphorus Compounds
  R. F. Hudson
AUTHOR INDEX—SUBJECT INDEX

VOLUME 6

Complexes of the Transition Metals with Phosphines, Arsines, and Stibines
  G. Booth
Anhydrous Metal Nitrates
  C. C. Addison and N. Logan
Chemical Reactions in Electric Discharges
  Adli S. Kan'an and John L. Margrave
The Chemistry of Astatine
  A. H. W. Aten, Jr.

The Chemistry of Silicon–Nitrogen Compounds
  U. Wannagat
Peroxy Compounds of Transition Metals
  J. A. Connor and E. A. V. Ebsworth
The Direct Synthesis of Organosilicon Compounds
  J. J. Zuckerman
The Mössbauer Effect and Its Application in Chemistry
  E. Fluck
AUTHOR INDEX—SUBJECT INDEX

VOLUME 7

Halides of Phosphorus, Arsenic, Antimony, and Bismuth
  L. Kolditz
The Phthalocynanines
  A. B. P. Lever
Hydride Complexes of the Transition Metals
  M. L. H. Green and D. L. Jones
Reactions of Chelated Organic Ligands
  Quintus Fernando
Organoaluminum Compounds
  Roland Köster and Paul Binger
Carbosilanes
  G. Fritz, J. Grobe, and D. Kummer
AUTHOR INDEX—SUBJECT INDEX

VOLUME 8

Substitution Products of the Group VIB Metal Carbonyls
  Gerard R. Dobson, Ingo W. Stolz, and Raymond K. Sheline
Transition Metal Cyanides and Their Complexes
  B. M. Chadwick and A. G. Sharpe
Perchloric Acid
  G. S. Pearson
Neutron Diffraction and Its Application in Inorganic Chemistry
  G. E. Bacon

Nuclear Quadrupole Resonance and Its Application in Inorganic Chemistry
Masaji Kubo and Daiyu Nakamura

The Chemistry of Complex Aluminohydrides
E. C. Ashby

AUTHOR INDEX—SUBJECT INDEX

VOLUME 9

Liquid–Liquid Extraction of Metal Ions
D. F. Peppard

Nitrides of Metals of the First Transition Series
R. Juza

Pseudohalides of Group IIIB and IVB Elements
M. F. Lappert and H. Pyszora

Stereoselectivity in Coordination Compounds
J. H. Dunlop and R. D. Gillard

Heterocations
A. A. Woolf

The Inorganic Chemistry of Tungsten
R. V. Parish

AUTHOR INDEX—SUBJECT INDEX

VOLUME 10

The Halides of Boron
A. G. Massey

Further Advances in the Study of Mechanisms of Redox Reactions
A. G. Sykes

Mixed Valence Chemistry—A Survey and Classification
Melvin B. Robin and Peter Day

AUTHOR INDEX—SUBJECT INDEX—
VOLUMES 1–10

VOLUME 11

Technetium
K. V. Kotegov, O. N. Pavlov, and V. P. Shvedov

Transition Metal Complexes with Group IVB Elements
J. F. Young

Metal Carbides
William A. Frad

Silicon Hydrides and Their Derivatives
B. J. Aylett

Some General Aspects of Mercury Chemistry
H. L. Roberts

Alkyl Derivatives of the Group II Metals
B. J. Wakefield

AUTHOR INDEX—SUBJECT INDEX

VOLUME 12

Some Recent Preparative Chemistry of Protactinium
D. Brown

Vibrational Spectra of Transition Metal Carbonyl Complexes
Linda M. Haines and M. H. Stiddard

The Chemistry of Complexes Containing 2,2'-Bipyridyl, 1,10-Phenanthroline, or 2,2',6',2''-Terpyridyl as Ligands
W. R. McWhinnie and J. D. Miller

Olefin Complexes of the Transition Metals
H. W. Quinn and J. H. Tsai

Cis and Trans Effects in Cobalt(III) Complexes
J. M. Pratt and R. G. Thorp

AUTHOR INDEX—SUBJECT INDEX

VOLUME 13

Zirconium and Hafnium Chemistry
E. M. Larsen

Electron Spin Resonance of Transition Metal Complexes
B. A. Goodman and J. B. Raynor

Recent Progress in the Chemistry of Fluorophosphines
John F. Nixon

Transition Metal Cluster with π-Acid
  Ligands
  R. D. Johnston
AUTHOR INDEX—SUBJECT INDEX

VOLUME 14

The Phosphazotrihalides
  M. Bermann
Low Temperature Condensation of High
  Temperature Species as a Synthetic
  Method
  P. L. Timms
Transition Metal Complexes Containing
  Bidentate Phosphine Ligands
  W. Levason and C. A. McAuliffe
Beryllium Halides and Pseudohalides
  N. A. Bell
Sulfur–Nitrogen–Fluorine Compounds
  O. Glemser and R. Mews
AUTHOR INDEX—SUBJECT INDEX

VOLUME 15

Secondary Bonding to Nonmetallic Elements
  N. W. Alcock
Mössbauer Spectra of Inorganic Compounds: Bonding and Structure
  G. M. Bancroft and R. H. Platt
Metal Alkoxides and Dialkylamides
  D. C. Bradley
Fluoroalicyclic Derivatives of Metals and
  Metalloids
  W. R. Cullen
The Sulfur Nitrides
  H. G. Heal
AUTHOR INDEX—SUBJECT INDEX

VOLUME 16

The Chemistry of Bis(trifluoromethyl)-
  amino Compounds
  H. G. Ang and Y. C. Syn

Vacuum Ultraviolet Photoelectron Spectroscopy of Inorganic Molecules
  R. L. DeKock and D. R. Lloyd
Fluorinated Peroxides
  Ronald A. De Marco and
  Jean'ne M. Shreeve
Fluorosulfuric Acid, Its Salts, and Derivatives
  Albert W. Jache
The Reaction Chemistry of Diborane
  L. H. Long
Lower Sulfur Fluorides
  F. Seel
AUTHOR INDEX—SUBJECT INDEX

VOLUME 17

Inorganic Compounds Containing the
  Trifluoroacetate Group
  C. D. Garner and B. Hughes
Homopolyatomic Cations of the Elements
  R. J. Gillespie and J. Passmore
Use of Radio-Frequency Plasma in
  Chemical Synthesis
  S. M. L. Hamblyn and B. G. Reuben
Copper(I) Complexes
  F. H. Jardine
Complexes of Open-Chain Tetradenate
  Ligands Containing Heavy Donor
  Atoms
  C. A. McAuliffe
The Functional Approach to Ionization
  Phenomena in Solutions
  U. Mayer and V. Gutmann
Coordination Chemistry of the Cyanate,
  Thiocyanate, and Selenocyanate Ions
  A. H. Norbury
SUBJECT INDEX

VOLUME 18

Structural and Bonding Patterns in
  Cluster Chemistry
  K. Wade

Coordination Number Pattern Recognition Theory of Carborane Structures
Robert E. Williams

Preparation and Reactions of Perfluorohalogenoorganosulfenyl Halides
A. Haas and U. Niemann

Correlations in Nuclear Magnetic Shielding. Part I
Joan Mason

Some Applications of Mass Spectroscopy in Inorganic and Organometallic Chemistry
Jack M. Miller and Gary L. Wilson

The Structures of Elemental Sulfur
Beat Meyer

Chlorine Oxyfluorides
K. O. Christe and C. J. Schack
SUBJECT INDEX

VOLUME 19

Recent Chemistry and Structure Investigation of Nitrogen Triiodide, Tribromide, Trichloride, and Related Compounds
Jochen Jander

Aspects of Organo-Transition-Metal Photochemistry and Their Biological Implications
Ernst A. Koerner von Gustorf, Luc H. G. Leenders, Ingrid Fischler, and Robin N. Perutz

Nitrogen–Sulfur-Fluorine Ions
R. Mews

Isopolymolybdates and Isopolytungstates
Karl-Heinz Tytko and Oskar Glemser
SUBJECT INDEX

VOLUME 20

Recent Advances in the Chemistry of the Less-Common Oxidation States of the Lanthanide Elements
D. A. Johnson

Ferrimagnetic Fluorides
Alain Tressaud and Jean Michel Dance

Hydride Complexes of Ruthenium, Rhodium, and Iridium
G. L. Geoffroy and J. R. Lehman

Structures and Physical Properties of Polynuclear Carboxylates
Janet Catterick and Peter Thornton
SUBJECT INDEX

VOLUME 21

Template Reactions
Maria De Sousa Healy and Anthony J. Rest

Cyclophosphazenes
S. S. Krishnamurthy, A. C. Sau, and M. Woods

A New Look at Structure and Bonding in Transition Metal Complexes
Jeremy K. Burdett

Adducts of the Mixed Trihalides of Boron
J. Stephen Hartman and Jack M. Miller

Reorganization Energies of Optical Electron Transfer Processes
R. D. Cannon

Vibrational Spectra of the Binary Fluorides of the Main Group Elements
N. R. Smyrl and Gleb Mamantov

The Mössbauer Effect in Supported Microcrystallites
Frank J. Berry
SUBJECT INDEX

VOLUME 22

Lattice Energies and Thermochemistry of Hexahalometallate(IV) Complexes, A_2MX_6, which Possess the Antifluorite Structure
H. Donald B. Jenkins and Kenneth F. Pratt

Reaction Mechanisms of Inorganic Nitrogen Compounds
G. Stedman

Thio-, Seleno-, and Tellurohalides of the Transition Metals
M. J. Atherton and J. H. Holloway

Correlations in Nuclear Magnetic Shielding, Part II
Joan Mason

Cyclic Sulfur–Nitrogen Compounds
H. W. Roesky

1,2-Dithiolene Complexes of Transition Metals
R. P. Burns and C. A McAuliffe

Some Aspects of the Bioinorganic Chemistry of Zinc
Reg H. Prince

SUBJECT INDEX

VOLUME 23

Recent Advances in Organotin Chemistry
Alwyn G. Davies and Peter J. Smith

Transition Metal Vapor Cryochemistry
William J. Power and Geoffrey A. Ozin

New Methods for the Synthesis of Trifluoromethyl Organometallic Compounds
Richard J. Lagow and John A. Morrison

1,1-Dithiolato Complexes of the Transition Elements
R. P. Burns, F. P. McCullough, and C. A. McAuliffe

Graphite Intercalation Compounds
Henry Selig and Lawrence B. Ebert

Solid-State Chemistry of Thio-, Seleno-, and Tellurohalides of Representative and Transition Elements
J. Fenner, A. Rabenau, and G. Trageser

SUBJECT INDEX

VOLUME 24

Thermochemistry of Inorganic Fluorine Compounds
A. A. Woolf

Lanthanide, Yttrium, and Scandium Trihalides: Preparation of Anhydrous Materials and Solution Thermochemistry
J. Burgess and J. Kijowski

The Coordination Chemistry of Sulfoxides with Transition Metals
J. A. Davies

Selenium and Tellurium Fluorides
A. Engelbrecht and F. Sladky

Transition-Metal Molecular Clusters
B. F. G. Johnson and J. Lewis

INDEX

VOLUME 25

Some Aspects of Silicon–Transition-Metal Chemistry
B. J. Aylett

The Electronic Properties of Metal Solutions in Liquid Ammonia and Related Solvents
Peter P. Edwards

Metal Borates
J. B. Farmer

Compounds of Gold in Unusual Oxidation States
Hubert Schmidbaur and Kailash C. Dash

Hydride Compounds of the Titanium and Vanadium Group Elements
G. E. Toogood and M. G. H. Wallbridge

INDEX

VOLUME 26

The Subhalides of Boron
A. G. Massey

Carbon-Rich Carboranes and Their Metal Derivatives
Russell N. Grimes

Fluorinated Hypofluorites and Hypochlorites
Jean'ne M. Shreeve

The Chemistry of the Halogen Azides
K. Dehnicke

Gaseous Chloride Complexes Containing Halogen Bridges
Harald Schäfer

One-Dimensional Inorganic Platinum-Chain Electrical Conductors
Jack M. Williams

Transition-Metal Alkoxides
R. C. Mehrotra

Transition-Metal Thionitrosyl and Related Complexes
H. W. Roesky and K. K. Pandey

INDEX

VOLUME 27

Alkali and Alkaline Earth Metal Cryptates
David Parker

Electron-Density Distributions in Inorganic Compounds
Koshiro Toriumi and Yoshihiko Saito

Solid State Structures of the Binary Fluorides of the Transition Metals
A. J. Edwards

Structural Organogermanium Chemistry
K. C. Molloy and J. J. Zuckerman

Preparations and Reactions of Inorganic Main-Group Oxide Fluorides
John H. Holloway and David Laycock

The Chemistry of Nitrogen Fixation and Models for the Reactions of Nitrogenase
Richard A. Henderson, G. Jeffery Leigh, and Christopher J. Pickett

Trifluoromethyl Derivatives of the Transition Metal Elements
John A. Morrison

INDEX